# Lecture Notes in Computer Science 13538

More information about this series at https://link.springer.com/bookseries/558

Parinya Chalermsook · Bundit Laekhanukit (Eds.)

# Approximation and Online Algorithms

20th International Workshop, WAOA 2022
Potsdam, Germany, September 8–9, 2022
Proceedings

 Springer

*Editors*
Parinya Chalermsook
Aalto University
Espoo, Finland

Bundit Laekhanukit
Shanghai University of Finance
and Economics
Shanghai, China

ISSN 0302-9743 ISSN 1611-3349 (electronic)
Lecture Notes in Computer Science
ISBN 978-3-031-18366-9 ISBN 978-3-031-18367-6 (eBook)
https://doi.org/10.1007/978-3-031-18367-6

This Springer imprint is published by the registered company Springer Nature Switzerland AG
The registered company address is: Gewerbestrasse 11, 6330 Cham, Switzerland

# Preface

WAOA 2022 received 21 submissions which were distributed among the program committee members. Each of the submissions was initially reviewed by three referees. However, due to a short review timeframe, almost all reviews were written by the PC members themselves. During the discussion phase, additional opinions were solicited in some difficult cases. The EasyChair system was used to manage the paper selection process, from submissions to notifications. This year is the first time WAOA has adopted a lightweight double-blind reviewing process. In our opinion, it worked out relatively well.

We would like to thank all the authors who submitted their papers to WAOA 2022, the invited keynote speaker, Tobias Mömke, who agreed to deliver the talk in Potsdam, the PC members who did an excellent job despite such a short timeframe, and the organizers from the Hasso Plattner Institute.

We are looking forward to seeing everyone again next year.

August 2022

Parinya Chalermsook
Bundit Laekhanukit

# Organization

## Program Committee Chairs

Parinya Chalermsook      Aalto University, Finland
Bundit Laekhanukit      Shanghai University of Finance and Economics, China

## Steering Committee

Evripidis Bampis      Sorbonne Université, France
Thomas Erlebach      Durham University, UK
Christos Kaklamanis      University of Patras, Greece
Nicole Megow      Universität Bremen, Germany
Laura Sanita      Bocconi University, Italy
Martin Skutella      Technische Universität Berlin, Germany
Roberto Solis-Oba      University of Western Ontario, Canada
Klaus Jansen      Christian-Albrechts-Universität zu Kiel, Germany

## Program Commitee

Hyung-Chan An      Yonsei University, South Korea
Amey Bhangale      University of California, Riverside, USA
Umang Bhaskar      Tata Institute of Fundamental Research, India
Parinya Chalermsook      Aalto University, Finland
Karthekeyan Chandrasekaran      University of Illinois Urbana-Champaign, USA
Karthik CS      Rutgers University, USA
Syamantak Das      Indraprastha Institute of Information Technology, India
Franziska Eberle      London School of Economics and Political Science, UK
Jittat Fakcharoenphol      Kasetsart University, Thailand
Zachary Friggstad      University of Alberta, Canada
Takuro Fukunaga      Chuo University, Japan
Waldo Gálvez      Universidad de O'Higgins, Chile
Chien-Chung Huang      École Normale Supérieure, France
Łukasz Jeż      University of Wrocław, Poland
Bundit Laekhanukit      Shanghai University of Finance and Economics, China
Pasin Manurangsi      Google Research, USA

| Eunjin Oh | Pohang University of Science and Technology, South Korea |
| Hanna Sumita | Tokyo Institute of Technology, Japan |
| Zhihao Tang | Shanghai University of Finance and Economics |
| Seeun William Umboh | The University of Sydney, Australia |
| Yuhao Zhang | Shanghai Jiao Tong University, China |

## Additional Referees

Kelin Luo
Mohsen Rezapour
Kamyar Khodamoradi
Ramin Mousavi
Zhihao Jiang

## Local Organization

| Tobias Friedrich | Hasso Plattner Institute, University of Potsdam, Germany |
| Simon Krogmann | Hasso Plattner Institute, University of Potsdam, Germany |
| Timo Kötzing | Hasso Plattner Institute, University of Potsdam, Germany |
| Gregor Lagodzinski | Hasso Plattner Institute, University of Potsdam, Germany |
| Pascal Lenzner | Hasso Plattner Institute, University of Potsdam, Germany |

# A PTAS for Unsplittable Flow on a Path (Invited Talk)

Tobias Mömke

University of Ausburg

**Abstract.** In the Unsplittable Flow on a Path problem (UFP) we are given a path with edge capacities, and a set of tasks where each task is characterized by a subpath, a demand, and a weight. The goal is to select a subset of tasks of maximum total weight such that the total demand of the selected tasks using each edge $e$ is at most the capacity of $e$. The problem admits a QPTAS [Bansal, Chakrabarti, Epstein, Schieber, STOC'06; Batra, Garg, Kumar, M ömke, Wiese, SODA'15]. After a long sequence of improvements [Bansal, Friggstad, Khandekar, Salavatipour, SODA'09; Bonsma, Schulz, Wiese, FOCS'11; Anagnostopoulos, Grandoni, Leonardi, Wiese, SODA'14; Grandoni, Mömke, Wiese, Zhou, STOC'18], the best known polynomial time approximation algorithm for UFP has an approximation ratio of $1+1/(e+1)+\epsilon < 1.269$ [Grandoni, Mömke, Wiese, SODA'22]. It has been an open question whether this problem admits a PTAS. We solve this open question and present a polynomial time $(1 + \epsilon)$-approximation algorithm for UFP.

# Contents

# Locating Service and Charging Stations

Rajni Dabas[1], Naveen Garg[2]($\boxtimes$), Neelima Gupta[1], and Dilpreet Kaur[2]

[1] Department of Computer Science, University of Delhi, New Delhi, India
{rajni,ngupta}@cs.du.ac.in
[2] Indian Institute of Technology Delhi, New Delhi, India
{naveen,dilpreet}@cse.iitd.ac.in

**Abstract.** In this paper, we consider the problem of locating service and charging stations to serve commuters. In the service station location problem we are given the paths followed by $m$ clients and wish to locate $k$ service stations, from a set of feasible locations, such that the maximum detour that a client has to take is minimized. We give a solution that has a maximum detour $3\mathtt{OPT} + L$ where $L$ is the length of the longest client-path.

Electric vehicles have a limited range and charging stations need to be located so that a client can drive from the source to destination without running out of charge. We consider two variants of the problem. In the first, we are given only the source and destination of each client and have to locate facilities at some subset of locations such that every client has a feasible path. In the second variant, we are also given the path that a client wishes to take and have to locate the facilities such that each client can follow its path without any detours. For both problems, our objective is to minimize the number of charging stations.

For all three problems, when the underlying graph is a tree and the facility can be located at any vertex on the tree, we show that the problem can be solved in polynomial time. On general graphs, the problem with source-destination pairs is at least as hard as the node-weighted group-Steiner tree but if we allow vehicles in our solution to have a range of 4 times that of the optimum then we obtain an 8-approximation. For the problem with specified paths, we show that finding an $o(\log m)$ approximation is hard even when vehicles of our solution have a range that is a constant times larger than that of the optimum solution.

**Keywords:** Facility location · Approximation algorithms

## 1 Introduction

In this paper, we consider facility location problems where clients correspond to paths in a network. This, for instance, is the case when we are given paths along which commuters in a city drive daily and wish to locate gas stations so

Rajni Dabas was supported by a UGC-JRF and Naveen Garg was supported by the Janaki and KA Iyer Chair.

P. Chalermsook and B. Laekhanukit (Eds.): WAOA 2022, LNCS 13538, pp. 1–19, 2022.
https://doi.org/10.1007/978-3-031-18367-6_1

that each user can fill gas without taking a significant detour from their daily commute. We assume that the number of facilities (gas/service stations) is fixed, say $k$. If for some client no gas station is located on her commute path then the client has to take a detour to the closest gas station from her path.

We first observe that the distance of a path from a facility does not define a metric and this makes facility location problems where clients are paths, very different from traditional facility location problems where clients are considered as vertices in a graph/points in a metric space. Facility location problems have been considered under different objective functions; in this paper, we will be extending the $k$-center and $k$-supplier problems to the setting of paths. Recall that in the $k$-center problem our objective is to locate $k$ facilities so as to minimize the maximum distance of a client from its nearest facility. The $k$-supplier problem is identical to the $k$-center problem with the only difference that the centres can be chosen only from a given set of possible locations. Unless P=NP, the best approximation ratio known and possible for $k$-center and the $k$-supplier problems are 2, 3 respectively [11]. For the $k$-center problem with paths as clients, we obtain a polynomial-time algorithm on trees and show that the problem is NP-hard if the facilities can be opened only on a subset of the vertices. On general graphs, we show how to open $k$ facilities from a given set of locations such that every path is within distance $3\mathtt{OPT} + L$ of an open facility where $L$ is the length of the longest path and $\mathtt{OPT}$ is the cost of the optimum solution.

We also consider the setting in which facilities are charging stations for Electric Vehicles (EV). An EV has a certain fixed range, say $B$, on a full charge and we should locate charging stations in a manner such that every vehicle can travel from its source to destination without running out of charge. We assume charging stations can only be located on a subset of vertices[1], $F$, and vehicles are allowed to use all of their charge to reach their destination. We consider two variants of the problem. In the first variant, we do not specify the path that a vehicle is required to take but only its source and destination. In the second variant, a path between the source and destination is specified and the vehicle is not allowed to deviate from it. For both these variants, our objective is to minimize the number of charging stations needed.

For both variants of the problem on trees, we assume that charging stations can be located at any vertex and give exact polynomial-time algorithms. However, on general graphs (and assuming charging stations can only be on a subset of vertices) these two variants behave rather differently in their approximability. When the route of each vehicle is fixed, then a simple approximation preserving reduction from the set cover problem establishes that one cannot approximate the problem better than $c \ln m$, where $c$ is a constant and $m$ is the number of clients. Allowing the vehicles of our solution to have a range $\alpha$ times that of the vehicles in the optimum solution does not provide any advantage and the same reduction can be extended to show that an approximation better than $c\alpha^{-1} \ln m$ is not possible unless P=NP.

---

[1] In a setting where the source of a client is her home, while it is reasonable to assume that the client charges her own vehicle, it would be unreasonable to put a charging station for other vehicles at her home.

When we are only specified the source-destination for each vehicle and are free to choose a route the problem is equivalent to finding a minimum node-weighted group Steiner tree and cannot be approximated better than $\Omega(\log^2 m)$, where $m$ is the number of clients. However, we can in polynomial time, find a solution which opens at most $O(\log m)\mathsf{OPT}$ charging stations where $\mathsf{OPT}$ is the minimum number of charging stations to service vehicles which have range half of the range of the vehicles in our solution. Further relaxation of the range of vehicles in our solution to 4 times the range of vehicles in the optimum solution leads to an 8-approximation; this result uses an interesting connection to the $k$-supplier problem [11].

Although our exact polynomial time algorithm on trees are primarily greedy algorithms, the correctness arguments require careful and delicate analysis. Our approximation guarantee for the service station location problem on general graphs is best possible. It is hard to obtain constant approximation algorithms for the two variants of charging station location considered in this paper, but we show that this is possible with some resource augmentation (assuming vehicles have larger range) for one of the variants.

## 2    Preliminaries and Related Work

Let $G = (V, E)$ be an undirected graph, $l : E \to \mathbb{R}^+$ a length function on the edges and $Z \subseteq V$ a set of *terminals*. Let $|V| = n$ and $C \subseteq Z \times Z$ be the source-destination pairs of a set of $m$ clients. Let $s_j \in Z$ be the source and $t_j \in Z$ the destination of client $j \in [m]$. For some of the problems considered, we also specify, for every client $j$, a path $P_j$ between $s_j$ and $t_j$ in $G$. We call such a path a *client-path* and let $\mathcal{P} = \{P_j, j \in [m]\}$. Let $F \subseteq V$ denote the set of vertices on which we can open the facilities - the charging and service stations.

The distance of a client-path $P_j$ from a facility at vertex $v_i$ is the length of the shortest path from $v_i$ to a vertex on $P_j$; we denote this quantity by $d(i, j)$. It is important to note that $d$ is not a metric. In particular, it could be the case that for client-paths $P_{j_1}, P_{j_2}$ and facility locations $v_{i_1}, v_{i_2}$ we have that $d(i_1, j_1) > d(i_2, j_1) + d(i_2, j_2) + d(i_1, j_2)$.

For the setting of electric vehicles, we assume that every vehicle starts with a full charge from its source and can arrive at its destination fully discharged. The *range* of an EV is the maximum distance it can travel on a full charge. For client $j$, a path $P$ from $s_j$ to $t_j$ is *feasible* with respect to a certain set of charging stations if the client can follow $P$ without running out of charge.

We now define the various problems considered in this paper.

**Service Station Location.** Given graph $G$, edge lengths $l$, client-paths $\mathcal{P}$, and an integer $k$, we wish to locate $k$ service stations on vertices in $F$ such that the maximum distance of a client-path from the nearest service station is minimized. We denote this problem by $\mathsf{SvSP}(G, l, \mathcal{P}, k)$.

**Charging Station Location for source-destination pairs.** Given graph $G$, edge lengths $l$, source-destination pairs for clients $C$, and range $B$, we wish to locate the smallest number of charging stations on vertices in $F$ such

that there is a feasible path for every client. We denote this problem by
ChSD$(G, l, C, B)$.

**Charging Station Location with specified paths.** Given a graph $G$, edge
lengths $l$, client-paths $\mathcal{P}$, and range $B$, we wish to locate the smallest number of charging stations on vertices in $F$ such that every client-path in $\mathcal{P}$ is
feasible. We denote this problem by ChSP$(G, l, \mathcal{P}, B)$.

**Related Work:** Facility location problems in which there is a path associated
with each client and the objective is to locate $k$ facilities so as to cover the
maximum number of paths (with bounded detours and without detours) have
been extensively studied [1,2,12,17,18]. Hodgson [12] and Berman et al. [2] were
the first ones to study the facility location problem in which clients are paths.
They consider the problem in which the client paths are specified and clients
are not allowed to take detours; they identify optimum locations for $k$ facilities
so as to cover the maximum number of clients (a client is said to be covered
if at least one facility is opened on its path). Mitra et al. [17] considered the
same problem in the setting of cellular networks. Berman et al. [1] and Mitra
et al. [18] generalised the problem to allow the clients to take small (bounded)
detours. The problems considered in these papers are different from ours as
they aim to maximise the number of clients covered whereas our objective is
to minimise either the detour that a client may have to take (SvSP()) with $k$
facilities or the number of facilities to be opened (ChSD(), ChSP()) with limited
range of electric vehicles.

Kuby and Lim [15] and Kuby et al. [16] consider the problem to capture
limited range of vehicles. A client is now said to be covered if sufficient charging
stations are opened on its path (i.e. no detours) so that it can follow the path
without running out of charge. Several heuristics have been provided in these
papers without any approximation guarantee.

Storandt and Funke [20] consider a problem similar to ChSD() and give an
$O(\log n)$ approximation via an algorithm for strongly connected dominating sets
(SCDS). Although their results also apply to directed graphs, they crucially
assume that charging stations can be located at any vertex of the graph. We
believe that restricting the location of charging stations to a subset of facilities
makes the problem much harder (see for example Claim 3.2). Funke et al. [6,7]
consider a problem similar to ChSP() in which the charging stations should be
located such that the *shortest path* (with and without detours) between every
pair of vertices in the graph is feasible. They observe that the problem of minimizing the number of charging stations needed can be formulated as a hitting
set problem and obtain an $O(\log n)$ approximation.

## 3   Locating Service Stations

In this section, we consider how to locate $k$ service stations so as to minimize
the maximum distance of any client path from the nearest station. When the
client-path is a single vertex this is the $k$-center problem which is known to be
NP-hard and for which the best possible approximation is 2.

## 3.1   Linear Program and the Integrality Gap

For a given instance $\text{SvSP}(G, l, \mathcal{P}, k)$, let $D$ be a guess on the optimum value. Let $y_i$ be an indicator variable which is 1 iff a facility is opened at vertex $v_i \in F$. Thus,

$$\sum_{i:v_i \in F} y_i = k$$

For a client $j$ and vertex $v_i \in F$ such that $d(i, j) \le D$, let $x_{ij} = 1$ iff $j$ is served by facility opened at $v_i$. Clearly,

$$i : v_i \in F, j \in [m] \qquad x_{ij} \le y_i$$
$$j \in [m] \sum_{i:v_i \in F} x_{ij} \ge 1$$

If there is a integer feasible solution, $y_i \in \{0, 1\}$, to these set of constraints, then the given $\text{SvSP}(G, l, \mathcal{P}, k)$ instance has a solution of value $D$.

The integrality constraints on the variables can be relaxed to obtain a linear program. However, this LP has an unbounded integrality gap. Let $G$ be a complete graph on $n$ vertices each edge of which has length 1. We associate a client with every pair of vertices $v_i, v_j \in V$ and let the edge $(v_i, v_j)$ be the path corresponding to this client. From our choice of client paths, it follows that the optimum solution to this instance is 0 iff there is a vertex cover of size $k$.

Let $k = n/2$ and $F = V$. Since any choice of $n/2$ vertices does not cover all edges of the complete graph, the optimum solution has value 1. However, the LP is feasible for $D = 0$, since we can assign $y_i = 1/2, i \in [n]$.

## 3.2   The Graph is a Tree

For this section we assume that $G$ is a tree, $T$. Then the path $P_j$ for client $j$ is the unique path in $T$ between $s_j, t_j$.

*Claim.* If we are only allowed to open service stations at a subset of vertices ($F \subset V$) then the problem $\text{SvSP}()$ is NP-hard even on a tree.

*Proof.* Let $H = (V', E')$ be an undirected graph and consider the problem of determining if $H$ has a vertex cover of size $k$. Let $T$ be a one-level tree with root $r$ and let the leaves of $T$ correspond to vertices in $V'$; let $\phi(v)$ be the leaf corresponding to vertex $v \in V'$. For each edge $e = (u, v) \in E'$ we add a client whose source/destination are the leaves $\phi(u), \phi(v)$. Let $k$ be the number of facilities we wish to open and let all vertices in $T$ other than $r$ be in $F$.

If $H$ has a vertex cover $S$ of size $k$, then opening service stations at the leaves $\phi(v), v \in S$, yields a solution in which there is a station on every client-path and hence $\text{OPT} = 0$. Conversely, a solution in which no client needs a detour corresponds to a vertex cover in $H$. Thus a polynomial time algorithm for $\text{SvSP}(T, l, \mathcal{P}, k)$ implies a polynomial time algorithm for vertex cover.

Hence for the rest of this section, we assume that $F = V$. Let $D$ be our guess on the optimum value. We root the tree $T$ at an arbitrary vertex, say $r$. For client $j$, let $r_j$ be the least common ancestor (LCA) of $s_j, t_j$ in $T$. Let $j^*$ be such that $r_{j^*}$ is furthest from the root; add $j^*$ to $S$. Let $f_{j^*}$ be the ancestor of $r_{j^*}$ closest to the root but within distance $D$ of $r_{j^*}$. We open a facility at $f_{j^*}$ thus covering all clients which are within distance $D$ of $f_{j^*}$. The clients whose paths are covered are removed from $C$ and the process is repeated until all clients are covered. Note that the number of facilities opened by the algorithm is $|S|$.

---

**Algorithm 1:** Locating service stations on trees

---

    **Input**   : $(T, l, \mathcal{P}, k, D)$
    **Output:** Set $F'$ of open service stations
1  $S \leftarrow \phi$, $F' \leftarrow \phi$, $\mathcal{P}' \leftarrow \mathcal{P}$
2  Let $r_j = LCA(s_j, t_j) \; \forall j \in \mathcal{P}$
3  **while** $\mathcal{P}'$ *not empty* **do**
4      |  Let $j^* \in \mathcal{P}'$ be such that $r_{j^*}$ is furthest from root
5      |  Let $f_{j^*} \in F$ be the ancestor of $r_{j^*}$ closest to $r$ within distance $D$ of $r_{j^*}$
6      |  $S \leftarrow S \cup \{j^*\}, F' \leftarrow F' \cup \{f_{j^*}\}, \mathcal{P}' \leftarrow \mathcal{P}' \setminus \{\text{Clients covered by } f_{j^*}\}$
7  **end**
8  **return** $F'$

---

**Lemma 1.** *If $D = \mathsf{OPT}$ then the Algorithm 1 opens at most $k$ facilities.*

*Proof.* Let $j_1, j_2 \in S$ and suppose $j_1$ was added to $S$ before $j_2$. We show that there is no vertex $v_i \in F$ such that $d(i, j_1) \leq D$ and $d(i, j_2) \leq D$. For contradiction, assume that there exist some such vertex $v_i$.

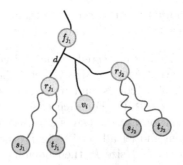

**Fig. 1.** Proof of Lemma 1.

First observe that $v_i$ is a descendant of $f_{j_1}$ since otherwise $d(i, j_1) > D$ (see Fig. 1). We next argue that $r_{j_2}$ is a descendant of $f_{j_1}$. If this is not the case then the distance of path $P_{j_2}$ from $v_i$ is at least as large as its distance from $f_{j_1}$. This implies that distance of $P_{j_2}$ from $f_{j_1}$ is at most $D$ which in turn implies that client $j_2$ is removed from the instance when $j_1$ is added to $S$.

Thus both $r_{j_1}, r_{j_2}$ are descendants of $f_{j_1}$. Since $j_1$ was added before $j_2$ to $S$ the distance of $r_{j_2}$ from $f_{j_1}$ is at most the distance of $r_{j_1}$ from $f_{j_1}$ which is at most $D$. This implies that $j_2$ would have been removed when $j_1$ was added to $S$ yielding a contradiction.

Thus no vertex in $T$ can cover more than one client of $S$. Hence any solution needs at least $|S|$ facilities to cover all clients within distance $D$. If the optimum solution opens $k$ facilities and has value $D$ then $|S| \leq k$ which implies that the number of facilities opened by our algorithm is at most $k$.

**Theorem 1.** *The problem SvSP($T, l, \mathcal{P}, k$) can be solved in polynomial time when $T$ is a tree and $F = V$.*

*Proof.* Since OPT is the distance between two vertices in $T$, we can run Algorithm 1 for $O(n^2)$ choices of $D$. The optimum is the smallest value of $D$ for which the algorithm opens at most $k$ service stations.

## 3.3   General Graphs

We next consider the case when $G$ is an arbitrary graph. Recall that we wish to open $k$ facilities at vertices in $F \subseteq V$, so as to minimize the maximum distance of any client-path $P_j, j \in [m]$ from its nearest open facility.

**Theorem 2.** *There is a polynomial-time algorithm that computes a solution to the problem SvSP($G, l, \mathcal{P}, k$) of cost at most $3\mathrm{OPT} + L$ where $L$ is the length of the longest path in $\mathcal{P}$ and OPT, is the cost of the optimum solution.*

*Proof.* Let $l(P_j)$ be the total length of edges on the path $P_j$; then $L = \max_{P_j \in \mathcal{P}} l(P_j)$. We follow the 3-approximation algorithm for $k$-supplier [11] and first guess the optimum value, say $D$. Define the distance between two paths $P_i, P_j$ as the minimum distance between two vertices one of which is on $P_i$ and the other on $P_j$.

Pick an arbitrary client-path, say $P_j$, and open a facility at a vertex $v_j \in F$ which is closest to $P_j$. Add $j$ to $S$ and remove all client-paths which are within distance $2D$ of $P_j$. Continue this process until no client paths remain.

**Lemma 2.** *If $D = \mathrm{OPT}$ then the above algorithm picks at most $k$ facilities.*

*Proof.* Note that the algorithm opens $|S|$ facilities. Since the distance between $P_i, P_j, i, j \in S$ is more than $2D$, there is no vertex $v \in F$ which is within distance $D$ from both $P_i$ and $P_j$. Thus any solution which covers the paths $P_i, i \in S$ within distance $D$ needs to open at least $|S|$ facilities. If the optimum solution, which opens $k$ facilities, has value $D$ then $|S| \leq k$.

We now argue that every path is within distance $3\mathrm{OPT} + L$ of an open facility. Suppose $P_i$ was removed in the step when path $P_j$ was picked and a facility opened at $v_j$. Since distance between $P_i, P_j$ is at most $2\mathrm{OPT}$, distance between

$P_j$ and $v_j$ is at most OPT and the length of $P_j$ is at most $L$, by triangle inequality it follows that the distance between $P_i$ and $v_j$ is at most $3\text{OPT} + L$.

We next show that our result is best possible unless P=NP.

**Theorem 3.** *Any polynomial time algorithm for SvSP$(G, l, \mathcal{P}, k)$ which returns a solution of value $\alpha\text{OPT} + \beta$ would have $\alpha \geq 3$ and $\beta \geq L$ unless P=NP.*

*Proof.* Let $H = (V, E)$ be an undirected graph. We construct an instance of SvSP(), by choosing $G = H$ and assigning every edge in $E$ a length $L$. Every pair of adjacent vertices $u, v$, in $G$ is the source-destination for a client, and the path corresponding to this client is the edge $(u, v)$. Let $k$ be the size of the minimum vertex cover in $H$.

The vertices in $G$ corresponding to a vertex cover in $H$ cover all client-paths within distance 0 and a solution which does not correspond to a vertex cover in $H$ has value at least $L$. Thus an algorithm which returns a solution of value at most $\alpha\text{OPT} + \beta$, $\beta < L$ must return a minimum vertex cover in $H$.

To argue that $\alpha \geq 3$, we consider an instance of $k$-supplier problem which is specified by a metric $d : V \times V \to \mathbb{R}^+$ and the set of facility locations, $F \subseteq V$. Recall that the $k$-supplier problem is to open $k$ facilities on vertices in $F$ such that the maximum distance of any client (a point in $C = V \setminus F$) from the nearest open facility is minimised. [11] show that a $3 - \epsilon$ approximation for this problem is NP-hard.

We construct an instance of SvSP() as follows.

1. For every $v \in C$ we include vertices $v^1, v^2$ in $G$. Let $V^1 = \cup_{u \in C} u^1$ and $V^2 = \cup_{u \in C} u^2$. We also include a vertex in $G$ for each $v \in F$ and let $F$ be this set of vertices.
2. For pairs $u, v^1$ and $u, v^2$, $u \in F$ we include edges $(u, v_1), (u, v_2)$ in $G$ of length $d(u, v)$.
3. For every pair $v_1, v_2$ we include an edge $v_1, v_2$ in $G$ of length 0. These vertices are also the source-destination for a client and the corresponding client-path is the edge $(v_1, v_2)$. Thus, $L = 0$.
4. $F$ is the set of vertices in $G$ where we can open at most $k$ service stations.

There is a one-to-one correspondence between solutions for the $k$-supplier problem and solutions to the SvSP() instance constructed above and the corresponding solutions have the same value. Consider a polynomial time algorithm which gives a solution of value at most $\alpha\text{OPT} + \beta$, $\alpha = 3 - \epsilon$, $\epsilon > 0$ for the SvSP() instance constructed. Since OPT can be scaled by scaling the metric $d$, the additive term $\beta$ in the approximation guarantee can be made negligible compared to the term $\alpha\text{OPT}$. Thus this algorithm yields a $3 - \epsilon$ approximation to the $k$-supplier problem which is not possible unless P=NP

## 4   Locating Charging Stations for Given Source-Destination Pairs

We next consider the setting of electric vehicles (EVs) which can only travel a certain distance, say $B$, on a full charge. In this section, we consider the variant where we are given only the source and destination for each client ($\text{ChSD}(G, l, C, B)$). We wish to find the smallest number of charging stations such that each client has a feasible path.

## 4.1 The Graph is a Tree

We first consider the case when $G$ is a tree $T = (V, E)$. If charging stations can only be located on a subset of vertices, the proof of Claim 3.2 can be modified as follows to show that the problem is NP-hard. We subdivide every edge of the tree considered in Claim 3.2 into two edges; the edge incident to the root has length 0 while the edge incident to the leaf has length $B$. It is easy to see that a vertex cover in $H$ corresponds to a solution to the $\text{ChSD}()$ problem in this instance.

We therefore assume in this section that $F = V$; thus charging stations can be located at any vertex of the tree. We build on the Algorithm 1 to obtain a polynomial-time algorithm for minimizing the number of charging stations for this setting.

**Structure of Feasible Paths.** For a client $j$, let $r_j$ be the least common ancestor of $s_j, t_j$ and let $T_j$ be the subtree rooted at $r_j$. Let $o_1, o_2, \ldots o_k$ be the facility locations in an optimum solution, $O$, in order of decreasing distance from the root.

*Claim.* There is a feasible path from $s_j$ to $t_j$ with respect to facilities in $O$, that uses at most one facility which is not in $T_j$.

*Proof.* Let $P$ be a minimal feasible path from $s_j$ to $t_j$ with respect to facilities in $O$ and let $o_a, o_b, a \neq b$ be the first and last facilities on $P$ that are not in $T_j$. Let $o_c, o_d$ be facilities on $P$ that precede and succeed $o_a$ and $o_b$. Then $l(o_c, o_a) \leq B$ and $l(o_b, o_d) \leq B$. Suppose $l(o_a, r_j) \leq l(o_b, r_j)$. Since $l(o_b, o_d) = l(o_b, r_j) + l(r_j, o_d)$, it follows that $l(o_a, o_d) = l(o_a, r_j) + l(r_j, o_d) \leq l(o_b, o_d) \leq B$ which implies that path $P$ needs only visit vertex $o_a$ outside $T_j$.

If $l(o_a, r_j) > l(o_c, r_j)$ then $l(o_a, r_j) + l(r_j, o_d) > l(o_c, r_j) + l(r_j, o_d) > B$ which is a contradiction. Thus $l(o_a, r_j) \leq l(o_c, r_j)$ and similarly $l(o_a, r_j) \leq l(o_d, r_j)$. This implies that $o_a$ is closer to the root than both $o_c, o_d$.

Let $P$ be a minimal feasible path and $o_{i_1}, o_{i_2}, \ldots, o_{i_a}$ be the facilities in order on the part of $P$ from $s_j$ to $r_j$.

*Claim.* $i_1 < i_2 < \cdots < i_a$

*Proof.* We prove this by contradiction. Let $b$ be the smallest index such that $i_b > i_{b+1}$. Then $l(r, o_{i_{b-1}}) > l(r, o_{i_b})$ and $l(r, o_{i_{b+1}}) > l(r, o_{i_b})$. Let $u_1$ be the least common ancestor of $o_{i_{b-1}}, o_{i_b}$ and $u_2$ the least common ancestor of $o_{i_{b+1}}, o_{i_b}$. Note that $l(o_{i_{b-1}}, o_{i_b}) \leq B$ and $l(o_{i_{b+1}}, o_{i_b}) \leq B$. We consider 3 cases:

1. $u_2$ is on the path from $u_1$ to $o_{i_{b-1}}$: Then $u_2$ is identical to $u_1$ and we view this as if $u_2$ is on the path from $u_1$ to $r$.

2. $u_2$ is on the path from $u_1$ to $o_{i_b}$: Then $l(u_1, o_{i_b}) > l(u_1, o_{i_{b+1}})$ which implies that $l(o_{i_{b-1}}, o_{i_{b+1}}) \leq B$.
3. $u_2$ is on the path from $u_1$ to $r$: Then $l(u_2, o_{i_b}) > l(u_1, o_{i_{b-1}})$ which implies that $l(o_{i_{b-1}}, o_{i_{b+1}}) \leq B$.

Since in all cases we have argued that $l(o_{i_{b-1}}, o_{i_{b+1}}) \leq B$, path $P$ can skip vertex $o_{i_b}$ while remaining feasible. Hence it is not minimal yielding a contradiction.

One can similarly argue that as we traverse $P$ from $t_j$ to $r_j$ we encounter facilities in decreasing order of their distance from the root. Thus, the distance from the root of the facilities encountered in any minimal feasible path, $P$, from $s_j$ to $t_j$ is a monotonically decreasing sequence followed by a monotonically increasing sequence. The facility on $P$ which is closest to the root is the only facility that may not belong to $T_j$.

**The Algorithm.** The structure of feasible paths suggests the following greedy algorithm for determining the minimum number of facilities we need to open. Once again, we root $T$ at an arbitrary vertex $r$. As before we would be removing clients as the algorithm progresses. Let $C'$ be the set of clients which have not yet been removed and let $Z' = \cup_{j \in C'} \{s_j, t_j\}$ be the corresponding set of terminal vertices. Note that initially $C' = C$ and $Z' = Z$. For each vertex, $v_i \in Z'$, let $a_i$ be the ancestor closest to the root and at most distance $B$ away from $v_i$. Let $X = \{a_i : v_i \in Z'\}$ and $x \in X$ be the vertex furthest from the root. Open a facility at $x$.

We now move sources/destinations of some clients to $x$. If both $s_j, t_j$ are within distance $B$ of $x$ for some client $j \in C'$ then we remove $j$ from $C'$. If $x$ is in $T_j$ and is within distance $B$ of $s_j$ (resp. $t_j$) then we move $s_j$ (resp. $t_j$) to $x$. The algorithm continues until $C'$ is empty. Algorithm 2 gives the algorithm as a pseudo-code.

**The Analysis.** Note that the source/destination of clients get closer to the root as the algorithm proceeds. If client $j$ is alive (has not been removed) at some step of the algorithm then $s_j, t_j$ denotes its location at that step. By our procedure for moving clients it follows that $s_j, t_j$ will always remain in the subtree $T_j$. Let $\hat{s}_j, \hat{t}_j$ denote the source and destination of client $j$ at the start of the algorithm. We begin with an observation about Algorithm 2.

**Lemma 3.** *When a facility is opened at vertex $x$, clients in $C'$ which have*

1. *either their source or destination (but not both) in the subtree rooted at $x$ will have that endpoint moved to $x$.*
2. *both their source and destination in the subtree rooted at $x$ will be removed from $C'$.*

*Proof.* Consider a client $j$ such that $s_j$ is in the subtree rooted at $x$ and $t_j$ is not in the subtree rooted at $x$. This implies $x \in T_j$. Since $x$ is an ancestor of $s_j$ it is closer to the root than $s_j$. $x$ must be within distance $B$ of $s_j$ or else we

---

**Algorithm 2:** Locating Charging Stations for source-destination pairs on trees

---

**Input** : $(T, l, C, B)$
**Output:** Set $F'$ of opened charging stations

1  $F' \leftarrow \phi$ , $C' \leftarrow C$, $Z' \leftarrow Z$
2  **while** $C'$ *not empty* **do**
3  $\quad Z' \leftarrow \cup_{j \in C'} \{s_j, t_j\}$
4  $\quad \forall v_i \in Z'$, let $a_i$ be the ancestor closest to $r$ within distance $B$ of $v_i$
5  $\quad X \leftarrow \{a_i : v_i \in Z'\}$. Let $x \in X$ be furthest from $r$
6  $\quad F' \leftarrow F' \cup \{x\}$
7  $\quad$ **for** $j \in C'$ **do**
8  $\quad\quad$ Let $r_j$ be the LCA of $s_j, t_j$
9  $\quad\quad$ **if**$(l(s_j, x) \leq B$ and $l(x, t_j) \leq B)$ **then** $C' \leftarrow C' \setminus \{j\}$; continue
10  $\quad\quad$ **if**$(l(s_j, x) \leq B$ and $x$ is a descendent of $r_j)$ **then** move $s_j$ to $x$
11  $\quad\quad$ **if**$(l(t_j, x) \leq B$ and $x$ is a descendent of $r_j)$ **then** move $t_j$ to $x$
12  $\quad$ **end**
13  **end**
14  **return** $F'$

---

would have picked a different facility location. Hence we would move $s_j$ to $x$ at this step. A similar argument can be made when $t_j$ is in the subtree rooted at $X$ and $s_j$ is not.

Now consider a client $j$ such that $s_j, t_j$ are in the subtree rooted at $x$. Again $x$ is within distance $B$ of both $s_j, t_j$ and hence we would remove $j$ from $C'$.

**Theorem 4.** *The problem ChSD$(T, l, C, B)$ can be solved in polynomial time when $T$ is a tree and $F = V$.*

*Proof.* Suppose our algorithm opens $p$ facilities and at step $i$, it opens a facility at location $x_i$. The algorithm only moves terminals closer to the root and this allows us to claim that $l(r, x_i) \geq l(r, x_{i+1})$. Let $i$ be the first index where the 2 sequences $x_1, x_2, \ldots x_p$ and $o_1, o_2, \ldots o_k$ differ. Let $T'$ be the subtree rooted at $x_i$ and let $Q = \{y_1, y_2, \ldots y_q\}$ be the facility locations (other than $x_i$) of our algorithm in $T'$. Since $Q \subseteq \{x_1, x_2 \ldots x_{i-1}\}$ the optimum solution opens facilities at exactly these locations in $T'$.

At step $i$ of our algorithm there was a client $j$ such that $x_i$ was the vertex closest to the root and within distance $B$ of $s_j$ (or $t_j$, we assume $s_j$). We now argue that if the optimum solution does not open a facility at another location (besides the set $Q$) in $T'$ then it has no feasible path for client $j$.

We first argue that $\hat{s}_j$ is as close to $x_i$ as it can possibly be at step $i$.

**Lemma 4.** *Let $y_s \in Q$ be the facility in $T'$, closest to $x_i$, to which there is a feasible path from $\hat{s}_j$ using facilities of $Q$. At step $i$, $s_j$ is at facility $y_s$.*

*Proof.* Let $P$ be the sequence of facilities (from $Q$) on a feasible path from $\hat{s}_j$ to $y_s$ and $P'$ be the sequence of facilities (from $Q$) on the path followed by $s_j$

in our algorithm. We claim that $P$ is a subsequence of $P'$. For contradiction assume this is not the case and let $y_a$ be the first location on $P$ that is not on $P'$. Let $y_b$ be the location preceding $y_a$ on $P$; note that $y_b$ is also on $P'$. Let $z_1, z_2, \ldots$ be the facilities on $P'$ in $Q$ following $y_b$. The facts that $l(y_a, y_b) \leq B$ and $l(y_b, z_1) \leq B$ and that $z_1, y_a$ is closer to the root than $y_b$ let us conclude that $l(z_1, y_a) \leq B$. This in turn lets us conclude that $l(z_2, y_a) \leq B$ and we can continue with locations on $P'$ arguing that they are within distance $B$ of $y_a$. Hence when $y_a$ was opened then $s_j$ would have been moved to that location by our algorithm which implies $y_a$ is on $P'$.

Since $P$ is a subsequence of $P'$, $s_j$ can only be closer to $r_j$ than $y_s$. However since $P'$ is also a feasible path from $s_j$ that only uses facilities in $Q$, $P'$ should also end at $y_s$.

We are now ready to complete the proof by considering the two cases when $r_j$ is in $T'$ and when it is not.

We first consider the case when $r_j$ is in $T'$. This implies that $s_j, t_j$ are in $T'$ and suppose they are at locations $y_s, y_t \in Q$ at step $i$.

1. If $l(y_s, y_t) \leq B$ we would have dropped client $j$ and it would not be in $C'$ at step $i$. Hence $l(y_s, y_t) > B$.
2. By our algorithm it follows that $y_s, y_t$ are in $T_j$ at step $i$.
3. By Lemma 4, there is no feasible path that using only facilities of $Q$ brings $s_j$ (resp $t_j$) any closer to $x_i$ - and hence to $r_j$ - than $y_s$ (resp. $y_t$). Hence any feasible path for client $j$ must use a facility, say $z$, not in $T_j$.
4. $z$ has to be within distance $B$ of both $y_s, y_t$ and should be closer to the root than either of $y_s, y_t$. From our definition of $y_s, y_t$ it follows that such a facility does not belong to $Q$.
5. $z$ therefore has to lie outside $T'$. From our choice of $x_i$ it follows that vertices not in $T'$ are at distance greater than $B$ from $s_j$ ($=y_s$).

We thus arrive at a contradiction.

Now consider the case when $r_j \notin T'$.

1. By Lemma 4, there is no feasible path that using only facilities of $Q$ brings $s_j$ any closer to $x_i$ than $y_s$.
2. The facility, say $z$ following $y_s$ on a feasible path should be within distance $B$ of $y_s$ and closer to the root than $y_s$. From our definition of $y_s$, it follows that such a facility does not belong to $Q$.
3. $z$ therefore has to lie outside $T'$. From our choice of $x_i$, it follows that vertices not in $T'$ are at a distance greater than $B$ from $s_j$ ($=y_s$).

We thus arrive at another contradiction.

Hence we conclude that the optimum solution opens at least one other location in $T'$ besides the locations in $Q$; let this be location $z$. We now claim that picking $x_i$ instead of $z$ does not make the optimum solution infeasible. From our proof of Lemma 3 it follows that this modified solution would have a feasible path for clients $j$ where $r_j \in T'$. For clients, $j$ for which only one of $s_j, t_j$ is in $T'$ (say $s_j$) the modified solution (by the proof of Lemma 3) would have a feasible path from $\hat{s}_j$ to $x_i$ which is only better than having a feasible path from $\hat{s}_j$ to a descendant of $x_i$ as provided by the original optimum solution.

In this manner, we can modify the optimum solution so that it corresponds exactly to the solution we picked. Hence the solution picked by our algorithm is optimum.

## 4.2   General Graphs

For arbitrary graphs, the problem of minimizing the number of charging stations is NP-hard even when $F = V$. It is no loss of generality to assume that the set of facility locations $F$ and the set of terminals $Z$ are disjoint. If such is not the case and $v \in F \cap Z$ we introduce a new vertex $v'$ and edge $(v, v')$ of length 0. Vertex $v$ is included in $F$ and $v'$ in $Z$. Construct a graph $H = (F \cup Z, E')$ where $(u, v) \in E'$ iff the length of the shortest path between $u, v \in F \cup Z$ (under length function $l$) is at most $B$ and both $u, v$ are not in $Z$. We associate a weight 1 with nodes in $F$ and 0 with nodes in $Z$.

A set of locations, $X \subseteq F$, is *feasible* if for every client $j$ there is a path from $s_j$ to $t_j$ in $H$ which only uses vertices of $X$. Finding a feasible set of minimum cardinality is exactly the same as finding a minimum weight *full* Steiner forest in the auxiliary graph $H$ constructed above. Recall that a Steiner tree/forest is *full* if all terminals are the leaves of the tree/forest. The requirement that the Steiner forest is full arises since the path between a source-destination pair, $s_j, t_j$, cannot use any terminals other than $s_j, t_j$.

The problem of finding a minimum weight full Steiner tree for edge-weighted graphs was shown to be equivalent [3] to finding a minimum weight group Steiner tree. The group Steiner tree cannot be approximated better than $\Omega(\log^2 k)$ [10] and an $O(\log^2 n \log k)$ [8] approximation is known; here $k$ is the number of groups and $n$ the number of vertices. The equivalence between full Steiner trees and group Steiner trees extends to the scenario when nodes have weights and also to Steiner forests. However, no non-trivial polynomial-time approximation algorithm is known for group Steiner tree/forest with node weights and the only result known is an online randomized algorithm that is polylog competitive [19] but runs in quasi-polynomial time.

**Theorem 5.** *For any fixed $\epsilon > 0$, the ChSD$(G, l, C, B)$ problem admits no efficient $\log^{2-\epsilon} |C|$ unless NP has quasi-polynomial LasVegas algorithms.*

To get around this difficulty we assume vehicles have a range of $2B$ and compare our solution to the optimum solution for vehicles with a range of $B$.

**Theorem 6.** *There is a polynomial time algorithm that computes a solution to the problem ChSD$(G, l, C, 2B)$ of size at most $O(\log m) \cdot$ OPT where OPT is the size of the optimum solution to the problem ChSD$(G, l, C, B)$ and $m = |C|$ is the number of clients.*

*Proof.* We begin by finding a minimum (node) weight Steiner forest, $F'$ which is an $O(\log |Z|)$ approximation to the optimum [14]. Note that there is a path between every source-destination pair in $F'$, although this path may contain a terminal node.

Consider a tree $T$ in $F'$ which has internal nodes which are terminals. Root $T$ at an arbitrary leaf. Let $t \in Z$ be an internal node of this tree with parent $u_0$ and children $u_1, u_2, \ldots u_p$. Since $H$ does not have an edge between two terminals none of the vertices in $\{u_0, u_1, \ldots, u_p\}$ are terminals. For $1 \le i \le p$ replace edge $(t, u_i)$ in $T$ with the edge $(u_0, u_i)$. We do this for all those internal nodes in $T$ which are terminals to obtain a full tree and then repeat the process for all trees in $F'$.

Note that the edges we introduce in the above procedure correspond to paths of length 2 in $H$. Hence these edges can be traversed by a vehicle with a range of $2B$ without running out of charge. This then yields the theorem.

We now consider a further relaxation and give a pseudo-approximation algorithm which opens at most 8OPT charging stations where OPT is the number of charging stations needed when the range of the vehicles is one-quarter the range of the vehicles used in our solution. Note that these analysis are in the nature of "resource augmentation" that is a common approach for analysing online scheduling algorithms [13].

**Theorem 7.** *There is a polynomial-time algorithm that computes a solution to the problem* $ChSD(G, l, C, 4B)$ *of size at most* 8OPT *where* OPT *is the size of the optimum solution to the problem* $ChSD(G, l, C, B)$.

*Proof.* Let $O \subseteq F$ be the vertices in an optimum solution. Note that every vertex of $Z$ is adjacent to a vertex of $O$ in $H$. Assume every edge of $H$ has length 1. Viewing the vertices of $Z$ as clients and those of $F$ as suppliers, we conclude that there exist $|O|$ suppliers who can service all clients within distance 1. Hochbaum and Shmoys [11] gave an algorithm to find $k$ suppliers who can service all clients within distance 3. We use their algorithm below to pick such a subset $X \subseteq F$.

1. Pick a vertex $v \in F$, adjacent to a vertex in $Z$, and include it in $X$.
2. Let $N(v) \subseteq F$ be vertices which are within distance 2 of $v$. Remove $N(v) \cup \{v\}$ from $F$. Remove all vertices in $Z$ adjacent to $N(v) \cup \{v\}$.
3. Repeat above steps until $Z = \phi$.

*Claim.* *([11]).* $|X| \le |O|$ *and every vertex of* $Z$ *has a path of length at most 3 (in $H$) to a vertex in $X$.*

We now construct an instance of the (edge-weighted) Steiner forest problem on a graph $H' = (F, E')$. $E'$ has an edge $(u, v), u, v \in F$ if there is a path of length at most 4 between $u$ and $v$ in $H$. Suppose there is a path of length at most 3 from $s_j$ to $x_1 \in X$ and from $t_j$ to $x_2 \in X$. We designate $(x_1, x_2)$ the source-destination pair for client $j$.

Since $O$ is a feasible solution to $ChSD(G, l, C, B)$ for every $(s_j, t_j) \in C$ there exists a path $s_j, v_1, v_2, \ldots v_p, t_j$ all of whose edges are in $H$ and $v_1, v_2, \ldots, v_p \in O$. If $x_1, x_2$ are the source-destination for client $j$ in the Steiner forest instance we constructed above then note that there is a path of length at most 4 in $H$

between $x_1, v_1$ and $x_2, v_p$. Thus all edges of the path $x_1, v_1, \ldots v_p, x_2$ are in $H'$ and this implies that there is a solution to our Steiner forest problem in $H'$ containing at most $|X| + |O|$ edges. We compute a Steiner forest containing at most $2(|X| + |O|)$ edges using the 2-approximation algorithm of Goemans and Williamson [9]. Since the number of nodes in a tree is at most twice the number of edges, this forest contains at most $4(|X| + |O|) \le 8|O|$ vertices and these are the locations of the charging stations for our solution.

# 5    Locating Charging Stations When Routes are Specified

We next consider the variant of the Charging Station location problem when we are given a path, $P_j$ for each client $j$, and have to locate charging stations such that each client can travel from its source to destination without deviating from its path and running out of charge. Our objective is to minimize the number of Charging Stations.

## 5.1    The Graph is a Tree

For this section we assume that $G$ is a tree $T = (V, E)$. Then the path $P_j$ for client $j$ is the unique path in $T$ between $s_j, t_j$. Our reduction in the proof of Claim 3.2 and its subsequent modification in Sect. 4.1 is an approximation preserving reduction from vertex cover in graphs to the ChSP() problem on trees when we are allowed to open charging stations only at a subset of vertices, $F \subseteq V$.

We, therefore, assume that $F = V$ and modify Algorithm 2 to minimize the number of facilities needed in this setting.

**Theorem 8.** *The problem ChSP$(T, l, \mathcal{P}, B)$ can be solved in polynomial time when $T$ is a tree and $F = V$.*

*Proof.* As before we root the tree at an arbitrary vertex, $r$ and shall be moving terminals up the tree. The movement of terminals $s_j, t_j$ is such that they always remain on the path $P_j$ and we remove client $j$ when $s_j, t_j$ are within $B$ distance of each other. Let $\mathcal{P}'$ be the set of clients which have not yet been removed and $Z' = \cup_{j \in C'}\{s_j, t_j\}$ be the corresponding set of terminal vertices.

For a client $j$, let $r_j$ be the least common ancestor of $s_j, t_j$ and $T_j$ the subtree rooted at $r_j$. Let $a_j \in F$ (resp. $b_j \in F$) be the furthest ancestor of $s_j$ (resp. $t_j$) which is at most distance $B$ away from it and contained in $T_j$. Let $X = \cup_{j \in C'}\{a_j, b_j\}$ and $x \in X$ be the vertex furthest from the root $r$. We open a facility at location $x$.

Our procedure for moving terminals is now different from Algorithm 2. We move $s_j$ (resp. $t_j$) to $x$ if $x$ is on the path $P_j$ and within distance $B$ of $s_j$ (resp. $t_j$). Client $j$ is removed from $C'$ if $s_j, t_j$ are within distance $B$ of each other.

Let $x_i$ be the facility opened in step $i$ of the algorithm. Once again it is easy to see that $l(r, x_i) \ge l(r, x_{i+1})$. Let $o_1, o_2, \ldots, o_k$ be facility locations of the optimum solution, $O$, ordered by decreasing distance from the root. Let $i$ be the earliest index at which the sequence of facilities in $O$ differs from $x_1, x_2, \ldots x_p$. Let $T'$ be the subtree rooted at $x_i$ and $Q = \{y_1, y_2, \ldots, y_q\}$ be the set of facility locations opened in $T'$ by our algorithm. Note that the optimum solution also opens facilities at all locations in $Q$. Once again we will argue that the optimum solution opens a facility in $T'$ that is not in $Q$.

Suppose $x_i = a_j, j \in \mathcal{P}$. We first consider the case when $x_i = r_j$. This implies $T' = T_j$. At step $i$, $s_j, t_j$ were more than $B$ distance apart. Thus any feasible solution would include a facility on the path $P_j$ between $s_j$ and $t_j$. This implies the optimum solution would have another facility in $T'$ besides the facilities in $Q$.

We now assume $x_i \neq r_j$. Note that $r_j$ cannot be in the subtree $T'$ since $a_j$ is never an ancestor of $r_j$. Hence we only need to consider the case when $x_i$ is in $T_j$. Suppose at step $i$, terminal $s_j$ is at location $y_s$. This implies any feasible path for client $j$ that starts at $\hat{s}_j$ and uses only locations in $Q$ cannot reach any closer to $r_j$ than $y_s$. Note that $x_i$ is the furthest ancestor of $y_s$ which is within distance $B$. If the optimum solution does not open another facility in $T'$ there would be no feasible path from $\hat{s}_j$ to $r_j$.

Thus we have established that the optimum solution should have another facility in $T'$ besides the facilities in $Q$. As before one can argue that location $x_i$ dominates any other choice of location in $T'$ and hence we can modify the optimum solution so that it includes $x_i$. Continuing in this manner we will obtain an optimum solution identical to the solution obtained by our algorithm which implies that our algorithm computes an optimum solution.

## 5.2   General Graphs

Finding a minimum collection of charging stations that ensures every path in $\mathcal{P}$ is feasible is equivalent to finding a set of vertices $X \subseteq F$ which hits every subpath of the given paths of length at least $B$. Funke et al. [6] use this observation and an approximation algorithm for the hitting set problem to obtain an $O(\log n)$ approximation for $\mathrm{ChSP}(G, l, \mathcal{P}, B)$.

We show this approximation is best possible by providing an approximation preserving reduction from the set cover problem to this charging station location problem.

**Theorem 9.** *Let* OPT *be the size of an optimum solution to the problem* $\mathrm{ChSP}(G, l, \mathcal{P}, B)$. *There is no polynomial time algorithm which finds a solution of size* $o\left(\frac{\log n}{\alpha}\mathrm{OPT}\right)$ *for the problem* $\mathrm{ChSP}(G, l, \mathcal{P}, \alpha B)$ *unless* P=NP. *In particular (for $\alpha = 1$), there is no polynomial time algorithm which finds a solution of size* $o(\mathrm{OPT}\log n)$ *for the problem* $\mathrm{ChSP}(G, l, \mathcal{P}, B)$ *unless* P=NP.

*Proof.* Given an instance of the set cover problem with elements $e_1, e_2, \ldots, e_n$ and sets $S_1, S_2, \ldots, S_m$ we create a graph $G = (V, E)$ as follows.

1. For every set $S_i$ in the set cover instance add vertex $v_i$ to $V$; $v_i$ is also included in $F$.

2. For element $e_j$ in the set cover instance add vertices $v_j^1, v_j^2, \ldots, v_j^{q+1}$ to $V$ where $q$ is the number of sets containing $e_j$. Also add vertices $s_j, t_j$ which are source and destination for a client $j$. Note that $s_j, t_j$ will be included in $V$.
3. For client $j$ add zero length edges $(v_j^i, v_j^{i+1}), 1 \le i \le q-1$ and edges $(s_j, v_j^1)$ and $(v_j^{q+1}, t_j)$ of length $B$. The path for client $j$ is $P_j = (s_j, v_j^1, v_j^2, \ldots, v_j^{q+1}, t_j)$.
4. Order the sets including element $e_j$ arbitrarily. If $S_i$ is the $p^{\text{th}}$ set containing $e_j$ then replace edge $(v_j^p, v_j^{p+1})$ with edges $(v_j^p, v_i), (v_i, v_j^{p+1})$ and both these edges have length 0. Thus the path $P_j$ corresponding to element $e_j$ has $2q+2$ edges and a total length $2B$.

Consider a set cover of size $k$ and pick vertices corresponding to these sets as locations for charging stations. Since all elements are covered by these sets, every path $P_j$ visits one of the charging stations which implies that the set of charging stations picked is feasible.

Conversely, consider a feasible set of charging stations in $G$. Since these can only be located on vertices in $F$ this corresponds to picking a collection of sets. Every path $P_j$ has length $2B$ and must visit a charging station. Hence the collection of sets covers all elements and forms a set cover. The hardness of approximating set cover [4,5] better than $c \log n$ implies a similar hardness on approximating the minimum number of charging stations.

If vehicles have range $\alpha B$ we make $\alpha$ copies of graph $G$. Let $v[i]$ denote vertex $v$ in the $i^{\text{th}}$ copy. We combine the paths corresponding to element $e_j$ in the various copies of $G$ into one path, $P_j$, as follows. For $1 \le i \le \alpha - 1$ identify vertices $v_j^{q+1}[i]$ and $s_j[i+1]$ and vertices $t_j[i]$ and $v_j^1[i+1]$. Let $G'$ be the resulting graph. Note that $P_j$ has $(2k+1)\alpha + 1$ edges and total length $(\alpha+1)B$ (Fig. 2).

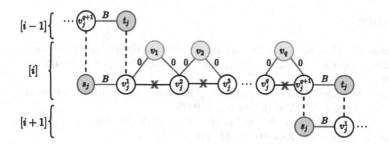

**Fig. 2.** Illustrating the reduction from set cover to $\texttt{ChSP}(G, l, \mathcal{P}, \alpha B)$.

Let $\texttt{OPT}$ be the number of sets in a minimum set cover. Picking all $\alpha$ copies of vertices corresponding to the sets in this minimum set cover yields a solution to the problem $\texttt{ChSP}(G, l, \mathcal{P}, B)$ of size $\alpha \texttt{OPT}$. Suppose by permitting vehicles to have range $\alpha B$ we could obtain an $o(\alpha^{-1} \log n)$-approximation in polynomial time. We could then use this algorithm to find a set of $o(\texttt{OPT} \log n)$ charging stations which would hit every path $P_j$ in $G'$. The sets corresponding to these charging stations cover all elements and form a set cover of size $o(\text{OPT} \log n)$. This would imply a $o(\log n)$ approximation for set cover in polynomial time which is not possible unless P=NP.

## 6    Conclusion

The problem of locating charging stations to build an efficient infrastructure for supporting EVs is of great relevance. Our paper shows that without assumptions on the nature of the road network it is not possible to obtain good approximation algorithms. It would therefore be natural to consider all these problems in the setting of planar graphs.

The two variants of charging station location that we considered seem to sit at two extremes. By specifying only source-destination for each client we might be reducing the number of charging stations needed but this likely comes at the expense of the clients having to take a long path. On the other hand, when we specify the path that the clients should follow and open charging stations to make these paths feasible, we are likely opening a large number of facilities. It would be interesting to consider the setting when client paths are specified but the clients may need to take small detours to avoid running out of charge. An interesting and likely challenging objective in this setting would be to minimize the maximum *total detour* that the clients have to make.

While we have been able to obtain exact algorithm on trees for all three problems considered it seems hard to extend these algorithms to graphs with bounded treewidth. We believe that even finding a set of $k$ vertices which would hit a given set of paths (the SvSP$(p)$ roblem with no detours) in a series-parallel graph is a challenging problem.

## References

1. Berman, O., Bertsimas, D., Larson, R.C.: Locating discretionary service facilities, II: maximizing market size, minimizing inconvenience. Oper. Res. **43**(4), 623–632 (1995)
2. Berman, O., Larson, R.C., Fouska, N.: Optimal location of discretionary service facilities. Transp. Sci. **26**(3), 201–211 (1992)
3. Biniaz, A., Maheshwari, A., Smid, M.H.M.: On the hardness of full steiner tree problems. J. Discrete Algorithms **34**, 118–127 (2015)
4. Dinur, I., Steurer, D.: Analytical approach to parallel repetition. In: Proceedings of the Forty-Sixth Annual ACM Symposium on Theory of Computing, STOC 2014, pp. 624–633. Association for Computing Machinery, New York (2014)
5. Feige, U.: A threshold of ln n for approximating set cover. J. ACM **45**(4), 634–652 (1998)
6. Funke, S., Nusser, A., Storandt, S.: Placement of loading stations for electric vehicles: no detours necessary! J. Artif. Intell. Res. **53**, 633–658 (2015)
7. Funke, S., Nusser, A., Storandt, S.: Placement of loading stations for electric vehicles: allowing small detours. In: Twenty-Sixth International Conference on Automated Planning and Scheduling (2016)

8. Garg, N., Konjevod, G., Ravi, R.: A polylogarithmic approximation algorithm for the group steiner tree problem. J. Algorithms **37**(1), 66–84 (2000)

9. Goemans, M.X., Williamson, D.P.: A general approximation technique for constrained forest problems. SIAM J. Comput. **24**(2), 296–317 (1995)

10. Halperin, E., Krauthgamer, R.: Polylogarithmic inapproximability. In: Proceedings of the Thirty-Fifth Annual ACM Symposium on Theory of Computing, STOC 2003, pp. 585–594. Association for Computing Machinery, New York (2003)

11. Hochbaum, D.S., Shmoys, D.B.: A best possible heuristic for the k-center problem. Math. Oper. Res. **10**(2), 180–184 (1985)

12. Hodgson, M.J.: A flow-capturing location-allocation model. Geogr. Anal. **22**(3), 270–279 (1990)

13. Kalyanasundaram, B., Pruhs, K.: Speed is as powerful as clairvoyance. J. ACM (JACM) **47**(4), 617–643 (2000)

14. Klein, P., Ravi, R.: A nearly best-possible approximation algorithm for node-weighted steiner trees. J. Algorithms **19**, 02 (2000)

15. Kuby, M., Lim, S.: The flow-refueling location problem for alternative-fuel vehicles. Socioecon. Plann. Sci. **39**(2), 125–145 (2005)

16. Kuby, M., Lines, L., Schultz, R., Xie, Z., Kim, J.-G., Lim, S.: Optimization of hydrogen stations in Florida using the flow-refueling location model. Int. J. Hydrogen Energy **34**(15), 6045–6064 (2009)

17. Mitra, S., et al.: Trajectory aware macro-cell planning for mobile users. In: 2015 IEEE Conference on Computer Communications (INFOCOM), pp. 792–800. IEEE (2015)

18. Mitra, S., Saraf, P., Sharma, R., Bhattacharya, A., Ranu, S., Bhandari, H.: Netclus: a scalable framework for locating top-k sites for placement of trajectory-aware services. In: 2017 IEEE 33rd International Conference on Data Engineering (ICDE), pp. 87–90. IEEE (2017)

19. Naor, J., Panigrahi, D., Singh, M.: Online node-weighted steiner tree and related problems. In: Ostrovsky, R. (ed.) IEEE 52nd Annual Symposium on Foundations of Computer Science, FOCS 2011, Palm Springs, CA, USA, 22–25 October 2011, pp. 210–219. IEEE Computer Society (2011)

20. Storandt, S., Funke, S.: Enabling e-mobility: facility location for battery loading stations. In: Twenty-Seventh AAAI Conference on Artificial Intelligence (2013)

# Graph Burning and Non-uniform
# $k$-centers for Small Treewidth

Matej Lieskovský[ID] and Jiří Sgall[(✉)][ID]

Faculty of Mathematics and Physics,
Computer Science Institute of Charles University, Prague, Czechia
{ml,sgall}@iuuk.mff.cuni.cz

**Abstract.** We study the graph burning problem and give a polynomial-time approximation scheme (PTAS) for arbitrary graphs of constant treewidth. This significantly extends the previous results, as a PTAS was known only for disjoint union of paths.

As a building block, we give an algorithm that proves the non-uniform $k$-center problem to be in XP when parameterized by the number of different radii and the treewidth of the graph. This extends the known exactly solvable cases of the non-uniform $k$-center problem; in particular this also solves the $k$-center with outliers on graphs of small treewidth exactly.

**Keywords:** Graph algorithms · Approximation algorithms · PTAS · $k$-center problem

## 1  Introduction

We study the problem of graph burning originally introduced in [3]. In this problem, we are given an undirected graph. In each round, we choose one node to be set on fire. In addition, in every time step fire spreads to all neighbors of all previously burning nodes. The objective is to minimize the number of steps to spread the fire to all the nodes. More precisely, the burning number $g$ of a graph is defined as the minimum number of time steps needed to achieve the state with all nodes burning. A 3-approximation algorithm for computing $g$ has been given in [2], recently a minor improvement to the ratio $3 - 2/OPT$ was given in [10], but still no $(3 - \epsilon)$-approximation for general graphs is known.

We approach the graph burning problem by restricting the graphs to be burned. Our main result is a PTAS for graph burning of all graphs of a constant treewidth, with a slight generalization that allows edge lengths. This result significantly improves previous results, as no PTAS was known for graph burning all tree graphs. Previous results applied only to linear forests (unions of disjoint paths), where both a PTAS and NP-hardness were known [2,4].

To design the PTAS, we study the non-uniform $k$-center (NUkC) problem, which generalizes both the graph burning problem and the classical and well-studied (uniform) $k$-center problem.

P. Chalermsook and B. Laekhanukit (Eds.): WAOA 2022, LNCS 13538, pp. 20–35, 2022.
https://doi.org/10.1007/978-3-031-18367-6_2

In an instance of the (uniform) $k$-center problem we are given a set of $n$ nodes, a metric $d$ defining distances between the nodes, and parameters $k$ and $r$ that tell us how many centers we are allowed to use and the radius of every center's reach, respectively. We must then decide if a set of $k$ nodes can be selected as centers so that every node will be within distance $r$ of some center. The $k$-center decision problem, first formulated in 1964 by Hakimi [11], is known to be strongly NP-hard.

In the optimization version of the $k$-center problem, one minimizes $r$ for a fixed $k$. Hochbaum and Shmoys [12] gave a 2-approximation algorithm and proved it to be optimal unless P=NP.

The non-uniform $k$-center (NU$k$C) problem is a generalization where different centers use given, possibly different, radii, see Definition 2.1 below. This problem was first introduced in [5] although a special case had already been analyzed in [6].

We can view the (decision variant of the) graph burning problem as a special case of the NU$k$C problem for graphs with unit-length edges and parameter $g$, where the goal is to cover the graph using $g$ centers, each having a unique integer radius from 0 to $g-1$. The locations of the centers then correspond to the nodes set on fire in each step, starting from the largest diameter.

To design our PTAS for graph burning, we study the NU$k$C problem with a constant number of different radii. We show that the non-uniform $k$-center problem is polynomial if both the number of different radii and treewidth are constant. More precisely, we give an XP-algorithm when the NU$k$C problem is parameterized by the number of different radii and treewidth.

To make this algorithm work, we design a concept of "tidy" assignments of centers to vertices, as our main technical contribution. This allows us to process the graph and check partial solutions in the ordering given by the tree decomposition. We then give a recursive—or, equivalently, dynamic programming—algorithm for the problem. The techniques are fairly standard for graphs of bounded treewidth and the existence of such an algorithm can be deduced also from the extension of Courcelle's theorem in [1].

Nevertheless, this algorithm significantly extends the known exactly solvable cases of the non-uniform $k$-center problem. In particular, for the previously studied problem of the $k$-center with outliers, which allows us to use a given number of centers of radii $r$ (the usual centers) and 0 (the outliers), we give an exact algorithms on graphs of constant treewidth.

Finally, we use this XP algorithm for the NU$k$C problem to design a PTAS for burning graphs of a constant treewidth, using a simple reduction.

For the sake of succinctness, all our algorithms below are formulated to answer the decision variant of the problem; however, they can be modified by standard dynamic programming techniques to also return a solution if one exists.

**Other Related Work.** The $k$-center problem is widely studied also in parameterized complexity. Most related to our result, Katsikarelis et al. [15] gives an XP-algorithm for $k$-center parameterized by the cliquewidth (and thus also treewidth) using similar methods and running time as ours. However, our algorithm is able to handle general distances and distinct radii (precisely, the NUkC problem with constant number of radii). They also show the hardness of the $k$-center problem which implies that a better (FPT) running time is not possible for exact algorithms. Interestingly, to complement this hardness result, they also give an approximation scheme running in FPT time.

Previous work on FPT algorithms for $k$-center includes most notably an exact algorithm for planar graphs by Demaine et al. [8]. Feldmann [9] studied the hardness of approximation on graphs of small highway dimension, obtaining a 3/2-approximation in FPT time if parameterized also by $k$.

New results about FPT algorithms and hardness of exact algorithms for graph burning, together with an extensive survey of the literature is given in [17].

For the non-uniform $k$-center problem, there have been recent developments concerning approximation algorithms. Chakrabarty et al. [5] show that no $O(1)$-approximation can exist for NUkC problem if the number of radii is a part of the input (unless P = NP). The positive results thus focus on a small number of distinct radii. In [5], authors focus on two radii. The special case known as $k$-center with outliers where one of the radii is zero is solved with an approximation factor of 2, which is optimal unless P = NP. An approximation factor of $(1+\sqrt{5})$ is given otherwise. It has been conjectured that $O(1)$-approximation exists for any constant number of radii, and such algorithms were given so far for 3 and 4 radii [13,14].

## 2    Solving $k$-center Problem with $\ell$ Different Radii on Trees

We first define the NUkC problem formally for general graphs.

**The NUkC Problem with $\ell$ Different Radii.** In an instance of this problem we are given an undirected graph $G$ on $n$ nodes with edge lengths from $\mathbb{R}_0^+$, a list of $\ell$ radii $r_1, r_2, \ldots, r_\ell \in \mathbb{R}_0^+$, and a vector $(k_1, k_2, \ldots, k_\ell)$ where $k_i \in \mathbb{N}$ specifies how many centers of radius $r_i$ we are allowed to use.

**Definition 2.1.** *For a given instance of the NUkC problem the metric $d(u, v)$ is defined as the length of the shortest path from $u$ to $v$. We also define $K = V(G) \times \{1, \ldots, \ell\}$ to be the set of all possible centers and $K_v = \{(w, i) \in K \mid d(v, w) \leq r_i\}$ to be the set of all centers that will cover node $v$.*

*A solution to an instance of the NUkC problem is a subset $S \subseteq K$ such that it contains at least one center from each $K_v$.*

The TreeNUkC problem with $\ell$ different radii is the NUkC problem with $\ell$ different radii with instances restricted to graphs $G$ that are trees. In the rest of this section, we consider only this subproblem. Furthermore, we root the tree arbitrarily.

**Subtrees and Partial Solutions.** We shall use $T_v$ to denote the subtree of $G$ with root $v$ containing $v$ and all its descendants. A partial solution for $T_v$ is any subset of $K \cap (V(T_v) \times \{1, \ldots, \ell\})$; note that we do not require all vertices to be covered.

For a solution $S \subseteq K$ we define a partial solution $S_v = S \cap (V(T_v) \times \{1, \ldots, \ell\})$ to be its restriction to $T_v$.

For each partial solution we consider its cost vector, describing the number of centers used, and its coverage level, describing its quality. Our algorithm computes recursively for each subtree and cost vector the best solution.

**Cost Vectors.** The cost vector $\boldsymbol{x}$ of a partial solution is a vector of $\ell$ natural numbers $x_1, x_2, \ldots, x_\ell$ where $x_i$ is the number of centers with radius $r_i$ the partial solution contains. We can view $k_1, k_2, \ldots, k_\ell$ given by the instance as the target cost vector. For a given instance, we define $k = \sum k_i$ and the set of valid cost vectors as $\mathcal{C} = \{(y_1, y_2, \ldots, y_\ell) \in \mathbb{N}^\ell \mid (\forall i)(y_i \leqslant k_i)\}$. Since any instance with at least $n$ centers is trivial, we assume $k < n$, which ensures that $|\mathcal{C}| \leqslant n^\ell$.

**Coverage Level.** We define the coverage level of any partial solution for any given subtree $T_v$. A coverage level consists of a sign and a non-negative real number. A coverage level is $+c$ if the partial solution covers all nodes in $T_v$ and in addition any nodes outside $T_v$ within distance up to $c$ from the root $v$ (and not more). A coverage level is $-c$ if not all nodes in $T_v$ are covered by the partial solution and $c$ equals the maximal distance from $v$ to a node not covered. We also define a natural linear ordering on the coverage levels so that $-b \leqslant -a \leqslant +a \leqslant +b$ for all $a \leqslant b \in \mathbb{R}_0^+$. Note that coverage levels thus differ from real numbers by the existence of both a positive and a negative zero. A coverage level of $+0$ denotes that all nodes in $T_v$ are covered while a coverage level of $-0$ denotes that the root $v$ is not covered.

For clarity we shall use $\alpha, \beta, \gamma$ to denote coverage levels while $a, b, c$ will be used to denote the absolute value of the corresponding coverage level.

**Table of Configurations.** For every subtree we compute a table $t$ which contains the maximum coverage level achievable for any given cost vector.

**Lemma 2.2.** *For any solution $S$ to an instance of the TreeNUkC problem, if the corresponding partial solution $S_v = S \cap (V(T_v) \times \{1, \ldots, \ell\})$ for subtree $T_v$ has a negative coverage level, then all nodes outside of $T_v$ are covered by some center outside of $T_v$ in solution $S$.*

*Proof.* For a contradiction, assume that there is a node $u$ outside of $T_v$ that is covered by some center in $S_v$. Let $-c$ be the coverage level of $S_v$. Since the coverage level is negative, there exists a node $w$ in $T_v$ with $d(v, w) = c$ that is not covered by $S_v$. We also know that $d(v, w) > d(u, v)$ since otherwise $w$ would be covered by the same center as $u$. However, the center outside of $T_v$ that covers $w$ also covers all nodes within distance $c$ from $v$ and thus also covers $u$.    □

**Combining Partial Solutions.** Let us now describe how to compute the table of configurations for a given (sub)tree. Any partial solution for $T_v$ can be thought of as consisting of a partial solution for $T_w$ where $w$ is a child of $v$ and a partial solution for the remaining nodes of the tree, which we shall denote as $T_v \setminus T_w$. We define below the function $f$ which defines how coverage levels of the two partial solutions are combined. The three arguments $f$ takes are the coverage level of

$T_v \setminus T_w$, the coverage level of $T_w$, and the length of the edge $v, w$ respectively. We divide the cases according to the signs of the coverage levels.

$$f(+a, +b, d) = + \max(a, b - d)$$

$$f(+a, -b, d) = \begin{cases} +a & \text{if } a \geqslant b + d \\ -(b + d) & \text{otherwise} \end{cases}$$

$$f(-a, +b, d) = \begin{cases} -a & \text{if } a > b - d \\ +(b - d) & \text{otherwise} \end{cases}$$

$$f(-a, -b, d) = - \max(a, b + d)$$

The following two lemmata verify that the function $f$ has the desired properties.

**Lemma 2.3.** *Let $T$ be any rooted tree, $v$ be the root of $T$, $w$ be any child of $v$ and $S$ be the subtree of $T$ rooted in $w$. If $P$ is a partial solution for $T \setminus S$ with cost vector $\boldsymbol{x}$ and coverage level $\alpha$ and $Q$ is a partial solution for $S$ with cost vector $\boldsymbol{y}$ and coverage level $\beta$, then $P \cup Q$ is a partial solution for $T$ with cost vector $\boldsymbol{x} + \boldsymbol{y}$ and coverage level $f(\alpha, \beta, d(v, w))$.*

*Proof.* If $\alpha = +a$ and $\beta = +b$, then all nodes in $S$ are covered by the centers of $Q$ and all nodes in $T \setminus S$ are covered by the centers of $P$, ensuring that the resulting coverage level will be positive. Any node outside of $T$ at distance $r$ from $v$ is at distance $r + d(v, w)$ from $w$. Thus the centers of $P$ cover any nodes outside $T$ within distance $a$ of $v$ and centers of $Q$ cover any nodes outside $T$ within distance $b - d(v, w)$ of $v$. The resulting coverage level is thus $+ \max(a, b - d(v, w))$.

If $\alpha = +a$ and $\beta = -b$, the centers in $P$ cover all uncovered nodes in $S$ up to the distance $a - d(v, w)$ from $w$ (or none if negative). If this quantity is smaller than $b$, there is an uncovered node in $S$ at distance $b + d(v, w)$ from $v$ and it is the most distant uncovered node. Otherwise all nodes in $T$ are covered as well as all other nodes at distance up to $a$ from $v$; by Lemma 2.2 centers in $S$ do not contribute to covering additional nodes outside of $T$.

If $\alpha = -a$ and $\beta = +b$, the centers in $Q$ cover all uncovered nodes in $T \setminus S$ up to the distance $b - d(v, w)$ from $v$ (or none if negative). If this quantity is smaller than $a$, there is an uncovered node in $S$ at distance $a$ from $v$ and it is the most distant uncovered node. Otherwise all nodes in $T$ are covered as well as all other nodes at distance up to $b - d(u, v)$ from $v$; similarly as in Lemma 2.2, centers in $T \setminus S$ do not contribute to covering additional nodes outside of $T$.

If $\alpha = -a$ and $\beta = -b$ then we consider the node $p$ not covered in $P$ with $d(v, p) = a$ and the node $q$ not covered in $Q$ with $d(w, q) = b$; note that $d(v, p) = b + d(v, w)$. We claim that among $p$ and $q$, the one more distant from $v$ remains uncovered. Should this more distant node be covered by a center in the other subtree, then that center would cover all the nodes of the same or smaller distance from $v$ in both subtrees, including the other node among $p$ and $q$, which is a contradiction. As all the nodes more distant from $v$ than both $p$

and $q$ are covered by centers in its subtree, it follows that the coverage level of $T$ is $-\max(a, b + d(v, w))$. □

**Lemma 2.4.** *The function $f(\alpha, \beta, d)$ is non-decreasing with increasing $\alpha$ or $\beta$.*

*Proof.* If the value of $f$ is negative, it is equal to $\min(\alpha, \beta - d)$, which is non-decreasing in $\alpha$ and $\beta$.

If the value of $f$ is positive, it is equal to $\max(\alpha, \beta - d)$, which is non-decreasing in $\alpha$ and $\beta$.

It remains to consider the cases when the value of $f$ may change its sign. This can only happen when we move from one of the cases or subcases in the definition of $f$ to another one.

If $\alpha$ or $\beta$ changes from $-0$ to $+0$, we are either moving from the $f(-a, -b, d)$ case which is always negative or we are moving to the $f(+a, +b, d)$ case which is always positive. Thus we are either moving from negative to positive coverage level or the sign does not change.

If we are changing subcases in one of the cases where exactly one of $\alpha$ and $\beta$ is positive, one can check that the subcase condition is monotone, i.e., increasing either $\alpha$ or $\beta$ can only flip the subcases from the negative one to the positive one. □

The last two lemmata imply the following.

**Corollary 2.5.** *For any given tree and cost vector, the maximum achievable coverage level is computable using only the maximum coverage levels achievable for each subtree and cost vector.*

We start a calculation for each node $v$ from a table $t_0$, which represents the table of configurations for a leaf node. We then combine it with the tables of configurations of the children of $v$, computed recursively, using the previous observations. The algorithm follows.

**Lemma 2.6.** *Function PARTIAL_solutions($v$) computes a correct table $t$ for $T_v$, that is, the following conditions hold:*
*(a) If $t[z] = \gamma$, then there exists a partial solution for $T_v$ with cost vector $z$ and coverage level $\gamma$.*
*(b) If there exists a partial solution for $T_v$ with cost vector $z$ and coverage level $\gamma$, then $t[z] \geq \gamma$.*

*Proof.* We proceed inductively. As the basis of induction, observe that $t_0$ is the correct table of configurations for all single-node trees as putting all centers at the root is a valid solution.

Now for $i = 1, \ldots, degree(v)$, let $T'$ be the subtree of $T_v$ consisting of $v$ and the subtrees $T_u$ for the first $i - 1$ children $u$ of $v$ considered in the loop on line 5; let $w$ be the $i$-th child of $v$. By induction, we assume that $t_{old}$ and $t_{ch}$ are the correct tables for $T'$ and $T_w$, respectively. We prove that the table $t$ after the $i$-th iteration of the loop on line 5 is correct for the combined subtree $T' \cup T_w$.

---

**Algorithm** TREE-COVERAGE     Solver for non-uniform $k$-center on trees

---

1: **function** TREENU$k$C($tree, r_1, r_2, \ldots, r_\ell, k_1, k_2, \ldots, k_\ell, C$)

2:  $\quad t_0[\boldsymbol{x}] \leftarrow \begin{cases} -0 & \text{for } \boldsymbol{x} = \mathbf{0} \\ \max_{i|x_i>0}(r_i) & \text{for } \boldsymbol{x} \in C \setminus \{\mathbf{0}\} \end{cases}$

3:  $\quad$ **function** PARTIAL_SOLUTIONS($node$)

4:  $\quad\quad t \leftarrow t_0$

5:  $\quad\quad$ **for** $child \in node.children$ **do**

6:  $\quad\quad\quad t_{ch} \leftarrow$ PARTIAL_solutions($child$)

7:  $\quad\quad\quad t_{old} \leftarrow t$

8:  $\quad\quad\quad t[\boldsymbol{x}] \leftarrow -\infty \quad$ **for** $\boldsymbol{x} \in C$

9:  $\quad\quad\quad$ **for** $\boldsymbol{x} \in C$ **do**

10: $\quad\quad\quad\quad$ **for** $\boldsymbol{y} \in C$ **do**

11: $\quad\quad\quad\quad\quad$ **if** $(\boldsymbol{x} + \boldsymbol{y}) \in C$ **then**

12: $\quad\quad\quad\quad\quad\quad \gamma = f(t_{old}[\boldsymbol{x}], t_{ch}[\boldsymbol{y}], d(node, child))$

13: $\quad\quad\quad\quad\quad\quad t[\boldsymbol{x} + \boldsymbol{y}] \leftarrow \max(t[\boldsymbol{x} + \boldsymbol{y}], \gamma)$

14: $\quad\quad$ **return** $t$

15: $\quad t =$ PARTIAL_solutions($tree.root$)

16: $\quad$ **return** True if $t[k_1, k_2, \ldots, k_\ell] \geq +0$

---

(a) For each $\boldsymbol{z} \in C$, the final value of $t[\boldsymbol{z}] = \gamma$ is assigned in line 13 for some $\boldsymbol{x}$ and $\boldsymbol{y}$ with $\boldsymbol{x} + \boldsymbol{y} = \boldsymbol{z}$. By induction, there exist a partial solution $P$ for $T'$ with cost vector $\boldsymbol{x}$ and coverage level $t[\boldsymbol{x}]$ and a partial solution $Q$ for $T_w$ with cost vector $\boldsymbol{y}$ and coverage level $t[\boldsymbol{y}]$. By Lemma 2.3, $P \cup Q$ is a partial solution for $T' \cup T_w$ with cost vector $\boldsymbol{x} + \boldsymbol{y}$ and coverage level $f(t_{old}[\boldsymbol{x}], t_{ch}[\boldsymbol{y}], d(v, w))$ which is the value $\gamma$ assigned in line 13.

(b) Suppose that we have a partial solution $R$ for $T' \cup T_w$; let $P$ be the partial solution consisting of centers of $R$ in $T'$ and $Q$ be the partial solution consisting of centers of $R$ in $T_w$. Let $\boldsymbol{x}$ and $\boldsymbol{y}$ be the cost vectors of $P$ and $Q$; obviously $\boldsymbol{z} = \boldsymbol{x} + \boldsymbol{y}$. Let $\alpha$ and $\beta$ be the coverage levels of $P$ and $Q$. By Lemma 2.3, the coverage level of $R$ is $f(\alpha, \beta, d(v, w))$. By induction, $t_{old}[\boldsymbol{x}] \geq \alpha$ and $t_{ch}[\boldsymbol{y}] \geq \beta$. Now by Lemma 2.4, $f(t_{old}[\boldsymbol{x}], t_{ch}[\boldsymbol{y}], d(v, w))$ is at least the coverage level of $R$. Since this value is among the values $\gamma$ considered in line 13 for $t[\boldsymbol{z}]$, the inductive claim follows. $\qquad\square$

**Theorem 2.7.** *Algorithm* TREE-COVERAGE *solves the TreeNUkC problem in polynomial time, namely with* $\mathcal{O}(n^{2\ell+1})$ *arithmetic operations.*

*Proof.* The correctness of the algorithm follows from Lemma 2.6.

For the bound on complexity, note that we compute two auxiliary tables for each node: one when considered as a child and one when considered as a root of a subtree. Calculation of each table considers $|C|^2 = \mathcal{O}(n^{2\ell})$ pairs $\boldsymbol{x}, \boldsymbol{y}$ and a constant number of arithmetic operations for each of them. $\qquad\square$

# 3   Generalization to Graphs with Small Treewidth

We shall now demonstrate a generalization of the previous algorithm for trees to graphs $G$ of small treewidth.

## 3.1   Tree Decompositions

The concepts of treewidth and tree decompositions are intimately linked, as will be apparent from the definitions below. Our algorithm shall make use of the existence of a *nice tree decomposition*. This is a concept introduced by Kloks in [16], but we use the variation from Cygan et al. [7], further modified so that the root is always an empty forget bag.

**Definition 3.1.** *A tree decomposition of a graph $G$ is a tree $\mathbb{T}$ where every vertex $x \in \mathbb{T}$ is assigned a set $B_x \subseteq V(G)$, which we shall call a bag, and the following conditions are satisfied:*

- *If $\{u, v\} \in E(G)$ then there exists a vertex $x \in \mathbb{T}$ such that $\{u, v\} \subseteq B_x$.*
- *For all $v \in V(G)$ the subgraph induced by $\{x \in \mathbb{T} \mid v \in B_x\}$ is a tree.*

*The width of a tree decomposition $\mathbb{T}$ is the size of the largest bag minus one. The treewidth of a graph $G$ is the width of the tree decomposition of $G$ with the minimum width.*

**Definition 3.2.** *A nice tree decomposition is a rooted tree of bags that consists of five different types of bags.*

- *Leaf bag is empty and has no children.*
- *Introduce bag adds a new node compared to its only child*
- *Forget bag removes a node compared to its only child*
- *Join bag has two children bags with identical contents*
- *Edge bag is labeled with an edge it introduces and has identical contents to its child.*

*The bags of a nice tree decomposition must further satisfy the following:*

- *There is exactly one forget bag for every node*
- *For every edge $uv$ in the graph there exists exactly one bag labeled with that edge*
- *A bag labeled with edge $uv$ contains both $u$ and $v$*
- *The root of the tree of bags is an empty forget bag*

Cygan et al. [7] prove the following lemma about constructing a nice tree decomposition. This also holds for our modification, since to satisfy the additional requirement that the root is an empty forget bag, we can add in linear time a sequence of forget bags as predecessors of the original root.

**Lemma 3.3.** *Given a tree decomposition of graph $G$, a nice tree decomposition of $G$ with equal width can be found in polynomial time.*

## 3.2   Tidy Assignments

Since bags contain multiple nodes, there is no straightforward way to measure the quality of a partial solution as we did using the coverage level previously. Instead, we shall predict which center is going to cover a given node and remember this information for each possible partial solution for a bag.

**Definition 3.4.** *An assignment (of centers) is a function $A : V(G) \to K$ assigning every node $v \in V(G)$ a center from $K_v$.*

Observe that the image of $A$ is a solution with an associated cost vector (possibly non-valid). We thus extend the idea of cost vectors to assignments. We say that an assignment $A$ is an assignment of solution $S$ if the image of $A$ is a subset of $S$.

We will now focus on assignments where all nodes on any shortest path from a node to the assigned center are assigned the same center. We call such assignments tidy. Tidiness of an assignment can be checked edge by edge and also guarantees that each center is used for the node where it is located. These two properties together allow us to fomulate our recursive algorithm and account for the number of used centers correctly.

For the following definition and subsequent observations, we abuse notation so that distance or path to $A(u)$ denotes the distance or path to the node where the center is located, i.e., to the first component of $A(u)$. Thus, if $A(u) = (w, i)$, then $d(u, A(u))$ denotes $d(u, w)$ and a path to $A(u)$ means a path to $w$.

**Definition 3.5.** *For a given assignment $A$ and an edge $uv \in E(G)$, we say that $A$ is $uv$-tidy if either of the following holds:*

- *$A(u) = A(v)$*
- *$d(u, A(u)) < d(u, v) + d(v, A(u))$ and $d(v, A(v)) < d(v, u) + d(u, A(v))$*

*Furthermore, an assignment is tidy, if it is $uv$-tidy for every edge $uv \in E(G)$.*

**Lemma 3.6.** *Let $A$ be a tidy assignment. Then any node $v$ on a shortest path from $u$ to $A(u)$ has $A(v) = A(u)$. In particular, if the assignment uses some center $(v, i)$, then $A(v) = (v, i)$.*

*Proof.* We prove the claim for $v$ adjacent to $u$ on a shortest path from $u$ to $A(u)$, the lemma then follows inductively. Indeed, if $v$ is on the shortest path then $d(u, A(u)) = d(u, v) + d(v, A(u))$. Since $uv$ is an edge and the assignment is tidy, the first condition in the definition has to hold, thus $A(v) = A(u)$.  □

Now we show that we can restrict considered solutions to tidy assignments.

A greedy assignment of a solution to the NUkC problem is constructed as follows: Let us select an arbitrary linear order $\preccurlyeq$ on the centers of the solution. For every node $v$ of the graph and every center $(w, i)$ used by the solution we compute the excess coverage $x(v, (w, i)) = r_i - d(v, w)$. The greedy assignment then assigns to $v$ the center which maximizes the excess coverage. If multiple centers have equal excess coverage we select the minimal one of those in $\preccurlyeq$. Note that at least one greedy assignment exists for any solution.

**Lemma 3.7.** *Every greedy assignment is a tidy assignment.*

*Proof.* Suppose that a greedy assignment $A$ is not $uv$-tidy for an edge $uv$. Thus $A(u) \neq A(v)$ and, w.l.o.g., $d(u, A(u)) = d(u, v) + d(v, A(u))$, swapping $u$ and $v$ if needed. Observe that $x(u, A(u)) = x(v, A(u)) - d(u, v)$ and $x(u, A(v)) \geqslant x(v, A(v)) - d(u, v)$. Consider why $v$ was assigned $A(v)$ instead of $A(u)$. If $x(v, A(v)) > x(v, A(u))$, then $x(u, A(v)) > x(u, A(u))$. If $x(v, A(v)) = x(v, A(u))$ and $A(v) \preccurlyeq A(u)$ then $x(u, A(v)) \geqslant x(u, A(u))$ and $A(v) \preccurlyeq A(u)$. In either case $u$ would not be assigned $A(u)$ during the construction of a greedy assignment. $\square$

**Corollary 3.8.** *For every inclusion-wise minimal solution $S$ there exists a tidy assignment with image equal to $S$.*

## 3.3 The Algorithm

We conceptualize the algorithm as deciding the existence of a tidy assignment with a valid cost vector. Since the number of tidy assignments can be exponential, we will once again proceed via dynamic programming. We process $\mathbb{T}$ from the leaves towards the root, computing partial results for every bag.

**Subgraphs.** Let us define $G_B$ as the subgraph of $G$ consisting of the nodes and edges introduced by bags in the subtree of $\mathbb{T}$ rooted in $B$. Note that if $B$ is the root of $\mathbb{T}$, $G_B$ is equal to the entire graph $G$.

**Forgotten Nodes.** We define $F_B = V(G_B) \setminus B$. This is the set of forgotten nodes for bag $B$, which are the nodes forgotten by bags in the subtree of $\mathbb{T}$ rooted in $B$. Observe that any edge incident with a node in $F_B$ is in $E(G_B)$. Also note that the set of forgotten nodes for the root bag is the set of all nodes.

**Subtree Assignment.** The subtree assignment (of centers) for a bag $B$ is a function $A^{\text{tree}} : V(G_B) \to K$. Note that the range of a subtree assignment is not restricted to centers located in $G_B$. We say that a subtree assignment $A^{\text{tree}}$ for bag $B$ is $B$-tidy if and only if it is $uv$-tidy for all edges $uv \in E(G_B)$.

**Cost Vectors.** We extend the notion of cost vectors to subtree assignments. For any subtree assignment $A^{\text{tree}}$ for bag $B$, its cost vector is the number of centers of each type in the intersection of $F_B \times \{1, \ldots, \ell\}$ and the image of $A^{\text{tree}}$. Note that we count only the centers located at the forgotten nodes, excluding the nodes in the bag itself. Consequently, we increment the cost vector only when a center's node is forgotten making use of the fact that every node is forgotten exactly once by the definition of the tree decomposition. Lemma 3.6 implies that for any used center $(v, i)$ and a tidy assignment $A$ we have $A(v) = (v, i)$, which guarantees that any used center is eventually accounted for in the cost vector.

For manipulating cost vectors, we shall use the common notation where $e_i$ is the vector with $i$th entry set to 1 and all other entries set to 0. The algorithm uses the same set of all valid cost vectors $\mathcal{C}$ as defined for the TREE-COVERAGE algorithm.

**Bag Assignment.** For any bag $B$ the bag assignment (of centers) is a function $A^{\mathrm{bag}} : B \to K$. We extend the concept of $uv$-tidiness to bag assignments for edges $uv \in E(G)$ with $u, v \in B$. For convenience, when manipulating bag assignments, we also view $A^{\mathrm{bag}}$ as a set of pairs where $(v, (w, i)) \in A^{\mathrm{bag}}$ if and only if $A^{\mathrm{bag}}(v) = (w, i)$. When changing the bag, we sometimes need to restrict the domain of the function $A^{\mathrm{bag}}$ to a set $B'$; we use the standard notation $A^{\mathrm{bag}}|_{B'}$.

**Valid Configurations.** For every bag $B$ we compute the set of all pairs of a valid cost vector and a bag assignment for which there exists a corresponding $B$-tidy subtree assignment with the given cost vector. We call such pairs valid configurations.

**Lemma 3.9.** *A valid configuration with cost vector $x$ for the root bag of $\mathbb{T}$ exists if and only if there exists a solution with cost vector $x$ for the entire graph $G$.*

*Proof.* Suppose a valid configuration with cost vector $x$ exists for the root bag $B$. By definition, then there exists a subtree assignment $A^{\mathrm{tree}}$ which is $B$-tidy and has cost vector $x$. Since $B$ is the root of $\mathbb{T}$, a $B$-tidy subtree assignment for $B$ corresponds to a tidy assignment for $G$. The image of any tidy assignment for $G$ is a solution with the same cost vector. Thus a solution exists if a valid configuration exists.

The other direction follows from the fact that a greedy assignment exists for every solution and both has a valid cost vector and, by Lemma 3.7, is tidy.   □

**Algorithm Graph-Coverage.** We process the nice tree decomposition recursively, computing the set of valid configurations for each bag. Without loss of generality, we only consider configurations that have a valid cost vector and a bag assignment mapping each node $v \in B$ a center from $K_v$. We shall use $Config_B$ to denote this set of valid configurations for bag $B$. Observe that $|K| \leqslant n\ell$. If the graph has treewidth tw, $|B| \leqslant \mathrm{tw} + 1$ holds for any bag. Thus the number of possible bag assignments for $B$ is at most $(n\ell)^{\mathrm{tw}}$. Since $|\mathcal{C}| \leqslant n^\ell$, the number of valid configurations for any bag is at most $n^{\mathrm{tw}+\ell}\ell^{\mathrm{tw}}$. See the algorithm's pseudocode on the next page.

**Lemma 3.10.** *Function* PROCESS *computes the correct set of valid configurations for bag $B$, that is,* PROCESS$(B) = Config_B$.

*Proof.* We proceed inductively from the leaves of $\mathbb{T}$ which are all leaf bags. We distinguish cases by the type of bag. Thanks to the inductive process, we assume we know the correct set of valid configurations for any child bags.

**Leaf Bag.** Since $G_B$ for any leaf bag is an empty graph, there exists only a single (empty) subtree assignment for it with zero partial cost vector and an empty bag assignment. Thus $\{(\mathbf{0}, \varnothing)\}$ is the correct set of valid configurations for any leaf bag.

**Introduce Bag.** Observe that $A^{\mathrm{tree}}$ is a subtree assignment for $B$ if and only if $A^{\mathrm{tree}}(v) \in K_v$ and $(A^{\mathrm{tree}})'$ is a subtree assignment for $B'$, where $(A^{\mathrm{tree}})'$

| | |
|---|---|
| **Algorithm** GRAPH-COVERAGE | |

1: **function** TwNUkC($root\_bag, r_1, r_2, \ldots, r_\ell, k_1, k_2, \ldots, k_\ell, \mathcal{C}$)
2:     **function** PROCESS($B$)
3:         **switch** based on the type of $B$ **do**
4:             **case** leaf
5:                 **return** $\{(\mathbf{0}, \varnothing)\}$
6:             **case** introduce
7:                 let $v$ be the node introduced by $B$ and let $B'$ be the child bag of $B$
8:                 $P = $ PROCESS($B'$)
9:                 **return** $\{(\mathbf{x}, A^{\mathrm{bag}} \cup \{(v,c)\}) \mid (\mathbf{x}, A^{\mathrm{bag}}) \in P, c \in K_v\}$
10:             **case** join
11:                 let $B'$ and $B''$ be the child bags of $B$
12:                 $P = $ PROCESS($B'$)
13:                 $Q = $ PROCESS($B''$)
14:                 **return** $\{(\mathbf{x} + \mathbf{y}, A^{\mathrm{bag}}) \mid (\mathbf{x}, A^{\mathrm{bag}}) \in P, (\mathbf{y}, A^{\mathrm{bag}}) \in Q, \mathbf{x} + \mathbf{y} \in \mathcal{C}\}$
15:             **case** edge
16:                 let $uv$ be the edge introduced by $B$ and let $B'$ be the child bag of $B$
17:                 $P = $ PROCESS($B'$)
18:                 **return** $\{(\mathbf{x}, A^{\mathrm{bag}}) \in P \mid A^{\mathrm{bag}}$ is $uv$-tidy$\}$
19:             **case** forget
20:                 let $v$ be the node removed by $B$ and let $B'$ be the child bag of $B$
21:                 $t = \varnothing$
22:                 $P = $ PROCESS($B'$)
23:                 **for** $(\mathbf{x}, A^{\mathrm{bag}}) \in P$ **do**
24:                     **if** $A^{\mathrm{bag}}(v) = (v, i)$ for some $i$ **then**    $\triangleright$ $v$ is a center with radius $r_i$
25:                         **if** $\mathbf{x} + \mathbf{e}_i \in \mathcal{C}$ **then** $t = t \cup \{(\mathbf{x} + \mathbf{e}_i, A^{\mathrm{bag}}|_B)\}$
26:                     **else**                                      $\triangleright$ $v$ is not a center
27:                         $t = t \cup \{(\mathbf{x}, A^{\mathrm{bag}}|_B)\}$
28:                 **return** $t$
29:         $solutions = $ PROCESS($root\_bag$)
30:         **return** True if $solutions \neq \varnothing$

denotes the restriction of the function $A^{\mathrm{tree}}$ to the set of vertices of $G_{B'}$. Since $V(G_B) \backslash B = V(G_{B'}) \backslash B'$, the partial cost vector of $(A^{\mathrm{tree}})'$ is equal to the partial cost vector of $A^{\mathrm{tree}}$. We also already know that PROCESS($B'$) = $\mathit{Config}_{B'}$. Thus PROCESS($B$) = $\mathit{Config}_B$.

**Join Bag.** Observe that $G_{B'} \cup G_{B''} = G_B$. Thus $A^{\mathrm{tree}}$ is a tidy subtree assignment for $B$ if and only if $A^{\mathrm{tree}}|_{G_{B'}}$ and $A^{\mathrm{tree}}|_{G_{B''}}$ are tidy subtree assignments for $B'$ and $B''$ respectively. Since $V(G_{B'}) \cap V(G_{B''}) = B = B' = B''$, a subtree assignment for $B'$ can be combined with a subtree assignment for $B''$ if and only if their restrictions to $B$ are equal. Furthermore, the partial cost vector of $A^{\mathrm{tree}}$ is equal to the sum of the partial cost vectors of $A^{\mathrm{tree}}|_{G_{B'}}$ and $A^{\mathrm{tree}}|_{G_{B''}}$, as the sets of forgotten nodes are disjoint. Thus PROCESS($B$) = $\mathit{Config}_B$.

**Edge Bag.** $A^{\mathrm{tree}}$ is a subtree assignment of $B$ if and only if it is also a subtree assignment of $B'$. $A^{\mathrm{tree}}$ is a tidy subtree assignment of $B$ if and only if it is a tidy subtree assignment of $B'$ and is $vw$-tidy. Thus $(\boldsymbol{x}, A) \in Config_B$ if and only if $(\boldsymbol{x}, A) \in Config_{B'}$ and $A$ is $vw$-tidy and thus $\text{PROCESS}(B) = Config_B$.

**Forget Bag.** $A^{\text{tree}}$ is a subtree assignment for $B$ if and only if it is also a subtree assignment for $B'$. We have $B \cup \{v\} = B'$, $F_B = F_{B'} \cup \{v\}$, and if $(\boldsymbol{x}, A^{\text{bag}}) \in Config_{B'}$, then $A^{\text{bag}}$ is a valid assignment and thus it is a restriction of some tidy subtree assignment $A^{\text{tree}}$ for $B'$. Then $(\boldsymbol{y}, A^{\text{bag}}|_B) \in Config_B$ where $\boldsymbol{y} = \boldsymbol{x} + \boldsymbol{e}_i$ if $(v, i)$ is in the image of $A^{\text{tree}}$ and $\boldsymbol{y} = \boldsymbol{x}$ otherwise. Lemma 3.6 implies that $A^{\text{bag}}(v) = (v, i)$ if and only if $(v, i)$ is in the image of $A^{\text{tree}}$, regardless of the choice of $A^{\text{tree}}$, and thus $\text{PROCESS}(B) = Config_B$. $\qquad\square$

**Theorem 3.11.** *Algorithm* GRAPH-COVERAGE *solves the NUkC problem in polynomial time, namely with* $\mathcal{O}(|\mathbb{T}| n^{2\ell + \text{tw}} \ell^{2 \cdot \text{tw}})$ *arithmetic operations.*

*Proof.* By Lemma 3.10, $\text{PROCESS}(root\_bag) = Config_{root\_bag}$. By definition of valid configurations, a valid configuration exists for the root bag if and only if there exists a tidy assignment for the entire graph with a valid cost vector. The correctness of the algorithm follows.

For the bound on time complexity, we recall the fact that the number of valid configurations for any bag is at most $n^{\text{tw}+\ell} \ell^{\text{tw}}$. Thus the set of valid configurations for a bag can be computed in $\mathcal{O}(n^{2\ell + \text{tw}} \ell^{2 \cdot \text{tw}})$ if given the sets of valid configurations for each child. $\qquad\square$

## 4   A PTAS for Burning Graphs of Small Treewidth

As indicated in the introduction, we deviate from the scheduling view of graph burning and instead view it as a non-uniform $k$-center problem with centers of different radii.

An instance of the graph burning problem consists of a graph $G$ on $n$ nodes with edge lengths from $\mathbb{R}_0^+$. The distance $d(u, v)$ is defined as the length of the shortest path from $u$ to $v$. For the decision variant of the problem, we are also given a natural number $g$. The task is to determine if it is possible to select $g$ centers, each with a unique integer radius from 0 to $g - 1$, so that the centers cover all nodes of $G$. We can view this problem as a variation of NUkC problem where $\ell = g$ and $r_i = i - 1$. For the optimization problem, the objective is to minimize $g$. Note that even if the edge lengths are non-integral and/or large, the possible values of $g$ are restricted to integers $1, \ldots, n$.

Let us now demonstrate how the GRAPH-COVERAGE algorithm can be used to construct a PTAS for the graph burning on graphs of small treewidth. In particular, for every $\epsilon > 0$ there exists an algorithm that is in XP when parameterized by treewidth of $G$ and always either proves that no solution for $g$ exists or finds a solution for $\lfloor (1 + \epsilon)g \rfloor$. We shall once again assume that we are given a tree decomposition $\mathbb{T}$ for $G$ with $|B| \leq \text{tw} + 1$ for all bags.

**Modifying Instances.** When given an instance $I$ of the graph burning problem, we proceed by constructing an instance $I'$ of NUkC problem which has a solution if $I$ had a solution. We then show that if $I'$ has a solution then a solution exists for instance $I''$ of graph burning, where $I''$ differs from $I$ only in the parameter $g'' \leqslant (1 + 2\epsilon)g + 2$.

We construct $I'$ by increasing the radii of the centers in such a way as to reduce the number of distinct radii being used while not increasing any radius by more than $1 + g\epsilon$.

**Lemma 4.1.** *If $S$ is a solution for an instance $I = (G, (r_1, \ldots, r_\ell), (k_1, \ldots, k_\ell))$ of NUkC problem with a valid cost vector, and both $r'_i \geqslant r_i$ and $k'_i \geqslant k_i$ for all $i$, then $S$ is also a solution for an instance $I' = (G, (r'_1, \ldots, r'_\ell), (k'_1, \ldots, k'_\ell))$ of NUkC problem with a valid cost vector.*

*Proof.* By Definition 2.1 $K \subseteq V(G) \times \{1, \ldots, \ell\}$, $K_v = \{(w, i) \in K \mid d(v, w) \leqslant r_i\}$, and $S \subseteq K$ is a solution to the instance $I = (G, \boldsymbol{r}, \boldsymbol{x})$ if and only if $S \cap K_v \neq \varnothing$ for all $K_v$.

Since $r_i \leqslant r'_i$, we know that $K_v \subseteq K'_v$ and thus $S \cap K'_v \neq \varnothing$ if $S \cap K_v \neq \varnothing$. Thus $S$ is a solution to $I'$ if it is a solution to $I$.

Since $k_i \leqslant k'_i$, if a solution's cost vector is valid for $I$, it is also valid for $I'$. $\square$

**Lemma 4.2.** *Let $I = (G, \boldsymbol{r}, \boldsymbol{k})$ and $I' = (G, \boldsymbol{r'}, \boldsymbol{k'})$ be instances of NUkC problem where $\boldsymbol{r'}$ contains $k_i$ copies of $r_i$ and $\boldsymbol{k'}$ consists of $\sum k_i$ ones. Then the instances $I$ and $I'$ are equivalent, i.e., given a solution $S$ of $I$, one can construct a solution $S'$ of $I'$ and vice versa.*

*Proof.* Given $S$, we construct $S'$ by changing, for each $i$, the (at most) $k_i$ centers $(v, i)$ into centers $(v, j)$ with distinct $j$ such that $r'_j = r_i$; by definition, $k_i$ of such $j$'s are available. Conversely, given $S'$, we replace each center $(v, j)$ by a center $(v, i)$ use such that $r_i = r'_j$. For a given $r$, let $R = \{i \mid r_i = r\}$. Then $S'$ has at most $\sum_{i \in R} k_i$ centers with $r'_j = r$, so we can distribute them among the $i \in R$ to give a solution $S$. $\square$

---

**Algorithm** GRAPH-BURNING

---
1: **function** TwBURN(*root_bag*, $g$, $\epsilon$)
2:     $s = 1 + \lfloor g\epsilon \rfloor$
3:     $\ell = \lceil g/s \rceil$
4:     $(r'_1, r'_2, r'_3, \ldots, r'_\ell) = (s - 1, 2s - 1, 3s - 1, \ldots, \ell s - 1)$     $\triangleright$ $g \leqslant \ell s \leqslant 1 + (1 + \epsilon)g$
5:     $(k_1, k_2, k_3, \ldots, k_\ell) = (s, s, s, \ldots, s)$
6:     $\mathcal{C} = \{(y_1, y_2, \ldots, y_\ell) \in \mathbb{N}^\ell \mid (\forall i)(y_i \leqslant s)\}$
7:     **return** TwNUkC(*root_bag*, $r_1, r_2, \ldots, r_\ell, k_1, k_2, \ldots, k_\ell, \mathcal{C}$)

---

**Theorem 4.3.** *Algorithm* GRAPH-BURNING *gives a PTAS for the graph burning problem on graphs with a constant treewidth.*

*Proof.* We show that (i) for any positive instance, GRAPH-BURNING gives a positive answer and (ii) whenever GRAPH-BURNING gives a positive answer, then there exists a solution for the graph burning instance with $g$ increased to $(1 + 2\varepsilon)g$.

The instance $I$ of graph burning can also be viewed as an instance of NUkC problem with $\ell = g$, $r_i = i - 1$ and $\mathcal{C} = \{0, 1\}^\ell$. By Lemma 4.1, if a solution with a valid cost vector exists for $I$, there also exists a solution with a valid cost vector for instance where $r_i$ is increased to the nearest $ts - 1 \geq r_i$. The instance $I'$ is then created by grouping the sets of $s$ identical radii together and (possibly) adding up to $s - 1$ centers of the greatest radius. Thus, by Lemma 4.2, a solution for $I'$ exists if a solution for $I$ exists. This proves (i), as $I'$ is the instance submitted to TwNUkC at line 7 of the algorithm.

To show (ii), let $S'$ be a solution for instance $I'$ of NUkC problem. By Lemma 4.2, we can construct a solution for an instance where the identical radii are ungrouped. We can then replace the $s$ copies of radius $r'_i$ with $r'_i, r'_i + 1, \ldots, r'_i + s - 1$, thus getting an instance with $(1+\epsilon)g$ unique radii $s-1, s, s+1, \ldots, (\ell+1)s-2$ for which, by Lemma 4.1, we can construct a solution given a solution to $I'$. Since $(\ell+1)s = (\lceil g/s \rceil + 1)s < (g/s + 2)s = g + 2s$, we have $(\ell+1)s - 2 \leq g + 2s - 3 \leqslant g + 2g\epsilon - 1$. Thus the centers used by this instance are a subset of the centers available in an instance of graph burning $(G, \lfloor (1 + 2\epsilon)g \rfloor)$.

Time complexity is entirely dominated by the complexity of TwNUkC. Since $s = 1 + \lfloor g\epsilon \rfloor$, we know that $\ell = \lceil g/s \rceil \leqslant 1 + 1/\epsilon$ and thus we get $\mathcal{O}(|\mathbb{T}|n^{2+2/\epsilon+\text{tw}}(1 + 1/\epsilon)^{2 \cdot \text{tw}})$. $\square$

## 5   Conclusions

We presented an algorithm that puts the non-uniform $k$-center problem in XP when parameterized by the number of different radii and the treewidth of the graph. We then used this algorithm to construct a PTAS for the graph burning problem for graphs of small treewidth. This significantly extends the PTAS from [4], which only applied to linear forests.

**Acknowledgements.** Partially supported by project SVV-2020-260578 and GA ČR project 19-27871X. We are grateful to anonymous referees for many helpful comments and references.

## References

1. Arnborg, S., Lagergren, J., Seese, D.: Easy problems for tree-decomposable graphs. J. Algorithms **12**(2), 308–340 (1991). https://doi.org/10.1016/0196-6774(91)90006-K
2. Bessy, S., Bonato, A., Janssen, J., Rautenbach, D., Roshanbin, E.: Burning a graph is hard. Discrete Appl. Math. **232**, 73–87 (2017). https://doi.org/10.1016/j.dam.2017.07.016
3. Bonato, A., Janssen, J., Roshanbin, E.: How to burn a graph. Internet Math. **12**(1–2), 85–100 (2015). https://doi.org/10.1080/15427951.2015.1103339

4. Bonato, A., Kamali, S.: Approximation algorithms for graph burning. In: Gopal, T.V., Watada, J. (eds.) TAMC 2019. LNCS, vol. 11436, pp. 74–92. Springer, Cham (2019). https://doi.org/10.1007/978-3-030-14812-6_6

5. Chakrabarty, D., Goyal, P., Krishnaswamy, R.: The non-uniform k-center problem. ACM Trans. Algorithms 16(4) (2020). https://doi.org/10.1145/3392720

6. Charikar, M., Khuller, S., Mount, D.M., Narasimhan, G.: Algorithms for facility location problems with outliers. In: Kosaraju, S.R. (ed.) Proceedings of the 12th Annual Symposium on Discrete Algorithms (SODA), pp. 642–651. ACM/SIAM (2001). http://dl.acm.org/citation.cfm?id=365411.365555

7. Cygan, M., Nederlof, J., Pilipczuk, M., Pilipczuk, M., van Rooij, J., Wojtaszczyk, J.O.: Solving connectivity problems parameterized by treewidth in single exponential time (2011)

8. Demaine, E.D., Fomin, F.V., Hajiaghayi, M., Thilikos, D.M.: Fixed-parameter algorithms for (k, r)-center in planar graphs and map graphs. ACM Trans. Algorithms 1(1), 33–47 (2005). https://doi.org/10.1145/1077464.1077468

9. Feldmann, A.E.: Fixed-parameter approximations for k-center problems in low highway dimension graphs. Algorithmica 81(3), 1031–1052 (2019). https://doi.org/10.1007/s00453-018-0455-0

10. García-Díaz, J., Pérez-Sansalvador, J.C., Rodríguez-Henríquez, L.M.X., Cornejo-Acosta, J.A.: Burning graphs through farthest-first traversal. IEEE Access 10, 30395–30404 (2022). https://doi.org/10.1109/ACCESS.2022.3159695

11. Hakimi, S.L.: Optimum locations of switching centers and the absolute centers and medians of a graph. Oper. Res. 12(3), 450–459 (1964). https://doi.org/10.1287/opre.12.3.450

12. Hochbaum, D.S., Shmoys, D.B.: A best possible heuristic for the k-center problem. Math. Oper. Res. 10(2), 180–184 (1985). https://doi.org/10.1287/moor.10.2.180

13. Inamdar, T., Varadarajan, K.R.: Non-uniform k-center and greedy clustering. In: Czumaj, A., Xin, Q. (eds.) 18th Scandinavian Symposium and Workshops on Algorithm Theory, SWAT. LIPIcs, vol. 227, pp. 28:1–28:20. Schloss Dagstuhl - Leibniz-Zentrum für Informatik (2022). https://doi.org/10.4230/LIPIcs.SWAT.2022.28

14. Jia, X., Rohwedder, L., Sheth, K., Svensson, O.: Towards non-uniform k-center with constant types of radii. In: Bringmann, K., Chan, T. (eds.) 5th Symposium on Simplicity in Algorithms, SOSA@SODA 2022, pp. 228–237. SIAM (2022). https://doi.org/10.1137/1.9781611977066.16

15. Katsikarelis, I., Lampis, M., Paschos, V.T.: Structural parameters, tight bounds, and approximation for (k, r)-center. Discrete Appl. Math. 264, 90–117 (2019). https://doi.org/10.1016/j.dam.2018.11.002

16. Kloks, T. (ed.) Pathwidth of pathwidth-bounded graphs. In: Treewidth. LNCS, vol. 842, pp. 147–172. Springer, Heidelberg (1994). https://doi.org/10.1007/BFb0045388

17. Kobayashi, Y., Otachi, Y.: Parametrized complexity of graph burning. Algorithmica 84, 2379–2393 (2022). https://doi.org/10.1007/s00453-022-00962-8

# Scheduling with Machine Conflicts

Moritz Buchem[1], Linda Kleist[2],
and Daniel Schmidt genannt Waldschmidt[3]([✉])

[1] School of Business and Economics, Maastricht University,
Maastricht, The Netherlands
m.buchem@maastrichtuniversity.nl
[2] Department of Computer Science, TU Braunschweig, Braunschweig, Germany
kleist@ibr.cs.tu-bs.de
[3] Institute for Mathematics, TU Berlin, Berlin, Germany
dschmidt@math.tu-berlin.de

**Abstract.** We study the scheduling problem of makespan minimization with machine conflicts that arise in various settings, e.g., shared resources for pre- and post-processing of tasks or spatial restrictions. In this context, each job has a blocking time before and after its processing time, i.e., three parameters. Given a set of jobs, a set of machines, and a graph representing machine conflicts, the problem SCHEDULINGWITH-MACHINECONFLICTS (SMC), asks for a conflict-free schedule of minimum makespan in which the blocking times of no two jobs intersect on conflicting machines.

We show that, unless P = NP, SMC on $m$ machines does not allow for a $\mathcal{O}(m^{1-\varepsilon})$-approximation algorithm for any $\varepsilon > 0$, even in the case of identical jobs and every choice of fixed positive parameters, including the unit case. Complementary, we provide approximation algorithms when a suitable collection of independent sets is given. Finally, we present polynomial time algorithms to solve the problem for the case of unit jobs SMC-UNIT on special graph classes. As our main result, we solve SMC-UNIT for bipartite graphs by using structural insights for conflict graphs of star forests. As the set of active machines at each point in time induces a bipartite graph, the insights yield a local optimality criterion.

**Keywords:** Scheduling · Machine conflict · Approximation algorithm · NP-hard · Inapproximability · Star forest · Bipartite graph

## 1 Introduction

Distributing tasks smartly is a challenge we face in numerous settings, ranging from every day life to optimization of industrial processes. Often these assignments must satisfy additional requirements. In this work, we study a variant

D. Schmidt genannt Waldschmidt—was funded by the DFG under Germany's Excellence Strategy - The Berlin Mathematics Research Center MATH+ (EXC-2046/1, project ID: 390685689).

P. Chalermsook and B. Laekhanukit (Eds.): WAOA 2022, LNCS 13538, pp. 36–60, 2022.
https://doi.org/10.1007/978-3-031-18367-6_3

of the well-studied scheduling problem of makespan minimization when machine conflicts are present. These conflicts arise in various contexts as a result of shared resources or spatial constraints which prohibit machines to complete certain tasks simultaneously. We are particularly interested in situations when external pre- and post-processing of jobs is necessary immediately before and after jobs are internally processed by a machine.

Conflicts of pre- and post-processing may be due to shared resources or spatial constraints. Examples of shared resources arise in manufacturing and logistics, where a common server is used for loading and unloading jobs onto and from machines immediately before or after jobs can be processed. Specific examples mentioned in the literature include manufacturing systems served by a single robot which can only serve one machine at a time [22, 30] or steel production in which furnaces must be served non-preemptively before and after heating processes [41]. Another example of shared resources appears in computing problems, in which different processors may share different databases or external processors that must be accessed before and after executing tasks on the processor and can only be accessed by one processor at a time. An up-to-date example of spatial conflicts occurs in pandemics when schedulers are faced with potentially infectious jobs which should keep sufficient distance to each other, e.g., in testing or vaccination centers. Similarly, spatial conflicts play a crucial role when jobs may have private information or data that should not be shared; e.g., the interrogation of suspects in multiple rooms.

**The Problem.** SCHEDULINGWITHMACHINECONFLICTS (SMC) is a scheduling problem in which jobs on conflicting machines are processed such that certain blocking intervals of their processing time do not overlap. An instance of SMC is defined by a set of jobs $\mathcal{J}$ and a conflict graph $G = (V, E)$ on a set of machines $V$ where two machines $i$ and $i'$ are *in conflict* if and only if $\{i, i'\} \in E$. In contrast to classical scheduling problems, each job $j$ has three parameters $(\overleftarrow{b}_j, p_j, \overrightarrow{b}_j)$, where $\overleftarrow{b}_j$ and $\overrightarrow{b}_j$ denote the first and second *blocking time* of $j$, respectively, and $p_j$ denotes its *processing time*. Together they constitute the *system time* $q_j = \overleftarrow{b}_j + p_j + \overrightarrow{b}_j$; note that the order $\overleftarrow{b}_j, p_j, \overrightarrow{b}_j$ must be maintained. We seek schedules in which the blocking times of no two jobs on conflicting machines intersect. For an example consider Fig. 3. Formally, a *(conflict-free) schedule $\Pi$* is an assignment of jobs to machines and starting times such that

- for each point in time, every machine executes at most one job,
- for every edge $\{i, i'\} \in E$ and two jobs $j, j' \in \mathcal{J}$ assigned to machines $i$ and $i'$, respectively, the intervals of the blocking times of $j$ and $j'$ do not overlap interiorly in time.

Moreover, jobs cannot be interrupted, i.e., the starting and completions times of each job differ by the system time. In other words, all schedules are non-preemptive. The *makespan* $\|\Pi\|$ of a schedule $\Pi$ is defined as the earliest point in time when all jobs are completed. We seek for a schedule with minimum makespan. Throughout this paper, we use $n := |\mathcal{J}|$ and $m := |V|$ to refer to the number of jobs and machines, respectively.

## 1.1   Our Contribution and Organization

We first consider the problem SMC with identical jobs (SMC-ID). Identifying intrinsic connections to maximum independent sets, we show that even if $(\overleftarrow{b}, p, \overrightarrow{b})$ are fixed positive parameters for any $\epsilon > 0$, there is no $\mathcal{O}(m^{1-\varepsilon})$-approximation for SMC-ID, unless $P = NP$ (Theorem 1). However, when a suitable collection of maximum independent sets is given or can be found in polynomial time, we present approximation algorithms with performance guarantee better than 2.5 (Theorem 2). An approximation algorithm can also be obtained when approximate maximum independent sets are at hand (Theorem 3).

In Sect. 3, we consider SMC with unit jobs (SMC-UNIT), i.e., $\overleftarrow{b} = p = \overrightarrow{b} = 1$. Motivated by the inapproximability result for SMC-UNIT on general graphs (Theorem 1), we focus on special graph classes. As our main result, we present a polynomial time algorithm to compute optimal schedules on bipartite graphs. Bipartite graphs are of special interest, because for any conflict graph and for every point in time, the set of active machines induces a bipartite graph. Hence, our insights can be understood as *local optimality criteria* of schedules for all graphs. To solve the problem to optimality we develop a divide-and-conquer algorithm based on structural insights for stars. Moreover, we provide an efficient representation of schedules so that the running time of our algorithm is polynomial in the size of $G$ and $\log(n)$.

Full details of all proofs are presented in the appendix (see Sect. 5 and Sect. 6). A full version can be found at [11].

## 1.2   Related Work

The problem SMC generalizes the classical scheduling problem of *makespan minimization on parallel identical machines*, also denoted by $P||C_{\max}$ in the three-field notation [21]. In fact, $P||C_{\max}$ is equivalent to two special cases of SMC:

- if the blocking times of all jobs vanish, i.e., $\overleftarrow{b}_j = \overrightarrow{b}_j = 0$ for all $j \in \mathcal{J}$, and
- if the edge set of the conflict graph is the empty set.

For a constant number of machines, $P||C_{\max}$ is weakly NP-hard, while it is strongly NP-hard when $m$ is part of the input [18]. Graham [19,20] introduced list scheduling algorithms to obtain the first constant approximation algorithms for this problem. Improved approximation guarantees have been achieved by a fully polynomial time approximation scheme (FPTAS) when $m$ is constant [38] and a polynomial time approximation scheme (PTAS) when $m$ is part of the input [26]. In subsequent work, the latter has been improved to efficient polynomial time approximation schemes (EPTAS), we refer to [4,13,25,28,29].

*Scheduling with pre- and post-processing* has been considered in different models in the literature. One such model was introduced by Hall et al. [22] in which jobs must be scheduled non-preemptively on identical parallel machines but have to be pre-processed by a common server immediately. This model corresponds to SMC with $\overrightarrow{b}_j = 0$ for all jobs $j$ and a complete graph as the conflict graph. The special cases of unit first blocking times $\overleftarrow{b}_j = 1 \,\forall j$ and $m = 2$ [22]

and the case of identical processing times $p_j = p \, \forall j$ and $m = 2$ [10] were shown to be weakly NP-hard, while the cases of $\overleftarrow{b}_j = \overleftarrow{b} \, \forall j$ and $m = 2$ [22] and the case of $\overleftarrow{b}_j = 1 \, \forall j$ [35] were shown to be strongly NP-hard. On the positive side, if $\overleftarrow{b}_j = p_j = 1 \, \forall j$ the problem can be solved in time $\mathcal{O}(n)$ [22]. Moreover, Kravchenko and Werner [35] present a pseudo-poynomial algorithm for the case of $\overleftarrow{b}_j = 1 \, \forall j$ and $m = 2$. Xie et al. [41] extend this model to a single server used for pre- and post-processing. This problem corresponds to SMC with $K_m$ as the conflict graph. Xie et al. [41] and Jiang et al. [30] analyze the worst case performance of the list scheduling algorithms introduced by Graham [19,20]. Furthermore, heuristics and mixed-integer programming techniques were developed for several special cases [1–3,16,33]. Two other models are the master-and-slave problem introduced by Kern and Nawijn [31] and termed by Sahni [39] and the problem of scheduling jobs with segmented self-suspension introduced by Rajkumar et al. [37]. Chen et al. [12] present an approximation algorithm for the special case of a single suspension interval in which each job consists of three components.

The concept of *machine conflicts* has previously also been considered in the context of buffer minimization on multiprocessor systems. Chrobak et al. [14] show that there is no polynomial approximation ratio unless P = NP. For the online case, they present competitive algorithms for general graphs as well as special graph classes. Höhne and van Stee [27] develop competitive algorithms when the conflict graph is a path.

Scheduling with conflict graphs has also been investigated in presence of *job conflicts*. While in one model, the conflicting jobs cannot be scheduled on the same machine [7,9,15], in a second model, the conflicting jobs may not be processed concurrently on different machines [5,6,8,17,23,40]. In these works, the complexity and approximability has been investigated for special classes of conflict graphs in both settings.

## 2    Identical Jobs

In this section, we consider SMC with identical jobs, denoted by SMC-ID. More specifically, SMC-ID$(G, n, (\overleftarrow{b}, p, \overrightarrow{b}))$ denotes an instance with a conflict graph $G$ and $n$ identical jobs with parameters $(\overleftarrow{b}, p, \overrightarrow{b})$. We present hardness and approximation results for any fixed choice of $\overleftarrow{b}, p, \overrightarrow{b}$. Using the fact that there exists no $O(m^{\varepsilon-1})$-approximation for computing maximum independent sets [24,36,42], we obtain an inapproximability result for all fixed positive constants $\overleftarrow{b}, p, \overrightarrow{b}$.

**Theorem 1.** *For any $\epsilon > 0$, there exists no $\mathcal{O}(m^{1-\varepsilon})$-approximation for SMC-ID, unless P = NP. This even holds for any fixed positive parameters $(\overleftarrow{b}, p, \overrightarrow{b})$.*

Note that this result holds even for the case when the running time depends polynomially on $n$ (instead of $\log(n)$).

*Proof-Sketch.* We distinguish the two cases when either all jobs have long blocking times, i.e., $\max\{\overleftarrow{b}, \overrightarrow{b}\} > p$ (Theorem 5), or all jobs have short blocking times, i.e., $\max\{\overleftarrow{b}, \overrightarrow{b}\} \leq p$ (Theorem 8). By symmetry, we may assume that $\overrightarrow{b} \leq \overleftarrow{b}$.

For long blocking times, we show that any schedule is *basic*, i.e., it uses only an independent set of machines (Lemma 8). The inapproximability result follows from the connection to the problem of finding a maximum independent set.

For short blocking times, we use a notion generalizing a maximum independent set. For a graph $G = (V, E)$ and $c \in \mathbb{N}_{\geq 1}$, a *maximum induced c-colorable subgraph*, or short *maximum c-IS*, of $G$ is a set of $c$ disjoint independent sets $\mathcal{I}_1, \ldots, \mathcal{I}_c \subseteq V$ whose union has maximum cardinality. Clearly, a 1-IS is an independent set. Any schedule induces a $c$-IS and it can be found in polynomial time (Lemma 9). Because the length of a schedule is related to the size of its induced $c$-IS (Lemma 12), an approximation algorithm yields an approximate maximum $c$-IS with a performance guarantee of the same order. However, for any $c \in \mathbb{N}_{\geq 1}$, a maximum $c$-ISs is inapproximable [36].

By the proof of Theorem 1, finding a maximum $c$-IS polynomially reduces to SMC-ID. One may wonder how the difficulty changes, when maximum $c$-ISs of the graph are at hand. We show that if we are given a suitable collection of maximum independent sets of the conflict graph, we obtain approximation algorithms with performance guarantee better than 2.5. To this end, we define a class of partial schedules using a collection of independent sets of machines.

**c-Patterns.** Consider an instance of SMC-ID on a conflict graph $G$. Let $c \in \mathbb{N}_{\geq 1}$ with $c \leq \lfloor p/\overleftarrow{b} \rfloor + 1$ and let $\mathcal{I} = (I_1, I_2, \ldots, I_c)$ be a $c$-tuple of disjoint independent sets of $G$. A partial schedule of length $q + (c - 1) \cdot \overleftarrow{b}$ starting at time $t$ is called a *c-pattern on $\mathcal{I}$* if on each machine $i$ in $I_\ell$ with $\ell \in \{1, \ldots, c\}$, there is one job starting at time $t + (\ell - 1)\overleftarrow{b}$, see also Fig. 1.

**Fig. 1.** A 3-pattern on three disjoint independent sets $I_1, I_2, I_3$.

**Theorem 2.** *Let $G$ be a graph and let $q = \overleftarrow{b} + p + \overrightarrow{b}$ be the system time. If $\max\{\overleftarrow{b}, \overrightarrow{b}\} > p$ and we are given a maximum 1-IS of $G$, then an optimal schedule of SMC-ID can be computed in polynomial time.*

*If $0 < \overleftarrow{b} \leq \overrightarrow{b} \leq p$ and we are given a maximum $(\lfloor p/\overleftarrow{b} \rfloor + 1)$-IS of $G$, then SMC-ID allows for a $(2 + (\lfloor p/\overleftarrow{b} \rfloor + 1)^{-1})$-approximation, where $(2 + (\lfloor p/\overleftarrow{b} \rfloor + 1)^{-1}) \leq 2.5$. If $p/\overleftarrow{b} \in \mathbb{N}$, it even allows for an $(1 + p/q)$-approximation, where $(1 + p/q) < 2$.*

*Proof-Sketch.* For long blocking times, any schedule is basic and hence, an optimal schedule uses the provided maximum independent set and can thus be computed in polynomial time (Theorem 6).

For short blocking times, we provide a lower bound on the optimal makespan. Specifically, we bound the number of jobs starting within an interval of length $\lambda$ for some specific $\lambda \leq q$. We then show that the number of jobs is bounded by $\alpha$, the size of a maximum $(\lfloor p/\overleftarrow{b} \rfloor + 1)$-IS. This yields a lower bound of $\lambda \cdot \lceil n/\alpha \rceil$

on the optimal makespan. To obtain an upper bound, we construct a schedule which repeatedly uses $(k+1)$-patterns, where $k = \lfloor p/\overleftarrow{b} \rfloor$.    □

Similarly, if we are given an approximate $c$-IS of the conflict graph for some suitable $c$, corresponding approximation results can be derived.

**Theorem 3.** *Let $G$ be a graph. If* $\max\{\overleftarrow{b}, \overrightarrow{b}\} > p$ *and we are given a* $1/\gamma$-*approximate 1-IS of $G$, then* SMC-ID *allows for a* $\lceil \gamma \rceil$-*approximation.*

*If* $0 < \overrightarrow{b} \le \overleftarrow{b} \le p$ *and we are given a* $1/\gamma$-*approximate* $(\lfloor p/\overleftarrow{b} \rfloor + 1)$-*IS of $G$, then* SMC-ID *allows for a* $5\gamma$-*approximation.*

## 3    Unit Jobs

Now, we turn our attention to SMC-UNIT in which we are given $n$ identical unit jobs where $\overleftarrow{b} = p = \overrightarrow{b} = 1$ for all jobs. On one hand, there exists no $\mathcal{O}(m^{1-\varepsilon})$-approximation for SMC-UNIT on general graphs with $m$ vertices, unless $\mathrm{P} = \mathrm{NP}$ (Theorem 1). On the other hand, Theorem 2 yields a $4/3$-approximation algorithm for SMC-UNIT if we are given a maximum 2-IS. Therefore, to improve this performance guarantee, we focus on special graph classes (for which maximum 1- and 2-IS can be computed in polynomial time).

Complete graphs play a special role in the context when machines share a single resource. We show that SMC-UNIT on complete graphs can be reduced to SMC-UNIT on a single edge and solved efficiently.

**Lemma 1.** *For every $n$, an optimal schedule for* SMC-UNIT$(K_m, n)$ *can be computed in time linear in* $\log n$. *In particular, for* $m \ge 2$, *it coincides with an optimal schedule for $K_2$ of makespan* $4\lfloor n/2 \rfloor + 3(n \mod 2)$.

Bipartite graphs constitute the arguably most interesting graph class in this context because for each schedule, the set of active machines at any point in time induces a bipartite subgraph. Therefore, optimal schedules on bipartite graphs can be understood as a local optimality criterion.

**Observation 2.** *Consider an instance* SMC-UNIT *on a graph $G$ and a feasible schedule $\Pi$. For every point in time $t$, the set of machines processing a job at time $t$ in $\Pi$ induces a bipartite subgraph of $G$.*

Note that a maximum 1-IS of a bipartite graph can be computed in polynomial time [34] and the maximum 2-IS is trivially the entire vertex set. In the remainder, we present a polynomial time algorithm to compute optimal schedules.

**Theorem 4.** *For every bipartite graph $G$ and every $n$, an optimal schedule for* SMC-UNIT$(G, n)$ *can be computed in polynomial time.*

Our algorithm is based on a divide-and-conquer technique. In a first step, we derive structural insights of optimal schedules on stars which allow to solve SMC-UNIT on stars in polynomial time (Sect. 3.1). In a second step, we show how to exploit these insights to find optimal schedules on general bipartite graphs by considering a subgraph whose components are stars with special properties (Sect. 3.2). Finally, we present a polynomial time algorithm to find an adequate subgraph (Sect. 3.3).

### 3.1 Stars

An essential step towards our polynomial time algorithm for bipartite graphs is to investigate the structure of optimal schedules on stars. A *star* is a complete bipartite graph $S_\ell := K_{1,\ell}$ on $\ell + 1 \geq 2$ vertices. For $\ell \geq 2$, $S_\ell$ has $\ell$ leaves and a unique *center* of degree $\ell$. For $\ell = 1$, either vertex can be seen as the center of $S_1$. We show that optimal schedules on stars can be obtained by using special types of 1- and 2-patterns (see definition of $c$-pattern for $c = 1, 2$). We define the special patterns as follows.

**A/B-Patterns.** An *A-pattern* on a graph $H$ is a 1-pattern on some maximum 1-IS of $H$. A *B-pattern* on $H$ is a 2-pattern on some maximum 2-IS of $H$. Note that the difference to 1- and 2-patterns is that we do not specify the 1- and 2-ISs. An *AB-schedule* on $H$ consists of A- and B-patterns only. Consider Fig. 2 for an example. Let $n \in \mathbb{N}$. We say SMC-UNIT$(H, n)$ admits an optimal AB-schedule if there exists an optimal schedule that can be transformed into an AB-schedule on $H$ by possibly adding more jobs without increasing the makespan.

Using these patterns we derive a structural property of optimal schedules for SMC-UNIT on stars that allows us to compute them in polynomial time.

**Lemma 3.** *For every star $S$ and every $n$, SMC-UNIT$(S, n)$ admits an optimal AB-schedule.*

*Proof.* The statement is obvious for $S_1$. For a star $S_\ell$ with $\ell \geq 2$, consider an optimal schedule and determine a leaf processing a maximum number of jobs. Changing all leaves to this schedule may only increase the number of processed

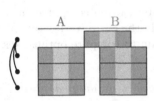

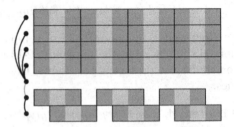

**Fig. 2.** An AB-schedule on $S_3$ consisting of one A-pattern followed by one B-pattern.

**Fig. 3.** Optimal schedule on a tree with 7 machines and 22 jobs with makespan 12.

jobs and yields a valid schedule in which all $\ell$ leaves have the same induced schedule. Finally, for each job $j$ processed on a leaf of $S_\ell$, there exist two cases: If no job is processed on the center vertex of the star during the system time (of length 3) of $j$, we obtain an A-pattern. Otherwise, the job on the center and the job on the leaves are shifted by exactly one time step and we obtain a B-pattern.

□

**Corollary 1.** *For every star $S$ and every $n$, an optimal schedule for* SMC-UNIT$(S, n)$ *can be computed in time linear in* $\log n$ *and* $|S|$.

*Specifically, for every $S_\ell$, there exists $X \in \{A, B\}$ such that an optimal schedule has at most 2 $X$-patterns, i.e., an optimal schedule has makespan*

$$\min_{k=0,1,2} \left\{ 4 \left\lceil \frac{n - k\ell}{\ell + 1} \right\rceil + 3k, 3 \left\lceil \frac{n - k(\ell + 1)}{\ell} \right\rceil + 4k \right\}, \tag{1}$$

*where $\lceil \cdot \rceil$ denotes the usual ceiling function; however, for negative reals it evaluates to 0.*

We later exploit the fact that for $S_3$, there exist two optimal AB-schedules which finish twelve jobs in time 12, namely 4 A-patterns as well as 3 B-patterns. This fact provides some flexibility for designing optimal schedules for general bipartite graphs.

## 3.2    Optimal Schedules for Bipartite Graphs for a Given Star Forest

While we can restrict our attention to AB-schedules for stars, this property does not generalize to all bipartite graphs. In fact, it does not even hold for trees as illustrated in Fig. 3.

**Observation 4.** *There exists a tree $T$ such that no optimal schedule for* SMC-UNIT$(T, n)$ *is an AB-schedule with respect to $T$.*

Interestingly, the optimal schedule shown in Fig. 3 is comprised of two AB-schedules on the two stars obtained by deleting the gray edge. Combining this insight with the optimality of AB-schedules for stars is the basis of our divide-and-conquer algorithm. The key idea is to find a spanning subgraph $H$ of $G$ for which optimal AB-schedules with respect to $H$ are among the optimal schedules for $G$. In particular, we identify subgraphs consisting of stars for which feasibility of AB-schedules can be encoded by certain vertex colorings of $G$. To do so, we introduce the following notions.

**Star Forests and I, II, III-Colorings.** A subgraph $H$ of a graph $G$ is a *star forest of $G$* if each component of $H$ is a star and $H$ contains all vertices of $G$.

We consider a star forest $H$ of a bipartite graph $G$. In particular, we want to use specific vertex subsets of $H$, denoted by $A_i$ and $B_i$, to process the jobs. The idea is to schedule A-patterns on the $A_i$'s and B-patterns on the $B_i$'s.

A vertex subset $A_1$ is a *I-coloring* of $(G, H)$ if it is a maximum independent set of both $G$ and $H$. A I-coloring allows to schedule an A-pattern on $H$ (by placing one job on each machine in $A_1$), yielding a valid schedule for $G$, see Fig. 4 (left).

Two disjoint vertex subsets $A_2, B_2$ are a *II-coloring* of $(G, H)$, if no vertex of $A_2$ is adjacent to another vertex from $A_2 \cup B_2$ in $G$ and the following properties hold: (i) for each $S = S_\ell$, $\ell \geq 3$, $A_2 \cap S$ is a maximum independent set of $S$, (ii) $B_2$ contains the vertices of each $S_1$, and (iii) for each $S_2$, either $A_2 \cap S_2$ is a maximum independent set of $S_2$ or $B_2$ contains the vertices of $S_2$. A II-coloring allows to schedule 3 A-patterns on stars with leaves in $A_2$ and 2 B-patterns on stars with vertices in $B_2$, see Fig. 4 (middle).

Two disjoint vertex subsets $A_3, B_3$ are a *III-coloring* of $(G, H)$, if no vertex of $A_3$ is adjacent to another vertex from $A_3 \cup B_3$ in $G$ and the following properties hold: (i) for each $S = S_\ell$, $\ell \geq 4$, $A_3 \cap S$ is a maximum independent set of $S$, (ii) $B_3$ contains the vertices of each $S_1$ and each $S_2$, and (iii) for each $S_3$, $A_3 \cap S_3$ is a maximum independent set of $S_3$ or $B_3$ contains the vertices of $S_3$. A III-coloring allows to schedule 4 A-patterns on stars with leaves in $A_3$ and 3 B-patterns on stars with vertices in $B_3$, see Fig. 4 (right).

Star forests and I, II, III-colorings help us to extend the structural insights on stars to general bipartite graphs. Particularly, we show that an optimal schedule on a star forest admitting I, II, III-colorings also yields a feasible schedule with respect to $G$ and is, therefore, also optimal.

**Lemma 5.** *Let $H$ be a star forest of a connected bipartite graph $G$ on at least two vertices. Given a I-coloring $A_1$, a II-coloring $(A_2, B_2)$ and a III-coloring $(A_3, B_3)$ of $(G, H)$, there exists a polynomial time algorithm to compute an optimal schedule for* SMC-UNIT$(G, n)$.

*Proof-Sketch.* Let $\Pi'$ be an optimal schedule for SMC-UNIT$(G, n)$. By Lemma 3, there exists an optimal AB-schedule $\Pi$ for SMC-UNIT$(H, n)$. Because $H$ is a subgraph of $G$, we have $\|\Pi\| \leq \|\Pi'\|$. We distinguish two cases.

If $\|\Pi\| \leq 20$, we show that there exists an optimal AB-schedule $\Pi^*$ on $H$ that is feasible for $G$. The schedule $\Pi^*$ is constructed as follows: Each A-pattern is scheduled on $A_1$, each B-pattern on $V$, 3 A-patterns on $A_2$, 2 B-patterns on $B_2$, 4 A-patterns on $A_3$, and 3 B-patterns on $B_3$.

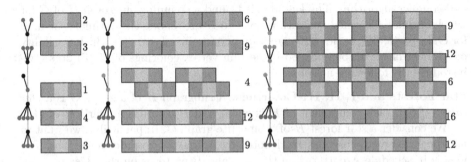

**Fig. 4.** A graph and a star forest with a I-, II-, and III-coloring and corresponding schedules. Vertices in $A_i$ are colored in asparagus and vertices in $B_i$ in blue. (Color figure online)

If $\|\Pi\| \geq 21$, we show that each star on $H$ contains either 4 A or 3 B patterns. Iteratively, shifting these to the front, we obtain a rest schedule of makespan at most 20. We thus need to compare 20 schedules to compute the optimal schedule.
$\square$

To complete the proof of Theorem 4, it remains to show how to find a star forest and the corresponding colorings.

### 3.3 Computing a Star Forest with I, II, III-Colorings

We compute such a star forest and corresponding I, II, III-colorings in four phases. First, we find an initial star forest $H$ admitting a feasible I-coloring. Then, we modify $H$ in two phases to ensure the existence of a II- and III-coloring of the star forest, respectively. Finally, we compute all colorings. Modifications of the initial star forest are necessary because of the possible appearance of so-called alternating paths.

**Alternating Paths.** Let $H = (V, E')$ be a star forest of a bipartite graph $G = (V, E)$. Let $C_1, \ldots, C_k$ be distinct stars of $H$ and $P$ be a path in $G$ on the vertices $v_1, v_2 \ldots, v_{2k-1}$ with the following properties:

- for even $i$, $v_i$ is a leaf of star $C_{i/2+1}$,
- for odd $i$, $v_i$ is the center of star $C_{(i+1)/2}$, and
- the edge $\{v_i, v_{i+1}\} \in E'$ if and only if $i$ is even.

We say $P$ is an *alternating path of type II* if $C_1 \simeq S_1$, $C_i \simeq S_2$ for all $i = 2, \ldots, k-1$ and $C_k \simeq S_\ell$ with $\ell \geq 3$. For an illustration of an alternating path of type II, see Fig. 5 (left).

**Fig. 5.** Alternating path of type II. Black edges belong to $H$, gray edges to $G \backslash H$. (Color figure online)

We say $P$ is an *alternating path of type III* if $C_1 \simeq S_2$, $C_i \simeq S_3$ for all $i = 2, \ldots, k-1$ and $C_k \simeq S_\ell$ with $\ell \geq 4$, see also Fig. 6 (left).

**Fig. 6.** Alternating path of type III. Black edges belong to $H$, gray edges to $G \backslash H$. (Color figure online)

If an alternating path of type II (III) exists in the star forest, then there is no feasible II-coloring $(A_2, B_2)$ (III-coloring $(A_3, B_3)$), as $B_2$ ($B_3$) must contain all nodes of the first star $C_1$, and hence also the nodes of all intermediate stars leading to adjacent vertices $v_{2k-3} \in B_2$ ($B_3$) and $v_{2k-2} \in A_2$ ($A_3$). However, an alternating path can be removed by swapping edges along it; this operation maintains the leaves of the star forest, see Figs. 5 and 6 (right).

**Observation 6.** *Let $H = (V, E')$ be a star forest of $G$ containing an alternating path $P$ of type II or III. Then, $H' := (V, E' \Delta P)$ is a star forest with the same set of leaves as $H$, where $\Delta$ denotes the symmetric difference.*

Algorithm 1 computes a star forest and I,II,III-colorings in polynomial time.

**Lemma 7.** *Algorithm 1 returns a star forest $H$ of $G$ and I-,II- and III- colorings $A_1, (A_2, B_2)$ and $(A_3, B_3)$ of $(G, H)$, respectively, in time polynomial in $G$.*

*Proof.* We first show that Algorithm 1 is well-defined (specifically line 7) and that the graph $H$, defined in line 16, is a star forest. Let $M$ be a maximum matching and $I$ be a maximum independent set of $G$. Both can be found in polynomial time using the maximum flow algorithm to find a maximum matching in bipartite graphs [34, Theorem 10.5]. Observe that the complement of a maximum independent set is a minimum vertex cover $U := V \backslash I$. By König's Theorem [32], it holds that $|U| = |M|$. In particular, every edge in $M$ contains exactly one vertex of $U$ and every vertex in $V' := V \backslash \bigcup_{e \in M} e$ is not contained in $U$. Therefore, for every $v \in V'$, there exists $u \in U$ such that $\{u, v\} \in E$, i.e., line 7 in Phase 1 is well-defined. By Lemma 6, modifying the star forest $(V, E')$ along an alternating path in Phase 2 and 3 results again in a star forest. It remains to show that $(V, E')$ is a star forest at the end of Phase 1. To this end, note that every edge of $E'$ is incident to exactly one vertex $u \in U$. Thus, every vertex in $U$ is the center of a star (on at least 2 vertices).

For the runtime, it is important to observe that in every iteration of Phase 2 (Phase 3), some $S_1$ ($S_2$) is removed and no new $S_1$ ($S_2$) is created. Therefore, the number of iterations in Phase 2 (Phase 3) is bounded by the number of $S_1$'s ($S_2$'s) before Phase 2 (Phase 3). An alternating path of type II (type III) can also be found in polynomial time by using BFS starting from a fixed $S_1$ ($S_2$). Because Phase 4 clearly runs in polynomial time, Algorithm 1 does as well.

Finally, we prove that Phase 4 computes valid colorings. Here, we exploit the fact that no alternating path of type II is created in Phase 3 (Lemma 16). Thus, after Phase 3, there exists no alternating path (of any type).

By definition, $A_1 := I$ is a maximum independent set of $G$. We argue that $A_1$ is also an maximum 1-IS of $H$. Note that while $U$ constitutes the centers of the stars, $V \backslash U = I = A_1$ are the leaves of the stars after Phase 1. Hence the claim is true after Phase 1. Lemma 6 ensures that this property is maintained in Phase 2 and 3. Hence, $A_1$ is a I-coloring.

For the type II-coloring, the algorithm ensures that $A_2$ contains the leaves of all $S_\ell$ with $\ell \geq 3$ and that $B_2$ contains all vertices of $S_1$. Hence, we only need to pay attention to $S_2$'s. The algorithm inserts the leaves of stars $S_2$ into $A_2$ if their center is adjacent to a vertex in $A_2$, as long as it is possible. The vertices of all remaining $S_2$'s are inserted into $B_2$. Thus, the properties (i), (ii) and (iii) of a II-coloring are fulfilled by $(A_2, B_2)$. It only remains to show that no vertex of $A_2$ is adjacent to another vertex of $A_2 \cup B_2$ in $G$. Because $A_2 \subset I$, no two vertices in $A_2$ are adjacent in $G$. Suppose there is a vertex $a \in A_2$ adjacent in $G$ to a vertex $b \in B_2$. Let $S^a$ and $S^b$ be the star containing $a$ and $b$, respectively. By construction $a$ is a leaf of $S^a$. Moreover, $b \in U$; otherwise $a, b \in A_1$, a contradiction to the fact that $A_1$ is an independent set. If $S^b \simeq S_2$, then the center $b$ is adjacent to $a \in A_2$, and the algorithm ensures that the leaves of $S^b$ are contained in $A_2$, and hence, $b \notin B_2$. Thus, $S^b \simeq S_1$.

If $S^a \simeq S_\ell, \ell \geq 3$, then there exists an alternating path of type II starting in $b$ and ending in the center of $S^a$, a contradiction.

If $S^a \simeq S_2$, the fact $a \in A_2$ implies that there exists a star $S_\ell, \ell \geq 2$, with a leaf adjacent to the center of $S^a$. If $\ell = 2$, there exists a further star whose leaf is adjacent to the considered star. Repeating this argument, the containment of $a \in A_2$ can be traced back to a star $S_\ell, \ell \geq 3$, and yields an alternating path of type II, a contradiction.

With arguments similar to the above, $(A_3, B_3)$ is a III-coloring; otherwise we find an alternating path of type III. Specifically, one can show that vertex $a$ belongs to a star $S_2$ and $b$ to a star $S_3$. □

Finally, Theorem 4 follows from Lemmas 1, 5 and 7.

# 4    Future Directions

Various interesting avenues remain open for future research. In particular, the investigation of graph classes capturing geometric information is of special interest for applications in which spatial proximity causes machines conflicts. Additionally, allowing for preemption constitutes an interesting direction.

---

**Algorithm 1.** Computing a star forest and I, II, III-colorings.

---

1: Input: Connected bipartite graph $G = (V, E)$ with $|V| \geq 2$.
2: Output: Star forest $H$ and I,II,III-colorings $A_1, (A_2, B_2), (A_3, B_3)$.

---

*Phase 1 – Initial star forest*

---

3: Compute a maximum matching $M$ and a maximum independent set $I$ of $G$
4: Set $U := V \backslash I$ (vertex cover).
5: Set $E' := M$ and $V' := V \setminus \bigcup_{e \in M} e$.
6: **while** $\exists\, v \in V'$ **do**
7:     Find $u \in U$ such that $\{u, v\} \in E$.
8:     Add $\{u, v\}$ to $E'$ and delete $v$ from $V'$.
9: **end while**

---

*Phase 2 – Removing alternating paths of type II*

---

10: **while** $\exists$ alternating path $P$ of type II **do**
11:     $E' = E' \Delta P$
12: **end while**

---

*Phase 3 – Removing alternating paths of type III*

---

13: **while** $\exists$ alternating path $P$ of type III **do**
14:     $E' = E' \Delta P$
15: **end while**

---

*Phase 4 – Computing the colorings*

---

16: $H := (V, E')$
17: $A_1 := I$
18: For each $S_\ell$ in $H$ with $\ell \geq 3$, add leaves of $S_\ell$ to $A_2$.
19: For each $S_1$ in $H$, add vertices of $S_1$ to $B_2$.
20: **while** $\exists\, S_2$ in $H$ such that its center is adjacent to a vertex of $A_2$ in $G$ **do**
21:     Add leaves of $S_2$ to $A_2$.
22: **end while**
23: For each $S_2$ in $H$ with $V(S_2) \cap A_2 = \emptyset$, add vertices of $S_2$ to $B_2$.
24: For each $S_\ell$ in $H$ with $\ell \geq 4$, add leaves of $S_\ell$ to $A_3$.
25: For each $S_\ell$ in $H$ with $\ell \in \{1, 2\}$, add vertices of $S_\ell$ to $B_3$.
26: **while** $\exists$ some $S_3$ in $H$ such that center $v$ of $S_3$ is adjacent to some $w \in A_3$ in $G$ **do**
27:     Add leaves of $S_3$ to $A_3$.
28: **end while**
29: For each $S_3$ in $H$ with $V(S_3) \cap A_3 = \emptyset$, add vertices of all $S_3$ to $B_3$.
30: **return** $H, A_1, (A_2, B_2), (A_3, B_3)$

---

# 5   Appendix I – Details for Sect. 2

All theorems in Sect. 2 are divided into two cases: either all jobs have long blocking times ($\max\{\overleftarrow{b}, \overrightarrow{b}\} > p$) or short blocking times ($\max\{\overleftarrow{b}, \overrightarrow{b}\} \leq p$). We split the proofs up according to the relation between the parameters and present the results in two subsections. Specifically, Theorem 1 follows from Theorem 5 and Theorem 8. Theorem 2 follows from Theorem 6 and Theorem 9. Theorem 3 follows from Theorem 7 and Theorem 10. Throughout the remainder, we denote the optimal makespan of a given instance by OPT.

## 5.1   Long Blocking Times

In this subsection, we consider SMC-ID where jobs have long blocking times, i.e., it holds $\max\{\overleftarrow{b}, \overrightarrow{b}\} > p$. These long blocking times lead to so-called *basic* schedules in which jobs on conflicting machines do not run in parallel.

**Basic Schedules.** A schedule $\Pi$ is *basic*, if for every edge $ii' \in E$ and for every pair of jobs $j$ and $j'$ assigned to $i$ and $i'$, respectively, their system times are non-overlapping, i.e., $S_j^\Pi \leq S_{j'}^\Pi$ implies $C_j^\Pi \leq S_{j'}^\Pi$.

The following lemma allows us to focus on basic schedules.

**Lemma 8.** *For an instance of* SMC-ID, *every schedule $\Pi$ is basic if*

$$(\max\{\overleftarrow{b}, \overrightarrow{b}\} > p).$$

*Proof.* Suppose for the sake of a contradiction that there exists two jobs $j$ and $j'$ assigned to machines $i$ and $i'$, respectively, with $ii' \in E$ such that $j'$ starts while $j$ is processed. Because their blocking times cannot overlap, we also know that $j'$ starts after the first blocking time of $j$ and the first blocking time of $j'$ ends before the second blocking time of $j$ starts, see Fig. 7. Consequently, $\overleftarrow{b}_{j'} \leq p_j$. Moreover, $p_{j'} \geq \overrightarrow{b}_j$ if $j'$ ends after $j$.

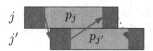

$$\begin{array}{ll} j \\ j' \end{array}$$

**Fig. 7.** Illustration of the proof of Lemma 8.

Because all jobs have the same parameters, we obtain the contradiction $\overleftarrow{b} = \overleftarrow{b}_{j'} \leq p_j = p$ and $p = p_{j'} \geq \overrightarrow{b}_j = \overrightarrow{b}$. $\qquad\qquad\square$

In the following, we present the proof of Theorem 1 for the case of long blocking times. To this end, for instances of SMC-ID we denote by $\alpha_1$ the cardinality of a maximum independent set of $G$.

**Theorem 5** *[Theorem 1 for long blocking times]. Unless $P = NP$,* SMC-ID *where $\overleftarrow{b}, p, \overrightarrow{b}$ are fixed and $\max\{\overleftarrow{b}, \overrightarrow{b}\} > p$ holds does not admit a $\mathcal{O}(m^{1-\varepsilon})$-approximation for any $\varepsilon > 0$.*

*Proof.* We restrict our attention to instances consisting of $n = \alpha_1$ jobs and for the sake of simplicity let $q = \overleftarrow{b} + p + \overrightarrow{b} = 1$ by scaling. By Lemma 8, schedules for SMC-ID with $\max\{\overleftarrow{b}, \overrightarrow{b}\} > p$ are basic. Clearly, the optimum makespan OPT is 1 because $q = 1$. Suppose for some constant $\kappa > 0$ there exists a $(\kappa m^{1-\varepsilon})$-approximation algorithm for $\varepsilon > 0$ and let $\Pi$ denote its schedule. Moreover, let $\beta$ denote the maximum number of machines processing jobs in parallel in $\Pi$ at any point in time. As we have at most $n$ distinct starting times in $\Pi$ we can compute $\beta$ in polynomial time. Because $\Pi$ is basic, we can transform $\Pi$ into a new schedule $\Pi'$ that processes all jobs non-idling on $\beta$ machines without increasing the makespan. Hence, we obtain $\|\Pi\| \geq \lceil n/\beta \rceil \geq n/\beta$. This fact together with the assumption $\|\Pi\| \leq \kappa m^{1-\varepsilon} \cdot \text{OPT} = \kappa m^{1-\varepsilon}$ yields $\beta \geq n/\|\Pi\| \geq n/\kappa m^{1-\varepsilon} = 1/\kappa m^{1-\varepsilon} \cdot \alpha_1$. In other words, the $(\kappa m^{1-\varepsilon})$-approximation algorithm implies an $(1/\kappa \cdot m^{\varepsilon-1})$-approximation algorithm for computing a maximum 1-IS for every graph $G$; a contradiction [24,42]. □

We now present the proofs of Theorems 2 and 3 for long blocking times, respectively.

**Theorem 6** *[Theorem 2 for long blocking times]. Let $G$ be a graph. If $\max\{\overleftarrow{b}, \overrightarrow{b}\} > p$ and we are given a 1-IS of $G$, then an optimal schedule of* SMC-ID *can be computed in polynomial time.*

*Proof.* Let $\Pi$ be an optimal schedule for SMC-ID with $\max\{\overleftarrow{b}, \overrightarrow{b}\} > p$. By Lemma 8, $\Pi$ is basic, hence, at any point in time at most $\alpha_1$ jobs are running. Therefore, we can modify the job-to-machine assignment of $\Pi$ such that all jobs are processed on a maximum 1-IS while maintaining the starting times of the jobs. As we did not increase the makespan, we have an optimal schedule where all jobs are assigned to a maximum 1-IS. Because all jobs are identical, evenly distributing all jobs over the machines yields an optimal schedule. □

**Theorem 7** *[Theorem 3 for long blocking times]. Let $G$ be a graph. If $\max\{\overleftarrow{b}, \overrightarrow{b}\} > p$ and we are given a $1/\gamma$-approximate 1-IS of $G$, then* SMC-ID *allows for a $\lceil \gamma \rceil$-approximation.*

*Proof.* Let $\mathcal{I}$ denote the given $1/\gamma$-approximate 1-IS, i.e., $|\mathcal{I}| \geq \alpha_1/\gamma$. Without loss of generality we assume $q = \overleftarrow{b} + p + \overrightarrow{b} = 1$. An optimal schedule distributes all jobs evenly over a maximum 1-IS, i.e., we have $\text{OPT} = \lceil n/\alpha_1 \rceil$ as $q = 1$. Since each system time is 1, we can find a schedule $\Pi$ with makespan $\|\Pi\| = \lceil n/|\mathcal{I}| \rceil \cdot q \leq \lceil \gamma \cdot n/\alpha_1 \rceil \leq \lceil \gamma \rceil \cdot \lceil n/\alpha_1 \rceil = \lceil \gamma \rceil \cdot \text{OPT}$ by distributing all jobs evenly on $\mathcal{I}$. □

### 5.2  Short Blocking Times

In this subsection, we consider SMC-ID where jobs have short blocking times, i.e., it holds $\max\{\overleftarrow{b}, \overrightarrow{b}\} \leq p$. By symmetry, we assume that $\overleftarrow{b} \geq \overrightarrow{b}$ and hence, we write $0 < \overrightarrow{b} \leq \overleftarrow{b} \leq p$. We first introduce a generalized notion of independent sets.

**(Maximum) Induced $c$-Colorable Subgraph.** For a graph $G = (V, E)$ and $c \in \mathbb{N}_{\geq 1}$, a *maximum induced $c$-colorable subgraph*, or short *maximum $c$-IS*, of $G$ is a set of $c$ disjoint independent sets $\mathcal{I}_1, \ldots, \mathcal{I}_c \subseteq V$ whose union has maximum cardinality. Clearly, a 1-IS is an independent set. Moreover, the cardinality of a maximum $c$-IS is defined as the cardinality of the union of $\mathcal{I}_1, \ldots, \mathcal{I}_c$. We denote the cardinality of a maximum $c$-IS by $\alpha_c$. These $c$-ISs can be used to obtain useful partial schedules.

**$c$-Pattern.** Let $c \in \mathbb{N}_{\geq 1}$ with $c \leq \lfloor p/\overleftarrow{b} \rfloor + 1$ and let $\mathcal{I} = (I_1, I_2, \ldots, I_c)$ be a $c$-tuple of disjoint independent sets of $G$. Recall that $q := \overleftarrow{b} + p + \overrightarrow{b}$ denotes the system time. A partial schedule of length $q + (c-1) \cdot \overleftarrow{b}$ starting at time $t$ is called a $c$-*pattern on* $\mathcal{I}$ if on each machine $i$ in $I_k$ with $k \in \{1, \ldots, c\}$, there is one job starting at time $t + (k-1)\overleftarrow{b}$.

In order to show connections between conflict-free schedules and $c$-ISs, we extract a $c$-IS from a conflict-free schedule. We say a job $j$ *blocks* a time $t$ in schedule $\Pi$ if one of its blocking times contains $t$, i.e., $t \in ((S_j^\Pi, S_j^\Pi + \overleftarrow{b}) \cup (C_j^\Pi - \overrightarrow{b}, C_j^\Pi))$. For a schedule $\Pi$, we define the quantity

$$\beta_c^\Pi := \max_{t_1 < \ldots < t_c} \left\{ \left| \bigcup_{k=1}^c \{i \in V : i \text{ processes a job in } \Pi \text{ which blocks time } t_k\} \right| \right\}.$$

Observe that for each time $t$ the machines which process a job blocking time $t$ form an 1-IS, because $\Pi$ is conflict-free. Hence, $\beta_c^\Pi$ corresponds to the cardinality of a $c$-IS, since machines are not counted more than once. It is easy to see that $\beta_c^\Pi$ can be computed in polynomial time for a constant $c$.

**Lemma 9.** *For a schedule $\Pi$ with $n$ jobs and a constant $c$, $\beta_c^\Pi$ can be computed in time polynomial in $n$.*

*Proof.* The schedule has $4n$ event times, namely the starting time of the blocking times and the processing time and its completion time for each job. For some point in time $t$ between every two consecutive event points, we count the number of machines processing a blocking interval. By definition of $\beta_c^\Pi$, it suffices to check $\mathcal{O}(n^c)$ tuples. $\qquad\square$

The definition of $\beta_c^\Pi$ helps us to bound the makespan of a schedule $\Pi$ from below stated in Lemma 12. To this end, we first give an upper bound on the number of jobs starting within an interval of length $\lambda$ which is defined as follows. For instances of SMC-ID with $0 < \overrightarrow{b} \leq \overleftarrow{b} \leq p$, we define

$$k := \lfloor p/\overleftarrow{b} \rfloor \text{ and } \lambda := (k+1)\overleftarrow{b} + \begin{cases} \overrightarrow{b} & \text{if } p/\overleftarrow{b} \in \mathbb{N} \\ 0 & \text{otherwise.} \end{cases}$$

**Lemma 10.** *For every schedule $\Pi$ of an instance SMC-ID with $0 < \overrightarrow{b} \leq \overleftarrow{b} \leq p$ and every time $t \geq 0$, the number of jobs starting within the interval $I := [t, t+\lambda)$ is at most $\beta_{k+1}^\Pi$.*

*Proof.* Because $I$ has length $\lambda \leq q$ and is half-open, at most one job starts on each machine within $I$. We partition the interval $I$ into an (possibly empty) interval $I_0$ (left-closed, right-open) of length $\lambda - (k+1)\overleftarrow{b}$ (evaluating to $\overrightarrow{b}$ if $p/\overleftarrow{b} \in \mathbb{N}$ and to $0$ otherwise) and $k+1$ disjoint (left-closed, right-open) intervals

$I_1, \ldots, I_{k+1}$, each of length $\overleftarrow{b}$. Clearly, the length of the intervals $I_0, \ldots, I_{k+1}$ add to $\lambda$. By $V_\ell$ we denote the set of machines that have a job starting in $I_\ell$. For each $\ell$, the jobs processed on a machine in $V_\ell$ block a point in time arbitrarily close to the right end of $I_\ell$, see Fig. 8 (left). Thus, all $V_\ell$'s are independent sets.

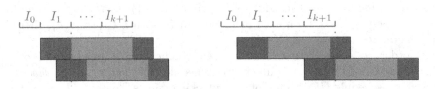

**Fig. 8.** Illustration for the proof of Lemma 10: (left) two jobs starting in $I_1$ and (right) a job starting in $I_0$ and a job starting in $I_{k+1}$.

Additionally, we show that $V_0 \cup V_{k+1}$ is an independent set. If $p/\overleftarrow{b} \notin \mathbb{N}$, then $V_0 = \emptyset$. It thus remains to consider the case $p/\overleftarrow{b} \in \mathbb{N}$. Let $t'$ be a point in time arbitrarily close to the right end of interval $I$. The second blocking time of job $j$ processed on a machine in $V_0$ starts in interval $[t + q - \overrightarrow{b}, t + q)$, and thus starts before $t + q$ and ends not before $t + q$. Hence $t'$ is blocked by the second blocking time of $j$. Additionally, observe that $t'$ is blocked by the first blocking time of every job processed on a machine in $V_{k+1}$, see also Fig. 8 (right). Consequently, each $V_\ell$ and $V_0 \cup V_{k+1}$ are independent sets and thus the number of jobs can be bounded from above by $\beta_{k+1}^\Pi$.                                                                □

Next, we show that an upper bound on the number of jobs starting within an interval of a specific lengths implies a lower bound on the makespan.

**Lemma 11.** *Let $\Pi$ be a schedule for an instance* SMC-ID *with $0 < \overrightarrow{b} \le \overleftarrow{b} \le p$ such that for some integer $\beta$, in every interval of length $L \le q$ at most $\beta$ jobs start. Then, for system time $q := \overleftarrow{b} + p + \overrightarrow{b}$, it holds that*

$$\|\Pi\| \ge L\lfloor n/\beta \rfloor + \begin{cases} 0 & \text{if } \beta \text{ divides } n \\ q & \text{otherwise.} \end{cases}$$

*Proof.* We divide $\Pi$ into intervals of length $L$ starting with 0. By assumption, at most $\beta$ jobs start in each interval. Hence, the number of these intervals (where some job is processed) is at least $\lfloor n/\beta \rfloor$. Moreover, if $\beta$ does not divide $n$, then at least some job (of length $q$) starts after time $L\lfloor n/\beta \rfloor + q$.                                                                □

Lemma 10 and 11 immediately imply a lower bound on the makespan of any schedule.

**Lemma 12.** *For every schedule $\Pi$ for an instance of* SMC-ID *with $0 < \overrightarrow{b} \le \overleftarrow{b} \le p$, the makespan is at least $\|\Pi\| \ge \lambda \cdot \lceil n/\beta_{k+1}^\Pi \rceil$. In particular, the optimal makespan is bounded from below by* OPT $\ge \lambda \cdot \lceil n/\alpha_{k+1} \rceil$.

Using Lemma 12, we prove Theorem 1 for short blocking times.

**Theorem 8** *[Theorem 1 for short blocking times]. Unless $P = NP$, SMC-ID where $\overleftarrow{b}, p, \overrightarrow{b}$ are fixed and $\max\{\overleftarrow{b}, \overrightarrow{b}\} \leq p$ holds does not admit a $\mathcal{O}(m^{1-\varepsilon})$-approximation for any $\varepsilon > 0$.*

*Proof.* Suppose that for some $\kappa > 0$ and some $\varepsilon > 0$ there exists a $(\kappa m^{1-\varepsilon})$-approximation algorithm $\mathcal{A}$. We assume $\overleftarrow{b} \geq \overrightarrow{b}$ and define $k := \lfloor p/\overleftarrow{b} \rfloor$ and $\lambda$ as above. For an instance of SMC-ID with short blocking times and $n \geq \alpha_{k+1}$, we consider the schedule $\Pi$ computed by $\mathcal{A}$. On the one hand, because $\mathcal{A}$ is a $(\kappa m^{1-\varepsilon})$-approximation, $\Pi$ has makespan

$$\|\Pi\| \leq (\kappa m^{1-\varepsilon}) \cdot \text{OPT}.$$

On the other hand, by Lemma 12, we have

$$\|\Pi\| \geq \lambda \cdot \lceil n/\beta_{k+1}^{\Pi} \rceil \geq \lambda n \cdot 1/\beta_{k+1}^{\Pi}.$$

Moreover, we obtain a feasible schedule by repeatedly using $(k+1)$-patterns on a maximum $(k+1)$-IS, while leaving the other machines idle. Recall that a $(k+1)$-pattern has length $(q + k\overleftarrow{b})$ and schedules $\alpha_{k+1}$ jobs. This yields the upper bound

$$\text{OPT} \leq (q + k\overleftarrow{b}) \cdot \lceil n/\alpha_{k+1} \rceil \leq 2(q + k\overleftarrow{b}) \cdot n/\alpha_{k+1},$$

where the last inequality uses the fact $n/\alpha_{k+1} \geq 1$ and thus $\lceil n/\alpha_{k+1} \rceil \leq 2(n/\alpha_{k+1})$. Altogether, we obtain

$$\beta_{k+1}^{\Pi} \geq \frac{\lambda n}{\|\Pi\|} \geq \frac{\lambda n}{\kappa m^{1-\varepsilon} \cdot \text{OPT}} \geq \frac{1}{\kappa m^{1-\varepsilon}} \cdot \frac{\lambda}{2(q + k\overleftarrow{b})} \cdot \alpha_{k+1},$$

where $\lambda/q + k\overleftarrow{b}$ is a constant $\leq 1$ since $\lambda \leq q$. By Lemma 9, we can compute a $(k+1)$-IS from $\Pi$ of size $\beta_{k+1}^{\Pi}$ in polynomial time and thus we obtain a $\mathcal{O}(m^{\varepsilon-1})$-approximation for computing a maximum $(k+1)$-IS; a contradiction [24, 36, 42]. $\square$

We now provide the proofs of Theorems 2 and 3 for short blocking times.

**Theorem 9** *[Theorem 2 for short blocking times]. If $0 < \overrightarrow{b} \leq \overleftarrow{b} \leq p$ and we are given a maximum $(\lfloor p/\overleftarrow{b} \rfloor + 1)$-IS of $G$, then SMC-ID allows for a $(q + k\overleftarrow{b})/\lambda$-approximation, with $(q + k\overleftarrow{b})/\lambda < 2 + 1/(k+1) < 2.5$ and if $p/\overleftarrow{b} \in \mathbb{N}$, we even get $(q + k\overleftarrow{b})/\lambda < 1 + p/q < 2$.*

*Proof.* We define a schedule $\Pi$ consisting of $\lceil n/\alpha_{k+1} \rceil$ many $(k+1)$-patterns on a maximum $(k+1)$-IS of $G$, while leaving all other machines idle. $\Pi$ has makespan $\|\Pi\| \leq (q + k\overleftarrow{b}) \cdot \lceil n/\alpha_{k+1} \rceil$. By Lemma 12, it holds that $\text{OPT} \geq \lambda \cdot \lceil n/\alpha_{k+1} \rceil$. We obtain

$$\frac{\|\Pi\|}{\text{OPT}} \leq \frac{q + k\overleftarrow{b}}{\lambda} = \begin{cases} \frac{q+p}{q} = 1 + \frac{p}{q} < 2 & \text{if } p/\overleftarrow{b} \in \mathbb{N}, \\ \frac{(k+1)\overleftarrow{b}+p+\overleftarrow{b}}{(k+1)\overleftarrow{b}} = 1 + \frac{p}{(k+1)\overleftarrow{b}} + \frac{\overleftarrow{b}}{(k+1)\overleftarrow{b}} < 2 + \frac{1}{k+1} \leq 2.5 & \text{if } p/\overleftarrow{b} \notin \mathbb{N}. \end{cases}$$

$\square$

**Theorem 10** *[Theorem 3 for short blocking times]. If $0 < \overrightarrow{b} \leq \overleftarrow{b} \leq p$ and we are given a $1/\gamma$-approximate $(\lfloor p/\overleftarrow{b} \rfloor + 1)$-IS of $G$, then* SMC-ID *allows for a $2\gamma/\lambda \cdot (q + k\overleftarrow{b})$-approximation, with $2\gamma/\lambda \cdot (q + k\overleftarrow{b}) < 5\gamma$. If $p/\overleftarrow{b} \in \mathbb{N}$, we get $2\gamma/\lambda \cdot (q + k\overleftarrow{b}) < 4\gamma$.*

*Proof.* Let $\mathcal{I}$ be the given $(k+1)$-IS of size $\beta \geq \alpha_{k+1}/\gamma$. We construct a schedule $\Pi$ by using $\lceil n/\beta \rceil$ many $(k+1)$-patterns on $\mathcal{I}$ which has makespan $\|\Pi\| \leq (q + k\overleftarrow{b}) \cdot \lceil n/\beta \rceil$. Moreover, Lemma 12 implies OPT $\geq \lambda \cdot \lceil n/\alpha_{k+1} \rceil \geq \lambda \cdot n/\alpha_{k+1}$. If $n \leq \beta$, then also $n \leq \alpha_{k+1}$, implying that $\lceil n/\beta \rceil = \lceil n/\alpha_{k+1} \rceil = 1$. Thus, we obtain

$$\frac{\|\Pi\|}{\text{OPT}} \leq \frac{(q + k\overleftarrow{b}) \cdot \lceil n/\beta \rceil}{\lambda \cdot \lceil n/\alpha_{k+1} \rceil} = \frac{(q + k\overleftarrow{b})}{\lambda}.$$

Hence, following the same steps as in the proof of Theorem 9, we obtain an upper bound on the performance guarantee of 2 if $p/\overleftarrow{b} \in \mathbb{N}$ and 2.5 if $p/\overleftarrow{b} \notin \mathbb{N}$. If $n > \beta$, we obtain $\lceil n/\beta \rceil \leq 2 \cdot n/\beta \leq 2\gamma \cdot n/\alpha_{k+1}$. Consequently, it holds

$$\frac{\|\Pi\|}{\text{OPT}} \leq \frac{(q + k\overleftarrow{b}) \cdot 2\gamma \cdot n/\alpha_{k+1}}{\lambda \cdot n/\alpha_{k+1}} \leq 2\gamma \cdot \frac{(q + k\overleftarrow{b})}{\lambda}.$$

Again, following the same steps as in the proof of Theorem 9, we obtain an upper bound on the performance guarantee of $4\gamma$ if $p/\overleftarrow{b} \in \mathbb{N}$ and $5\gamma$ if $p/\overleftarrow{b} \notin \mathbb{N}$.     $\square$

# 6   Appendix II – Details for Sect. 3

In the following we provide omitted proofs of Sect. 3.

**Lemma 1.** *For every $n$, an optimal schedule for* SMC-UNIT$(K_m, n)$ *can be computed in time linear in $\log n$. In particular, for $m \geq 2$, it coincides with an optimal schedule for $K_2$ of makespan $4\lfloor n/2 \rfloor + 3(n \mod 2)$.*

*Proof.* Let $\Pi$ be an optimal schedule of SMC-UNIT$(K_m, n)$. Note that for each point in time $t$, the set of machines processing a job at time $t$ induces a $K_1$ or $K_2$. Moreover, all vertices in $K_m$, $m \geq 2$, play the same role and hence we may shift all jobs to the same two vertices. Thus, we may reduce our attention to SMC-UNIT$(K_2, n)$. It is easy to check that two jobs are optimally processed in time 4. This implies the claim.     $\square$

The next lemma shows that we can restrict to schedules with integral starting times.

**Lemma 13.** *If $\overleftarrow{b}_j$, $p_j$, $\overrightarrow{b}_j$ are integral for all $j \in \mathcal{J}$, every feasible schedule $\Pi$ can be transformed into a feasible schedule $\Pi^*$ such that the starting time of each job is intergral and the makespan does not increase, i.e., $\|\Pi^*\| \leq \|\Pi\|$.*

*Proof.* Let $S_j^{\Pi}$ denote the starting time for each job $j \in J$. We define $\Pi^*$ by $S_j^{\Pi^*} := \lfloor S_j^{\Pi} \rfloor$ for each job $j$. It remains to show that $\Pi^*$ is feasible. Let $j$ and $j'$ be two jobs such that two of their blocking times $b$ and $b'$ intersect in $\Pi^*$. By integrality of $\Pi^*$ and the blocking times, they intersect in at least one time unit.

Without loss of generality, we assume that $b'$ does not start before $b$ in $\Pi$. With slight abuse of notation, $S_b^{\Pi}$ denotes the starting time of the blocking time $b$ in $\Pi$. Then, $S_{b'}^{\Pi} - S_b^{\Pi}$ and $S_{b'}^{\Pi^*} - S_b^{\Pi^*}$ differ by strictly less than 1. Hence, if they intersect in at least 1 time unit in $\Pi^*$, then they also intersect in $\Pi$. Therefore, by feasibility of $\Pi$, $j$ and $j'$ are scheduled on conflict-free machines. □

**Corollary 1.** *For every star $S$ and every $n$, an optimal schedule for* SMC-UNIT$(S, n)$ *can be computed in time linear in* $\log n$ *and* $|S|$.

*Specifically, for every $S_\ell$, there exists $X \in \{A, B\}$ such that an optimal schedule has at most 2 X-patterns, i.e., an optimal schedule has makespan*

$$\min_{k=0,1,2} \left\{ 4 \left\lceil \frac{n - k\ell}{\ell+1} \right\rceil + 3k, 3 \left\lceil \frac{n - k(\ell+1)}{\ell} \right\rceil + 4k \right\}, \tag{2}$$

*where $\lceil \cdot \rceil$ denotes the usual ceiling function; however, for negative reals it evaluates to 0.*

*Proof.* By Lemma 3, there exists an optimal AB-schedule. For a star $S_\ell$, an A-pattern processes $\ell$ jobs in time 3 and a B-pattern processes $(\ell+1)$ jobs in time 4. An AB-schedule for at least $n$ jobs with exactly $k$ A-patterns has a makespan of $4\lceil \frac{n-k\ell}{\ell+1} \rceil + 3k$. Similarly, an AB-schedule for at least $n$ jobs with exactly $k$ B-patterns has a makespan of $3\lceil \frac{n-k(\ell+1)}{\ell} \rceil + 4k$.

For $\ell \leq 2$, an optimal schedule contains at most 2 A-patterns. While 3 A-patterns process $3\ell$ jobs in time 9, 2 B-pattern process $2(\ell+1) \geq 3\ell$ jobs in time 8. Thus, any 3 A-patterns can be replaced by 2 B-patterns.

For $\ell \geq 3$, an optimal schedule contains at most 2 B-patterns: While 3 B-patterns process $3(\ell+1)$ jobs in time 12, 4 A-patterns finish $4\ell \geq 3(\ell+1)$ jobs in time 12. Thus, any 3 B-patterns can be replaced by 4 A patterns.

Therefore, for each star we only have to compare at most three different schedules with $k = 0, 1, 2$ A- or B-patterns, respectively. Taking one with minimum makespan induces an optimal solution. □

We next present the full proof of Lemma 5.

**Lemma 5.** *Let $H$ be a star forest of a connected bipartite graph $G$ on at least two vertices. Given a I-coloring $A_1$, a II-coloring $(A_2, B_2)$ and a III-coloring $(A_3, B_3)$ of $(G, H)$, there exists a polynomial time algorithm to compute an optimal schedule for* SMC-UNIT$(G, n)$.

*Proof.* Let $\Pi'$ be an optimal schedule for SMC-UNIT$(G, n)$. By Lemma 3, there exists an optimal AB-schedule $\Pi$ for SMC-UNIT$(H, n)$. Because $H$ is a subgraph of $G$, we have $\|\Pi\| \leq \|\Pi'\|$. First, we show how to solve SMC-UNIT$(G, n)$ for small optimal makespan values.

**Table 1.** AB-schedules on the stars of $H$ based on Corollary 1 and modification *. The number before A and B indicates the number of A- and B-patterns.

| $\|\Pi\|$ | $S_1$ | $S_2$ | $S_3$ | $S_\ell, \ell \geq 4$ |
|---|---|---|---|---|
| 1 | – | – | – | – |
| 2 | – | – | – | – |
| 3 | A | A | A | A |
| 4 | B | B | B | B |
| 5 | – | – | – | – |
| 6 | 2A* | 2A | 2A | 2A |
| 7 | A, B | A, B | A, B | A, B |
| 8 | 2B | 2B | 2B | 2B |
| 9 | 2B | 3A or 2B* | 3A | 3A |
| 10 | 2A, B* | 2A, B | 2A, B | 2A, B |
| 11 | A, 2B | A, 2B | A, 2B | A, 2B |
| 12 | 3B | 3B | 4A or 3B | 4A |
| 13 | B, 2B | B, (3A or 2B)* | B, 3A | B, 3A |
| 14 | 2A, 2B* | 2A, 2B | 2A, 2B | 2A, 2B |
| 15 | A, 3B | A, 3B | A, (4A or 3B) | A, 4A |
| 16 | B, 3B | B, 3B | B, (4A or 3B) | B, 4A |
| 17 | 2B, 2B | 2B, (3A or 2B)* | 2B, 3A | 2B, 3A |
| 18 | 2A, 3B* | 2A, 3B | 2A, (4A or 3B) | 2A, 4A |
| 19 | A, B, 3B | A, B, 3B | A, B, (4A or 3B) | A, B, 4A |
| 20 | 2B, 3B | 2B, 3B | 2B, (4A or 3B) | 2B, 4A |

**Claim 14.** *If $\|\Pi\| \leq 20$, then there exists an optimal AB-schedule $\Pi^*$ on $H$ that is feasible for $G$ (according to Table 1).*

To prove this claim, observe that $\|\Pi\| \geq q = 3$ and that there is no AB-schedule with $\|\Pi\| = 5$. We first show how to define a schedule $\Pi^*$ of makespan $\|\Pi\|$ according to Table 1 that has as many jobs as $\Pi$. Afterwards, we show that $\Pi^*$ is feasible with respect to $G$. To this end, we first concentrate on the four types of components in $H$. For $C \in \{S_1, S_2, S_3, S_{\ell \geq 4}\}$, let $L_C$ denote the set of all makespan (of at most 20) from schedules obtained by Corollary 1. For $C \in \{S_1, S_2, S_3, S_{\ell \geq 4}\}$ and $\|\Pi\| \in [20] \setminus \{1, 2, 5\}$, we use the optimal AB-schedule with makespan $\|\Pi\|$ (or the maximum value in $L_C$ that is $\leq \|\Pi\|$). However, we modify some schedules in order to guarantee feasibility later: modified entries are marked by an asterisk in Table 1.

For $S_1$ and for $\|\Pi\| \equiv 2 \pmod 4$, we use 2 A-patterns and $(\lfloor \|\Pi\|/4 \rfloor - 1)$ B-patterns instead of $\lfloor \|\Pi\|/4 \rfloor$ B-patterns. Since 2 A-patterns and 1 B-pattern differ in length 2, the modified schedule finishes within $\|\Pi\|$. It also schedules at least as many jobs as $\Pi$, because 1 A-pattern contains one job while 1 B-pattern contains two jobs.

For $S_2$ and for $\|\Pi\| \equiv 1 \pmod 4$, we also allow to schedule 3 A-patterns and $\lfloor\|\Pi\|/4\rfloor - 2$ B-patterns besides the optimal schedule using $\lfloor\|\Pi\|/4\rfloor$ B-patterns. As 3 A-patterns and 2 B-patterns differ in length 1, the modified schedule finishes within $\|\Pi\|$. Moreover, both 3 A-patterns and 2 B-patterns contain six jobs.

Table 1 displays the resulting patterns on the components. The schedule $\Pi^*$ is constructed as follows: Each A-pattern is scheduled on $A_1$, each B-pattern on $V$, 3 A-patterns on $A_2$, 2 B-patterns on $B_2$, 4 A-patterns on $A_3$, and 3 B-patterns on $B_3$. It remains to show that scheduling according to Table 1 is feasible with respect to $G$.

By definition of $A_1$, scheduling an A-pattern on $A_1$ yields a feasible schedule for SMC-UNIT$(G, n)$. This is used for the A-patterns in the cases where $\|\Pi\| \in \{3, 6, 10, 11, 14, 15, 18, 19\}$.

Because $G$ is bipartite, a B-pattern can be scheduled on $V$. This is used for the B-patterns in the cases where $\|\Pi\| \in \{4, 7, 8, 10, 11, 13, 14, 16, 17, 19, 20\}$.

Because $(A_2, B_2)$ is a II-coloring, scheduling 3 A-patterns on $A_2$ and 2 B-patterns on $B_2$ is feasible. This is used for the case where $\|\Pi\| \in \{9, 13, 17\}$.

Because $(A_3, B_3)$ is a III-coloring, scheduling 4 A-patterns on $A_3$ and 3 B-patterns on $B_3$ is feasible. We use this whenever $\|\Pi\| \in \{12, 15, 16, 18, 19, 20\}$. This proves Claim 14.

**Claim 15.** *There exists an optimal schedule that is comprised of blocks of length 12 following row 12 of Table 1 and one rest block of length at most 20 following Table 1.*

By Claim 14, we need to show Claim 15 for $\|\Pi\| \geq 21$. We may assume that for any two different components of $H$, their makespans in $\Pi$ differ by at most 3. Suppose there exists components $C_1$ and $C_2$ such that the makespan of $C_1$ exceeds the makespan of $C_2$ by at least 4. Deleting a last job from $C_1$ and inserting it after the last job of $C_2$, the makespan difference decreases. As a consequence, the makespan in $\Pi$ on each component is at least 18. This implies that $\Pi$ schedules at least 4 A-patterns or 3 B-patterns both of length 12 on each connected component: Let $C$ be a component of $H$. If $\Pi$ has at most 3 A-patterns and at most 2 B-patterns on $C$, then the makespan on $C$ is at most $3 \cdot 3 + 2 \cdot 4 = 17$. A contradiction. Therefore, if $\|\Pi\| \geq 21$, we modify $\Pi$ by scheduling the 4 A-patterns on $A_3$ and the 3 B-patterns on $B_3$. By definition of a III-coloring of $(A_3, B_3)$ this yields a feasible subschedule with respect to $G$. We repeat this procedure until the makespan of the remaining patterns is at most 20. This proves Claim 15.

We can compute the schedule obtained by Table 1 for each possible value $r \in [20] \setminus \{1, 2, 5\}$ on each connected component using the given I-,II- and III-colorings and filling it up with blocks of 12 following row 12 of Table 1. By Lemma 15, the schedule with minimum makespan is an optimal schedule.                    □

In the following, we complete the proof of Lemma 7 by showing that modifying a star forest which does not contain any alternating paths of type II along an alternating path of type III will create a new star forest which also does not contain any alternating path of type II.

**Lemma 16.** *Let $H = (V, E')$ be a star forest of $G$ without any alternating paths of type II. Let $P$ be an alternating path $P$ of type III. Then the star forest $H' = (V, E' \Delta P)$ contains no alternating paths of type II.*

*Proof.* Let $C_1, \ldots C_k$ be the stars of the alternating path $P$ in $H$ and let $C'_1, \ldots C'_k$ denote the corresponding stars in $H'$. Observe that $C'_i \simeq C_i \simeq S_3$ for all $i \in \{2, \ldots, k-1\}$. Moreover, $C'_1 = S_3$ and $C'_k = S_\ell$ for some $\ell \geq 3$. For the purpose of a contradiction, suppose that $H'$ contains an alternating path $P_2$ of type II. Clearly, $P_2$ and $P$ intersect; otherwise $P_2$ is also contained in $H$. Specifically, $P_2$ ends in some $C'_i$ with $i \in \{1, \ldots, k\}$. If $i > 1$, then $P_2$ and $P$ share exactly the center of $C'_i$ and thus $H$ contains $P_2$ as well. A contradiction. If $P_2$ ends in $C'_1$, then $H$ contains an alternating path of type II ending in $C_2$ that goes via $C_1 \simeq S_2$. Again, a contradiction. □

# References

1. Abdekhodaee, A.H., Wirth, A.: Scheduling parallel machines with a single server: some solvable cases and heuristics. Comput. Oper. Res. **29**(3), 295–315 (2002)
2. Abdekhodaee, A.H., Wirth, A., Gan, H.S.: Equal processing and equal setup time cases of scheduling parallel machines with a single server. Comput. Oper. Res. **31**(11), 1867–1889 (2004)
3. Abdekhodaee, A.H., Wirth, A., Gan, H.S.: Scheduling two parallel machines with a single server: the general case. Comput. Oper. Res. **33**(4), 994–1009 (2006)
4. Alon, N., Azar, Y., Woeginger, G.J., Yadid, T.: Approximation schemes for scheduling on parallel machines. J. Sched. **1**(1), 55–66 (1998)
5. Baker, B.S., Coffman, E.G., Jr.: Mutual exclusion scheduling. Theoret. Comput. Sci. **162**(2), 225–243 (1996)
6. Bodlaender, H.L., Fomin, F.V.: Equitable colorings of bounded treewidth graphs. Theoret. Comput. Sci. **349**(1), 22–30 (2005)
7. Bodlaender, H.L., Jansen, K.: On the complexity of scheduling incompatible jobs with unit-times. In: Borzyszkowski, A.M., Sokołowski, S. (eds.) MFCS 1993. LNCS, vol. 711, pp. 291–300. Springer, Heidelberg (1993). https://doi.org/10.1007/3-540-57182-5_21
8. Bodlaender, H.L., Jansen, K.: Restrictions of graph partition problems. Part I. Theor. Comput. Sci. **148**(1), 93–109 (1995)
9. Bodlaender, H.L., Jansen, K., Woeginger, G.J.: Scheduling with incompatible jobs. Discret. Appl. Math. **55**(3), 219–232 (1994)
10. Brucker, P., Dhaenens-Flipo, C., Knust, S., Kravchenko, S.A., Werner, F.: Complexity results for parallel machine problems with a single server. J. Sched. **5**(6), 429–457 (2002)
11. Buchem, M., Kleist, L., Schmidt genannt Waldschmidt, D.: Scheduling with machine conflicts. CoRR **abs/2102.08231** (2021). https://arxiv.org/abs/2102.08231
12. Chen, J.J., Hahn, T., Hoeksma, R., Megow, N., von der Brüggen, G.: Scheduling self-suspending tasks: new and old results. In: Proceedings of the 31st Euromicro Conference on Real-Time Systems (2019)
13. Chen, L., Jansen, K., Zhang, G.: On the optimality of approximation schemes for the classical scheduling problem. In: Proceedings of the 25th ACM-SIAM Symposium on Discrete Algorithms, pp. 657–668 (2013)

14. Chrobak, M., Csirik, J., Imreh, C., Noga, J., Sgall, J., Woeginger, G.J.: The buffer minimization problem for multiprocessor scheduling with conflicts. In: Orejas, F., Spirakis, P.G., van Leeuwen, J. (eds.) ICALP 2001. LNCS, vol. 2076, pp. 862–874. Springer, Heidelberg (2001). https://doi.org/10.1007/3-540-48224-5_70

15. Das, S., Wiese, A.: On minimizing the makespan when some jobs cannot be assigned on the same machine. In: Proceedings of the 25th Annual European Symposium on Algorithms (2017)

16. Gan, H.S., Wirth, A., Abdekhodaee, A.H.: A branch-and-price algorithm for the general case of scheduling parallel machines with a single server. Comput. Oper. Res. **39**(9), 2242–2247 (2012)

17. Gardi, F.: Mutual exclusion scheduling with interval graphs or related classes. Part I. Discret. Appl. Math. **157**(1), 19–35 (2009)

18. Garey, M.R., Johnson, D.S.: Computers and intractability: a guide to the theory of NP-completeness (1979)

19. Graham, R.L.: Bounds for certain multiprocessing anomalies. Bell Syst. Tech. J. **45**(9), 1563–1581 (1966)

20. Graham, R.L.: Bounds on multiprocessing timing anomalies. SIAM J. Appl. Math. **17**(2), 416–429 (1969)

21. Graham, R.L., Lawler, E.L., Lenstra, J.K., Rinnooy Kan, A.H.: Optimization and approximation in deterministic sequencing and scheduling: a survey. Ann. Discret. Math. **5**, 287–326 (1979)

22. Hall, N.G., Potts, C.N., Sriskandarajah, C.: Parallel machine scheduling with a common server. Discret. Appl. Math. **102**(3), 223–243 (2000)

23. Hansen, P., Hertz, A., Kuplinsky, J.: Bounded vertex colorings of graphs. Discret. Math. **111**(1–3), 305–312 (1993)

24. Håstad, J.: Clique is hard to approximate within 1- $\varepsilon$. Acta Math. **182**(1), 105–142 (1999)

25. Hochbaum, D.S.: Various notions of approximations: good, better, best and more. Approx. Algorithms NP-Hard Probl., 346–398 (1997)

26. Hochbaum, D.S., Shmoys, D.B.: Using dual approximation algorithms for scheduling problems theoretical and practical results. J. ACM **34**(1), 144–162 (1987)

27. Höhne, F., van Stee, R.: Buffer minimization with conflicts on a line. Theoret. Comput. Sci. **876**, 25–33 (2021)

28. Jansen, K.: An EPTAS for scheduling jobs on uniform processors: using an MILP relaxation with a constant number of integral variables. SIAM J. Discret. Math. **24**(2), 457–485 (2010)

29. Jansen, K., Klein, K.M., Verschae, J.: Closing the gap for makespan scheduling via sparsification techniques. Math. Oper. Res. **45**(4), 1371–1392 (2020)

30. Jiang, Y., Zhang, Q., Hu, J., Dong, J., Ji, M.: Single-server parallel-machine scheduling with loading and unloading times. J. Comb. Optim. **30**(2), 201–213 (2014). https://doi.org/10.1007/s10878-014-9727-z

31. Kern, W., Nawijn, W.N.: Scheduling multi-operation jobs with time lags on a single machine. In: Proceedings of the 2nd Twente Workshop on Graphs and Combinatorial Optimization (1991)

32. Kőnig, D.: Gráfok és mátrixok. Mat. Fizikai Lapok **38**, 116–119 (1931)

33. Kim, M.Y., Lee, Y.H.: MIP models and hybrid algorithm for minimizing the makespan of parallel machines scheduling problem with a single server. Comput. Oper. Res. **39**(11), 2457–2468 (2012)

34. Korte, B.H., Vygen, J.: Combinatorial Optimization, vol. 6. Springer, Heidelberg (2018)

35. Kravchenko, S.A., Werner, F.: Parallel machine scheduling problems with a single server. Math. Comput. Model. **26**(12), 1–11 (1997)
36. Lund, C., Yannakakis, M.: The approximation of maximum subgraph problems. In: Lingas, A., Karlsson, R., Carlsson, S. (eds.) ICALP 1993. LNCS, vol. 700, pp. 40–51. Springer, Heidelberg (1993). https://doi.org/10.1007/3-540-56939-1_60
37. Rajkumar, R., Sha, L., Lehoczky, J.P.: Real-time synchronization protocols for multiprocessors. In: Proceedings of the 9th IEEE Real-Time Systems Symposium, vol. 88, pp. 259–269 (1988)
38. Sahni, S.K.: Algorithms for scheduling independent tasks. J. ACM **23**(1), 116–127 (1976)
39. Sahni, S.K.: Scheduling master-slave multiprocessor systems. IEEE Trans. Comput. **45**(10), 1195–1199 (1996)
40. de Werra, D.: Restricted coloring models for timetabling. Discret. Math. **165**, 161–170 (1997)
41. Xie, X., Li, Y., Zhou, H., Zheng, Y.: Scheduling parallel machines with a single server. In: Proceedings of 2012 International Conference on Measurement, Information and Control, vol. 1, pp. 453–456. IEEE (2012)
42. Zuckerman, D.: Linear degree extractors and the inapproximability of max clique and chromatic number. In: Proceedings of the 38th Annual ACM Symposium on Theory of Computing, pp. 681–690 (2006)

# Knapsack Secretary Through Boosting

Andreas Abels[1], Leon Ladewig[2], Kevin Schewior[3(✉)],
and Moritz Stinzendörfer[4]

[1] Chair of Management Science, RWTH Aachen University, Aachen, Germany
andreas.abels@oms.rwth-aachen.de
[2] Munich, Germany
leonladewig@mail.de
[3] Department of Mathematics and Computer Science, University of Southern
Denmark, Odense, Denmark
kevs@sdu.dk
[4] Department of Mathematics, TU Kaiserslautern, Kaiserslautern, Germany
stinzendoerfer@mathematik.uni-kl.de

**Abstract.** We revisit the knapsack-secretary problem (Babaioff et al.;
APPROX 2007), a generalization of the classic secretary problem in
which items have different sizes and multiple items may be selected if
their total size does not exceed the capacity $B$ of a knapsack. Previous works show competitive ratios of $1/(10e)$ (Babaioff et al.), $1/8.06$
(Kesselheim et al.; STOC 2014), and $1/6.65$ (Albers, Khan, and Ladewig;
APPROX 2019) for the general problem but no definitive answers for the
achievable competitive ratio; the best known impossibility remains $1/e$
as inherited from the classic secretary problem. In an effort to make more
qualitative progress, we take an orthogonal approach and give definitive
answers for special cases.

Our main result is on the 1-2-knapsack secretary problem, the special case in which $B = 2$ and all items have sizes 1 or 2, arguably the
simplest meaningful generalization of the secretary problem towards the
knapsack secretary problem. Our algorithm is simple: It *boosts* the value
of size-1 items by a factor $\alpha > 1$ and then uses the size-oblivious approach by Albers, Khan, and Ladewig. We show by a nontrivial analysis that this algorithm achieves a competitive ratio of $1/e$ if and only
if $1.40 \lesssim \alpha \leq e/(e-1) \approx 1.58$.

Towards understanding the general case, we then consider the case
when sizes are 1 and $B$, and $B$ is large. While it remains unclear if $1/e$
can be achieved in that case, we show that algorithms based only on
the relative ranks of the item values can achieve precisely a competitive
ratio of $1/(e+1)$. To show the impossibility, we use a non-trivial generalization of the factor-revealing linear program for the secretary problem
(Buchbinder, Jain, and Singh; IPCO 2010).

Supported in part by the Independent Research Fund Denmark, Natural Sciences,
grant DFF-0135-00018B.

P. Chalermsook and B. Laekhanukit (Eds.): WAOA 2022, LNCS 13538, pp. 61–81, 2022.
https://doi.org/10.1007/978-3-031-18367-6_4

# 1   Introduction

In the classic secretary problem, there is a single position to be filled, and $n$ candidates arrive one by one in uniformly random order. Upon arrival of any candidate, they have to be rejected or accepted immediately and irrevocably only based on *ordinal* information on the candidates seen so far, that is, their relative ranks. The goal is to maximize the probability that the best candidate is selected. The origin of this problem is unclear; for a discussion, we refer to Ferguson's survey [16]. It is well known [13, 26] since the 1960s that a probability of $1/e$ can be achieved by selecting the first candidate that is better than the $n/e$ first candidates and that this is the best-possible probability under the typical assumption $n \to \infty$. Many extensions of this problem have since been considered, especially in recent years, partially due to relations to beyond-the-worst-case analyses of online algorithms (e.g., [1, 18, 20]) and to mechanism design (e.g., [4, 24]).

There is extensive work on multiple-choice variants of the secretary problem. Few of these works consider an ordinal setting [8, 19, 29]; the majority considers the *value* setting in which each arriving candidate (or item) $i$ is revealed along with a value $v_i \in \mathbb{R}_{\geq 0}$ and must be rejected or accepted immediately and irrevocably so that the set of accepted items obeys some combinatorial constraint. The goal is to obtain an algorithm with a (strong) competitive ratio $\rho$, i.e., that constructs a solution ALG such that $v(\mathsf{ALG})$, the sum of values of accepted items, is in expectation at least $\rho \cdot v(\mathsf{OPT})$ where OPT is the best solution that could have been constructed.

Whereas the results for the standard secretary problem carry over to the value setting, even relatively simple variants are not completely understood in that setting. This is arguably due to the sheer amount of conceivable strategies. For instance, the precise competitive ratio achievable in the 2-secretary problem, the variant in which two positions are to be filled, is *not* known—only that it is strictly larger than in the much-better-understood ordinal "counterpart", sometimes called the $(2, 2)$-secretary problem [8, 9].

The secretary variant that has probably received most attention is the matroid secretary problem [5], an extension of the $k$-secretary problem [24] (in which $k$ positions are to be filled) to any matroid constraint, see, e.g., the state-of-the-art result [15, 25] and the survey by Dinitz [12]. An orthogonal and also well-known extension of $k$-secretary is the *knapsack* secretary problem in which items additionally have sizes and the total size of accepted items must not exceed some given capacity $B$ [2, 4, 21, 23, 28]. While this line of work has improved the competitive ratio from $1/(10e)$ to $1/6.65$, no impossibility beyond $1/e$ has been found. For some secretary versions, e.g., the bipartite-matching variant [22], it is known that this ratio can in fact be matched.

Our paper may raise hope that the ratio of $1/e$ can in fact be matched for knapsack secretary. First, we consider the 1–2-knapsack problem. Here, items have sizes either 1 or 2 and the capacity $B$ is 2. We develop a $1/e$-competitive algorithm. To us, this result is both surprising and significant because the problem generalizes both the classic secretary problem, which severely restricts the set of candidate algorithms, and the not-entirely-understood 2-secretary problem. We also consider the problem with sizes either 1 or $B$ and $B$ large, for which we show initial results, namely that $1/(e+1) \pm o(1)$ is precisely the competitive ratio that can be achieved by *ordinal* algorithms. These are algorithms that only use the relative rank of the items and disregard the actual values.

## 1.1   Related Work

Kleinberg [24] first considers $k$-secretary as introduced above, gives an algorithm with competitive ratio $1 - \Theta(1/\sqrt{k})$, and shows that this ratio is asymptotically best possible. This result is reproduced by Kesselheim et al. [23] in the more general context of packing LPs. Buchbinder et al. [8] consider the $(j, k)$-secretary problem in the ordinal setting in which $j$ items can be selected and the goal is to maximize the expected ratio of elements selected from the top $k$ items. They also state the algorithm-design problems as linear programs, which they can only solve for small values of $j$ and $k$, but Chan et al. [9] can solve them for larger values. Any guarantee for the $(k, k)$-secretary problem carries over to the $k$-secretary problem, but Chan et al. [9] rule out the other direction. More specifically, Chan et al.'s results include an optimal algorithm for $(2, 2)$-secretary with guarantee approximately 0.489 and a (not necessarily optimal) algorithm for 2-secretary with guarantee approximately 0.492. Albers and Ladewig [3] revisit the problem and give simple algorithms with improved (albeit non-optimal) competitive ratios for many fixed values of $k$.

The knapsack secretary problem is introduced by Babaioff et al. [4] who give a $1/(10e)$-competitive algorithm, which was subsequently improved by Kesselheim et al. [23] to $1/8.06$ and by Albers, Khan, and Ladewig [2] to $1/6.65$. Essentially all known $\Omega(1)$-competitive algorithms for the knapsack secretary problem are somewhat wasteful in the competitive ratio, presumably at least partially for the sake of a simpler analysis, in that they randomize between different algorithms that are tailored to respective item sizes. It seems that qualitative progress can only be made by a more fine-grained analysis avoiding such case distinctions.

A variant of the knapsack secretary problem that has recently been considered is the fractional variant in which an item can also be packed fractionally, avoiding situations in which an arriving item cannot be selected at all, even when there is space. The currently best known achievable competitive ratio is $1/4.39$ [17], also achieved by a blended approach.

It is not difficult to see that no constant competitive ratio can be achieved when the items do not arrive in random but in adversarial order, even in the unit-value case [27]. Starting from this problem, problems in which other assumptions than the order are relaxed are considered as well. For instance, Zhou et al. [30] consider the version in which each item has a small size; Böckenhauer et al. [6] and Boyar et al. [7] introduce advice and untrusted predictions, respectively, to the problem.

Lower bounds for secretary problems in the value setting are rare. For some related problems [10,11,14], the rich class of strategies can be handled by, for any strategy, identifying an infinite set of values (using Ramsey theory) on which it is much better behaved. It is, however, not clear how such an approach could be applied, e.g., for knapsack secretary since it seems one would need to control how the values in the support are spread out, a property that is irrelevant in the other settings.

## 1.2   Our Contribution

The special case 1-2-knapsack is not only arguably the simplest special case that exhibits features of the knapsack problem distinguishing it from the matroid secretary problem. Since the problem generalizes both the standard secretary problem and 2-secretary, we believe that settling it in terms of the achievable competitive ratio is also interesting per se.

A good starting point for tackling 1-2-knapsack seems to be the extended secretary algorithm, which is $1/3.08$-competitive in the slightly more general case when all items have size larger than $B/3$ [2]. This algorithm simply ignores the item sizes, samples some prefix of length $cn$ for some optimized constant $c \in (0,1)$, and afterwards selects all items that surpass the largest value from the sampling phase and that can still be feasibly packed. It is, however, easy to see that this approach cannot achieve $1/e$: Achieving $1/e$ in an instance where the optimal solution consists of a large item requires setting $c = 1/e \pm o(1)$. The resulting algorithm will, however, not be $1/e$-competitive in an instance where the optimal solution consists of two small items of equal value, but there are many large items, each slightly more valuable than the individual small items, making sure that the small items are (almost) never selected by the algorithm. In this case, the competitive ratio of the algorithm will be essentially half the probability that the algorithm selects a (large) item, that is, $(1 - 1/e)/2 < 1/e$. We denote two instances of the above forms by $\mathcal{I}_1$ and $\mathcal{I}_2$, respectively, in the following. Clearly, it is possible to choose $c$ so as to balance between $\mathcal{I}_1$ and $\mathcal{I}_2$. As a small side result, we show that a ratio of approximately $0.353 < 1/e$ can be achieved that way.

The key observation leading to our $1/e$-competitive algorithm is that keeping $c = 1/e$ and internally multiplying (*boosting*) values of small items with a suitable constant factor $\alpha > 1$ prior to running the extended secretary algorithm may handle both $\mathcal{I}_1$ and $\mathcal{I}_2$: While this is clear for $\mathcal{I}_1$ when the ranking of values does not change through boosting, a small item may overtake the most valuable (large) item. This however means that this small item has relatively large (actual) value. Using that the algorithm also accepts the second-best item with a significant probability $(1/e^2)$, we can show that, with the right choice of $\alpha$, we still extract enough value from the small and large items to cover $1/e \cdot v(\mathsf{OPT})$. In $\mathcal{I}_2$, the small items would overtake the large items, significantly improving the expected value achieved by the algorithm; conversely, if they did not overtake, they would not have been harmfully valuable in the first place—again with the right choice of $\alpha$. To sum up, "$\mathcal{I}_1$ type" instances impose an upper bound on $\alpha$, and "$\mathcal{I}_2$ type" instances impose a lower bound on $\alpha$. We show that the algorithm is $1/e$-competitive if *and only if* $1.40 \lesssim \alpha \leq e/(e-1) \approx 1.58$ where the upper bound comes essentially from the above consideration for $\mathcal{I}_1$. Note that therefore, in particular, our boosting *is* different from ordering the items by their "bang for the buck" ratios.

We note that, while $\alpha$-boosting seems reminiscent of $\beta$-filtering [9] (for $\beta < 1$), applying $\beta$-filtering to the extended secretary algorithm will not yield a $1/e$-competitive algorithm. The extended secretary algorithm would be adapted by ignoring items with a value less than $\beta$ times the highest value seen so far. Note that indeed, a "$\mathcal{I}_1$ type" instance where all but the most valuable item have a similar small value, one would have to choose $c = 1/e \pm o(1)$ again, independently of $\beta$. But such an algorithm would again only be $(1 - 1/e)/2$-competitive on $\mathcal{I}_2$.

The crux of our analysis is distinguishing all possible cases beyond those covered by $\mathcal{I}_1$ and $\mathcal{I}_2$ in a smart way. To bound the algorithm's value in each of these cases, we precisely characterize the probabilities with which the algorithm selects an item depending on its size and its position in the (boosted) order of values, significantly extending observations made by Albers and Ladewig [3].

Before tackling the general case and understanding potentially complicated knapsack configurations, we propose considering a clean special case called 1-$B$-knapsack where items have sizes either 1 or $B$, and $B$ is large. One may be tempted to think that this special case is difficult in that selecting a small item early on may lead to a blocked knapsack and a horribly inefficient use of capacity, e.g., because all other items are large. On the other hand, when $B$ is large, one can easily avoid such situations by sampling. We do not give a conclusive answer on whether $1/e$ can be matched in this case, but we give some preliminary results.

Unfortunately, a competitive ratio of $1/e$ for 1-$B$-knapsack cannot be achieved with our boosting approach. The same consideration we made for $\mathcal{I}_1$ earlier (for 1-2-knapsack) to get an upper bound of $e/(e-1)$ on $\alpha$ still works; in contrast, a generalization of $\mathcal{I}_2$ rules out any constant boosting factor.

We then give another algorithm for 1-$B$ knapsack which can be viewed as a linear interpolation between the classic secretary algorithm and the algorithm by Kleinberg [24] for $k$-secretary. We show that it is $1/(e+1)$-competitive. This algorithm turns out to be *ordinal*, that is, its decisions only depend on the item sizes and the relative order of their values. Remarkably, we are able to show that $1/(e+1)$ is the best-possible guarantee such algorithms can achieve. We do so by generalizing the factor-revealing linear program due to Buchbinder et al. [8] by adding variables and constraints. Arguing that the LP indeed models our problem becomes more difficult because, in contrast to the setting of Buchbinder et al., at any time, even the size of the next item is random. We do so by showing reductions between our model and an auxiliary batched-arrival model.

## 2    Preliminaries

We use the following notation. Let $\mathcal{I} = \{1, \ldots, n\}$ be the set of items (also called *elements*), where each item $i \in \mathcal{I}$ is specified by a profit $v_i$ and a size $s_i$. Moreover, we are given a knapsack of capacity $B \in \mathbb{N}_{\geq 2}$. The goal is to find a maximum-profit packing, i.e., a subset of items $S$ such that $\sum_{i \in S} s_i \leq B$ and $\sum_{i \in S} v_i$ is maximized. Without loss of generality, we assume that all elements have distinct values and that $v_1 > v_2 > \ldots > v_n$. This way, the name of an item $i$ corresponds to the (global) *rank* in $\mathcal{I}$.

Throughout the following sections, an important subclass of the knapsack problem arises where each item has either size 1 or $B$.

**Definition 1 (1-$B$-knapsack).** *We call the special case of the knapsack problem where all items have size 1 or $B$ the 1-$B$-knapsack problem. Items of size 1 are called small and items of size $B$ are called large.*

Within the context of 1-$B$-knapsack, we use the following further notation. Let $\mathcal{I}_S$ be the set of small items. For any small item $i \in \mathcal{I}_S$, let $r_s(i)$ denote its rank among the small items. Note that $r_s(i)$ is at most the global rank $i$ of this item. Further, let $r'_g(a)$ denote the global rank of the small item $x$ that satisfies $r_s(x) = a$. When we use just the word "rank", we refer to the global rank.

Let OPT be an optimal offline algorithm. For any algorithm ALG, we overload the notation and use the same symbol also for the packing returned by the algorithm. Further, we denote by $v(\mathsf{ALG}) := \sum_{i \in \mathsf{ALG}} v_i$ the total profit of the packing returned by ALG. We are particularly interested in *online* algorithms, i.e., algorithms that are initially only given $n$ and are presented with the items one by one. Upon arrival of any item, an online algorithm has to irrevocably decide whether it includes the item or not. A special class of algorithms we consider are *ordinal* algorithms. These algorithms only have access to the item sizes and the *relative order* of item values.

We say that an online algorithm ALG is $\rho$-competitive if $\mathbb{E}[v(\mathsf{ALG})] \geq \rho \cdot v(\mathsf{OPT})$ for all instances, where the expectation is taken over a uniformly random arrival order (and possibly internal randomization that the algorithm uses). In general, we assume $n \to \infty$ for our bounds. Note that, for a fixed number of items, we can achieve a guarantee that is arbitrarily close to the guarantee for $n \to \infty$ by adding a sufficient amount of virtual dummy items.

Finally, throughout the paper, we use the notation $[k] := \{1, \ldots, k\}$ for any $k \in \mathbb{N}$.

## 3  Matching $1/e$ for 1–2-Knapsack

In this section, we develop an optimal algorithm for 1–2-knapsack. For this purpose, we first propose a natural algorithm for 1-$B$-knapsack, based on the size-oblivious approach from [2]. Here, items are accepted whenever their profit exceeds a certain threshold, similar to the optimal algorithm for the classic secretary problem. Therefore, we call it the *extended secretary algorithm*. From an initial sampling phase of length $cn$, where $c \in (0, 1)$ is a parameter of the algorithm, the best item is used as a reference element. Subsequently, any item beating the reference element is packed if it still fits. A formal description is given in Algorithm 1.

---

**Algorithm 1:** Extended secretary algorithm

---

**Input:** Instance of 1-$B$-knapsack arriving in uniformly random order, parameter $c \in (0,1)$.

**Output:** A knapsack packing.

**for** *round $\ell = 1$ to $n$* **do**

    **if** $\ell \leq cn$ **then**

        | Reject the current item;                                  `// sampling phase`

    **end**

    **if** $\ell > cn$ **then**

        Let $v^*$ be the highest profit seen up to round $\lfloor cn \rfloor$;

        Pack the current item if its profit exceeds $v^*$ and the remaining capacity is large enough;

    **end**

**end**

---

In the following, we denote Algorithm 1 by ALG and set

$$p_i(j) := \Pr[\text{ALG packs item } i \text{ as the } j\text{-th element}],$$
$$p_i := p_i(1),$$
$$P_i := \sum_{j=1}^{B} p_i(j).$$

Thus, $P_i$ is the probability that the algorithm packs item $i$ at all, while $p_i$ is the probability that it is packed as the first item. We first state some results on the values $p_i$, which have essentially been investigated in [3]. Indeed, the following results follow from that work and some simple observations.

**Lemma 1.** *For $i \in \mathbb{N}$, it holds that*

$$p_i = c \left( \ln \frac{1}{c} + \sum_{\ell=1}^{i-1} (-1)^{\ell+1} \binom{i-1}{\ell} \frac{c^\ell - 1}{\ell} \right) \pm o(1).$$

*Proof.* Let $i \in \mathbb{N}$. The extended secretary algorithm packs $i$ as the first item if and only if the SINGLE-REF algorithm from [3] with $r = 1$ and $k = i$ packs $i$ as the first item. Hence, the probability $p_i$ can be derived from [3] as follows: If $i = 1$, item $i$ is a dominating item in the terminology of [3] and Lemma 6 of [3] gives $p_1 = c \cdot \ln(1/c) - o(1)$. In the case $i \geq 2$, item $i$ is a non-dominating item in the terminology of [3]. Here, Lemma 4 of [3] gives $p_i = p_i(i)$ and Lemma 5 of [3] and gives $p_i(i) = p_1(i)$, that is, $p_i$ turns out to be the probability that the dominating item 1 is accepted as the $i$-th item by the SINGLE-REF algorithm. Again, the claim follows from Lemma 6 of [3]. □

Furthermore, observe that, since increasing the profit of an item cannot decrease its probability of being selected, we have $p_i \geq p_{i+1}$ for all $i \in [n-1]$. Note that ALG accepts no item if and only if the best item is in the sampling phase. Therefore, we have the following observation.

**Observation 1.** *It holds that*

$$\sum_{i=1}^{n} p_i = 1 - \Pr[\mathsf{ALG} \text{ accepts no item}]$$

$$= 1 - \Pr[\text{item 1 appears in sampling phase}] = 1 - c.$$

In the following subsection, we identify relations between the probabilities $P_i$ and $p_j$.

## 3.1 Structural Lemma

In this subsection, we show the following lemma connecting the probabilities $P_i$ to the probabilities $p_j$ from Lemma 1. The analysis showing the $1/e$-competitiveness of our algorithm is crucially based on this result. Note that we only use it for $B = 2$ but it holds for all $B$.

**Lemma 2.** *The probability that* $\mathsf{ALG}$ *packs element* $i \in \mathcal{I}$ *is*

$$P_i = \begin{cases} p_i & \text{if element } i \text{ is large,} & (1) \\ i_s^* \cdot p_i + \displaystyle\sum_{x=\mathrm{r_s}(i)+1}^{B^*} p_{\mathrm{r}_\mathrm{g}'(x)} & \text{if element } i \text{ is small,} & (2) \end{cases}$$

*with* $i_s^* := \min\{\mathrm{r_s}(i), B\}$ *and* $B^* := \min\{B, |\mathcal{I}_S|\}$.

Observe that (1) follows immediately: Any large element can only be packed when the knapsack is empty, i.e., as the first element. The proof of (2) requires a bit more work.

**Definition 2.** *Let* $E_{x,y}^{i,j}$ *be the event that the small elements* $i$ *and* $j$ *are packed as the* $x$-*th and* $y$-*th items, respectively.*

Note that the event that any item $i \in \mathcal{I}_S$ is packed as $x$-th item, where $x \geq 2$, can be partitioned according to the item packed first. Therefore, for any $i \in \mathcal{I}_S$ and $x \geq 2$,

$$p_i(x) = \sum_{j \in \mathcal{I}_S} \Pr\left[E_{1,x}^{j,i}\right]. \qquad (3)$$

We have the following technical lemmata.

**Lemma 3.** *Let* $i \in \mathcal{I}_S$ *be any small item and* $i_s^* = \min\{\mathrm{r_s}(i), B\}$. *For* $2 \leq \ell \leq i_s^*$, *it holds that* $\displaystyle\sum_{j \in \mathcal{I}_S} \Pr\left[E_{1,\ell}^{i,j}\right] = p_i$.

*Proof.* The first step is to show that at least $\ell$ elements are accepted in total, if element $i$ is accepted first. Since element $i$ has rank $\mathrm{r_s}(i)$ among the small elements, there are $\mathrm{r_s}(i) - 1$ small elements that are more valuable. Their position in the input sequence cannot be in the sampling phase, nor before element $i$ if it is packed first. So there are at least $i_s^*$ small elements that can be packed subsequently. Therefore, for $2 \leq \ell \leq i_s^*$, a small element is packed as $\ell$-th item. The claim follows by partitioning the event that $i$ is packed first according to the item $j \in \mathcal{I}_S$ packed as $\ell$-th item. □

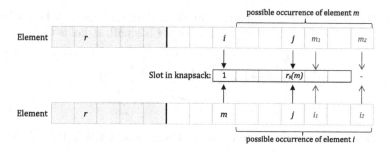

**Fig. 1.** Occurrence of element $i$ and $m$ in event $E^{m,j}_{1,\mathrm{r_s}(m)}$ and $E^{i,j}_{1,\mathrm{r_s}(m)}$

**Lemma 4.** *For any two small elements $i, j \in \mathcal{I}_S$ and any $x, y \in [B]$, we have* $\Pr\left[E^{i,j}_{x,y}\right] = \Pr\left[E^{j,i}_{x,y}\right].$

*Proof.* Consider any input sequence of $E^{i,j}_{x,y}$ and the sequence resulting from swapping the elements $i$ and $j$. Since both elements are not part of the sample, the reference element is not changed by the swap. Therefore, no element that was previously accepted will be rejected and none that was previously rejected will be accepted. Only the order of selection changes.                    □

**Lemma 5.** *For any small items $i, j, m \in \mathcal{I}_S$ with $\mathrm{r_s}(m) > 1$ and $\mathrm{r_s}(i) < \mathrm{r_s}(m)$, it holds that* $\Pr\left[E^{m,j}_{1,\mathrm{r_s}(m)}\right] = \Pr\left[E^{i,j}_{1,\mathrm{r_s}(m)}\right].$

*Proof.* Consider any input sequence from $E^{m,j}_{1,\mathrm{r_s}(m)}$. Since $\mathrm{r_s}(i) < \mathrm{r_s}(m)$ applies, element $i$ lies behind the element with rank $m$ in the sequence. If both are selected (see $i_1$ in Fig. 1), this also applies after they have been swapped (see Lemma 4 and $m_1$ in Fig. 1). If previously only element $m$ of the two is packed, only element $i$ (of the two) is selected after their swapping ($i_2$ and $m_2$ in Fig. 1), since in this case, nothing changes in the reference element either. Therefore $\Pr\left[E^{m,j}_{1,\mathrm{r_s}(m)}\right] \leq \Pr\left[E^{i,j}_{1,\mathrm{r_s}(m)}\right]$ applies.

Now consider any input sequence from $E^{i,j}_{1,\mathrm{r_s}(m)}$. We show that the element $m$ lies behind the element $i$ in the sequence since an element is packed as $z$-th item, where $z = \mathrm{r_s}(m)$. Assuming this did not apply and $m$ is in the sample, then there would be at most $\mathrm{r_s}(m) - 1$ small elements that can be packed.

In the case that it occurs in the sequence after the sampling phase, but before element $i$, there must be a more valuable element in the sample (because $m$ was not packed) and therefore there are again at most $\mathrm{r_s}(m) - 1$ small elements that can be selected. In particular, in both cases, no element is packed as $z$-th item for $\mathrm{r_s}(m)$. This is a contradiction to the fact that we consider an input sequence in $E^{i,j}_{1,\mathrm{r_s}(m)}$. Now, using the same argumentation as in the first case, it follows that $\Pr\left[E^{m,j}_{1,\mathrm{r_s}(m)}\right] \geq \Pr\left[E^{i,j}_{1,\mathrm{r_s}(m)}\right]$, which completes the proof.                    □

Using Lemmas 3, 4 and 5, we are now able to prove Lemma 2.

*Proof (Lemma 2).* Let $i \in \mathcal{I}$ be any item. If $i$ is large, it can only be packed as the first item, thus $P_i = p_i$. Now, assume that $i$ is small. It holds that

$$P_i = \sum_{x=1}^{B^*} p_i(x) = \underbrace{\sum_{x=1}^{i_s^*} p_i(x)}_{(*)} + \underbrace{\sum_{x=r_s(i)+1}^{B^*} p_i(x)}_{(**)}.$$

We next simplify both starred terms using Lemmas 3, 4 and 5. For $(*)$, it holds that

$$\sum_{x=1}^{i_s^*} p_i(x) = p_i(1) + \sum_{x=2}^{i_s^*} \sum_{j \in \mathcal{I}_S} \Pr\left[E_{1,x}^{j,i}\right] \qquad \text{(Equation (3))}$$

$$= p_i + \sum_{x=2}^{i_s^*} \sum_{j \in \mathcal{I}_S} \Pr\left[E_{1,x}^{i,j}\right] \qquad \text{(Lemma 4)}$$

$$= p_i + \sum_{x=2}^{i_s^*} p_i(1) \qquad \text{(Lemma 3)}$$

$$= i_s^* \cdot p_i.$$

For $(**)$, we obtain

$$\sum_{x=r_s(i)+1}^{B^*} p_i(x) = \sum_{x=r_s(i)+1}^{B^*} \sum_{j \in \mathcal{I}_S} \Pr\left[E_{1,x}^{j,i}\right] \qquad \text{(Equation (3))}$$

$$= \sum_{x=r_s(i)+1}^{B^*} \sum_{j \in \mathcal{I}_S} \Pr\left[E_{1,x}^{i,j}\right] \qquad \text{(Lemma 4)}$$

$$= \sum_{x=r_s(i)+1}^{B^*} \sum_{j \in \mathcal{I}_S} \Pr\left[E_{1,x}^{r_g'(x),j}\right] \qquad \text{(Lemma 5, } r_s(i) < x\text{)}$$

$$= \sum_{x=r_s(i)+1}^{B^*} p_{r_g'(x)}, \qquad \text{(Lemma 3, } 2 \le x = r_s(r_g'(x)) \le B\text{)}$$

which completes the proof. $\qquad\qquad\square$

The following corollary is an immediate consequence of Lemma 2 for $B = 2$.

**Corollary 1.** *For $B = 2$, the probability that* ALG *packs element $i \in \mathcal{I}$ is*

$$P_i = \begin{cases} p_i & \text{if } i \text{ is large,} \\ p_i + p_{r_g'(2)} & \text{if } i \text{ is small and } r_s(i) = 1, \\ 2p_i & \text{if } i \text{ is small and } r_s(i) > 1, \end{cases}$$

*where, if the second most valuable small item does not exist, we set $p_{r_g'(2)} = 0$.*

## 3.2 First Approach: Without Boosting

In this subsection, we study Algorithm 1 (as is) for 1–2-knapsack. Unfortunately, there are two instances such that it is impossible to choose the parameter $c$ so that Algorithm 1 is $(1/e)$-competitive on both instances.

**Lemma 6.** *For 1-2-knapsack, the competitive ratio of* ALG *is at most* 0.35767, *assuming* $n \to \infty$.

*Proof.* Let $1 > \varepsilon > 0$ be a constant. We define two instances $\mathcal{I}_1$ and $\mathcal{I}_2$. In the first instance $\mathcal{I}_1$, all items are large and only one item has substantial profit. Formally, let $v_1 = 1$, $v_i = \varepsilon^i$ for $2 \leq i \leq n$, and $s_i = 2$ for all $1 \leq i \leq n$. Then, for instance $\mathcal{I}_1$,

$$\lim_{\varepsilon \to 0} \mathbb{E}[v(\mathsf{ALG})] = P_1 \cdot v_1 = p_1 \cdot v(\mathsf{OPT}). \tag{4}$$

In the second instance $\mathcal{I}_2$, most items are large and essentially of the same profit. However, the optimal packing contains two small items that appear at ranks $n - 1$ and $n$. Formally, set $s_i = 2$ for $1 \leq i \leq n - 2$, $s_{n-1} = s_n = 1$, and $v_i = 1 + \varepsilon^i$ for all $i \in \{1, \ldots, n\}$. As item $n$ never beats any reference item, we have $P_n = 0$. Hence, the algorithm selects only items from $\{1, \ldots, n-1\}$ with positive probability, and always at most one item. For instance $\mathcal{I}_2$, we get

$$\lim_{\varepsilon \to 0} \mathbb{E}[v(\mathsf{ALG})] = \lim_{\varepsilon \to 0} \sum_{i=1}^{n} (P_i \cdot (1 + \varepsilon^i)) = \sum_{i=1}^{n} p_i$$

$$\overset{\text{Obs. (1)}}{=} 1 - c \leq \frac{1-c}{2} \cdot v(\mathsf{OPT}). \tag{5}$$

Overall, by Eqs. (4) and (5), the competitive ratio as $n \to \infty$ of ALG is bounded from above by

$$\max_{c \in (0,1)} \min\left\{p_1, \frac{1-c}{2}\right\} = \max_{c \in (0,1)} \min\left\{c \cdot \ln\frac{1}{c}, \frac{1-c}{2}\right\} \leq 0.35767.$$

This completes the proof. $\qquad\qquad\qquad\qquad\qquad\qquad\qquad\qquad\qquad\qquad$ $\square$

As a small side result, we show that this bound is almost tight. The techniques are similar to those used for our main result and presented in the full version of the paper.

**Proposition 1.** *For 1-2-knapsack, the competitive ratio of* ALG *is* $0.35317 - o(1)$, *setting* $c = 0.26888$ *and assuming* $n \to \infty$.

## 3.3 Optimal Algorithm Through $\alpha$-Boosting

The proof of Lemma 6 reveals the bottleneck of Algorithm 1: If the optimal solution consists of two elements having a high rank, the probability of selecting those items is small. This problem can be resolved by the concept of $\alpha$-*boosting*.

**Definition 3 ($\alpha$-boosting).** *Let $\alpha \geq 1$ be the boosting factor. For any item $i \in \mathcal{I}$, we define its boosted profit to be*

$$v_i' = \begin{cases} \alpha \cdot v_i & \text{if } i \text{ is small,} \\ v_i & \text{otherwise.} \end{cases}$$

In the following, we investigate Algorithm 1 enhanced by the concept of $\alpha$-boosting, denoted by $\mathsf{ALG}_\alpha$. This algorithm works exactly as given in the description of Algorithm 1, but works with the boosted profit $v_i'$ instead of the actual profit $v_i$ for any item $i \in \mathcal{I}$. Note that the unboosted algorithm analyzed in Proposition 1 is $\mathsf{ALG}_1$. For the remainder of this subsection, we fix $c = 1/e$. In particular, this implies $p_1 = 1/e \pm o(1)$ and $p_2 = 1/e^2 \pm o(1)$ according to Lemma 1.

So far, we did not specify the boosting factor $\alpha$. However, the following intuitive reasoning already shows that $\alpha$ should be bounded from above and below: If $\alpha$ is too large, we risk that $\mathsf{ALG}_\alpha$ packs small items with high probability, even when they are not part of the optimal packing. On the other hand, by the result of Proposition 1 we know that $\mathsf{ALG}_1$ cannot achieve an optimal competitive ratio. The following theorem provides lower and upper bounds on $\alpha$ such that $\mathsf{ALG}_\alpha$ is $(1/e)$-competitive.

**Theorem 1.** *For 1–2-knapsack, algorithm $\mathsf{ALG}_\alpha$ is $(1/e - o(1))$-competitive if and only if $1.400382 \lesssim \alpha \leq e/(e-1)$ and $c = 1/e$, assuming $n \to \infty$.*

*Proof.* For any item $x \in \mathcal{I}$, let $\rho(x)$ denote the global rank of $x$ after boosting. On a high level, we need to consider two cases.

In the first case, the optimal packing contains a single item $x$. If $\rho(x) = 1$, we immediately obtain $\mathbb{E}[v(\mathsf{ALG}_\alpha)] \geq p_1 v_x = (1/e) \cdot v(\mathsf{OPT})$. Now, suppose $\rho(x) \geq 2$. Let $a$ and $b$ be the items such that $\rho(a) = 1$ and $\rho(b) = 2$, respectively. Hence,

$$v_a' > v_b' \geq v_x' \geq v(\mathsf{OPT}).$$

We note that $a$ is small, as otherwise $v_a = v_a' > v(\mathsf{OPT})$. Moreover, for $\alpha < 2$, item $b$ is large: If $b$ was small, it would follow that $v_b' = \alpha \cdot v_b$ and therefore $v_a + v_b = v_a'/\alpha + v_b'/\alpha > (2/\alpha) \cdot v(\mathsf{OPT}) > v(\mathsf{OPT})$, contradicting the assumption that the optimal packing contains a single item. Therefore, $a$ is small and $b$ is large, implying $v_a = v_a'/\alpha > v(\mathsf{OPT})/\alpha$ and $v_b = v_b' \geq v(\mathsf{OPT})$. Hence,

$$\mathbb{E}[v(\mathsf{ALG}_\alpha)] \geq p_1 \cdot v_a + p_2 \cdot v_b \tag{6}$$

$$= \left(\frac{1}{e} \pm o(1)\right) \cdot \frac{v(\mathsf{OPT})}{\alpha} + \left(\frac{1}{e^2} \pm o(1)\right) \cdot v(\mathsf{OPT})$$

$$\geq \left(\frac{1}{e} \pm o(1)\right) \cdot v(\mathsf{OPT}),$$

where the latter inequality holds for $\alpha \leq e/(e-1)$. Note that, when $v_a' = 1$, $v_b' = 1 - \varepsilon$, and $v_z' = O(\varepsilon)$ for all other items $y$, Inequality (6) becomes satisfied with equality as $\varepsilon \to 0$. Therefore, $\mathsf{ALG}_\alpha$ is not $(1/e - o(1))$-competitive when $\alpha > e/(e-1)$.

In the remainder of the proof, we consider the case where the optimal packing contains two small items $x$ and $y$, where we assume $v_x > v_y$ without loss of generality. We set $j := \rho(x)$ and $k := \rho(y)$, where $1 \le j < k$. Now, let $a_1, \ldots, a_{j-1}$ and $b_{j+1}, \ldots, b_{k-1}$ denote the items appearing before $x$ and between $x$ and $y$, respectively, in the ordered sequence of boosted profits:

$$v'_{a_1} > \ldots > v'_{a_{j-1}} > v'_x > v'_{b_{j+1}} > \ldots > v'_{b_{k-1}} > v'_y.$$

We observe that neither $a$ items nor $b$ items can be small: Otherwise, the profit of such an item would be strictly larger than $v_y$, and as any two small items fit together, this item should be in the optimal packing instead of $y$. Therefore, we have $v_{a_i} = v'_{a_i} > v'_x = \alpha \cdot v_x$ for all $i \in \{1, \ldots, j-1\}$ and $v_{b_i} = v'_{b_i} > v'_y = \alpha \cdot v_y$ for all $i \in \{j+1, \ldots, k-1\}$.

Now, we can bound the expected profit of $\mathsf{ALG}_\alpha$ as follows:

$$\mathbb{E}[v(\mathsf{ALG}_\alpha)] \ge \left( \sum_{i=1}^{j-1} P_i \cdot \alpha \cdot v_x \right) + P_j \cdot v_x + \left( \sum_{i=j+1}^{k-1} P_i \cdot \alpha \cdot v_y \right) + P_k \cdot v_y \qquad (7)$$

$$= \left( \sum_{i=1}^{j-1} p_i \cdot \alpha \cdot v_x \right) + (p_j + p_k) \cdot v_x + \left( \sum_{i=j+1}^{k-1} p_i \cdot \alpha \cdot v_y \right) + 2p_k \cdot v_y$$

$$= \underbrace{\left( p_j + p_k + \alpha \cdot \sum_{i=1}^{j-1} p_i \right)}_{\lambda_x} \cdot v_x + \underbrace{\left( 2p_k + \alpha \cdot \sum_{i=j+1}^{k-1} p_i \right)}_{\lambda_y} \cdot v_y,$$

where we use Corollary 1 for the first equality.

If $\lambda_x < \lambda_y$ we immediately get $\lambda_x v_x + \lambda_y v_y > \lambda_x (v_x + v_y) \ge p_1 (v_x + v_y) = (1/e) \cdot v(\mathsf{OPT})$. Therefore, we assume $\lambda_x \ge \lambda_y$ in the following. By Chebyshev's sum inequality, it holds that $\lambda_x v_x + \lambda_y v_y \ge (1/2) \cdot (\lambda_x + \lambda_y) \cdot (v_x + v_y)$. Therefore, the competitive ratio is

$$\frac{\mathbb{E}[v(\mathsf{ALG}_\alpha)]}{v(\mathsf{OPT})} \ge \frac{\lambda_x + \lambda_y}{2} = \frac{1}{2} \cdot \left( (1-\alpha) \cdot p_j + 3p_k + \alpha \cdot \sum_{i=1}^{k-1} p_i \right). \qquad (8)$$

If $k = 2$, it follows that $j = 1$ and therefore Eq. (8) resolves to

$$\mathbb{E}[v(\mathsf{ALG}_\alpha)] \ge \frac{1}{2} \cdot (p_1 + 3p_2) \cdot v(\mathsf{OPT}) > \frac{1}{e} \cdot v(\mathsf{OPT}),$$

which holds independently of $\alpha$. For $k \ge 3$, $\mathsf{ALG}_\alpha$ is $(1/e - o(1))$-competitive by Eq. (8) if

$$\alpha \ge \frac{2/e - p_j - 3p_k}{\sum_{i=1}^{k-1} p_i - p_j} =: \theta_{j,k}.$$

**Table 1.** Upper bounds on $\theta_{j,k}$ for $3 \leq k \leq 10$ according to Eq. (9).

| $k$ | 3 | 4 | 5 | 6 | 7 | 8 | 9 | 10 |
|---|---|---|---|---|---|---|---|---|
| $\frac{1/e-3p_k}{\sum_{i=2}^{k-1} p_i}$ | 1.3475 | 1.3962 | **1.400382** | 1.3988 | 1.3968 | 1.3952 | 1.3941 | 1.3934 |

It remains to show $\theta_{j,k} \leq 1.400382$ for all $k \geq 3$ and $j$ with $1 \leq j < k$. For this purpose, we first show

$$\theta_{j,k} = \frac{2/e - p_j - 3p_k}{\sum_{i=1}^{k-1} p_i - p_j} \leq \frac{2/e - p_1 - 3p_k}{\sum_{i=1}^{k-1} p_i - p_1} = \frac{1/e - 3p_k}{\sum_{i=2}^{k-1} p_i} \pm o(1) \qquad \text{for any } k \geq 3.$$
(9)

Since $p_j$ is decreasing in $j$, the inequality in Eq. (9) follows immediately if we can show $2/e - 3p_k > \sum_{i=1}^{k-1} p_i$ for large-enough $n$. This inequality is easily verified for $k = 3$, as $2/e - 3p_3 > p_1 + p_2$, for large-enough $n$. For $k \geq 4$, note that $p_k < p_1 - 1/3$, again for large-enough $n$, which is equivalent to $2/e - 3p_k > 1 - p_1$. Using Observation 1, we obtain $\sum_{i=1}^{k-1} p_i < \sum_{i=1}^{n} p_i = 1 - c = 1 - p_1$. Combining both inequalities yields Eq. (9).

By computing the last term in Eq. (9) for $3 \leq k \leq 10$, we obtain the upper bounds on $\theta_{j,k}$ given in Table 1. Note that the maximum value is 1.400382. For $k \geq 11$, we obtain from Eq. (9) together with $p_i \geq 0$ for all $i \geq 11$ that

$$\theta_{j,k} \leq \frac{1/e - 3p_k}{\sum_{i=2}^{k-1} p_i} \leq \frac{1/e}{\sum_{i=2}^{11-1} p_i} < 1.398875 \pm o(1).$$

For the lower bound of approximately 1.400382 on $\alpha$, first note that for $j = 1$ and $k = 5$, it holds indeed that

$$\theta_{1,5} = \frac{2/e - p_1 - 3p_5}{\sum_{i=1}^{5-1} p_i - p_1} = \frac{1/e - 3p_5}{p_2 + p_3 + p_4} \pm o(1)$$

$$= -\frac{51}{16} + \frac{9}{4e} + \frac{75 - 522e + 486e^2}{16 - 96e + 288e^2 - 64e^3} \pm o(1) \approx 1.400382 \pm o(1).$$

Next, note that setting $v'_x, v'_{b_2}, v'_{b_3}, v'_{b_4}$, and $v'_y$ all equal to $1 + O(\varepsilon)$ and $v'(z) = O(\varepsilon)$ for all other items $z$ makes Inequality (7) as well as Inequaltiy (8) tight as $\varepsilon \to 0$. Therefore, the above arguments imply that $\alpha \geq \theta_{1,5}$ if and only if $\mathsf{ALG}_\alpha$ is $(1/e - o(1))$-competitive. This completes the proof. $\square$

## 4   Ordinal Algorithms for 1-$B$-Knapsack

In this section, we consider ordinal algorithms for 1-$B$-knapsack with $B$ large. Recall that ordinal algorithms have access to both item sizes and the relative order on item values (of previously arrived items) but not to the actual item values. We show the following theorem.

**Theorem 2.** *There is an ordinal $(1/(e+1) - o(1))$-competitive algorithm for the 1-B-knapsack problem, and every ordinal algorithm has a competitive ratio of at most $1/(e+1) + o(1)$ for this problem.*

We first discuss the lower bound, i.e., the algorithm. Note that, while the input is any combination of large and small items, the optimal solution still consists of either the single most valuable item $\mathsf{OPT}_L$ or of a set of up to $B$ small items $\mathsf{OPT}_S$. Our algorithm can be viewed as a linear combination of (near-)optimal algorithms $\mathsf{ALG}_L$ and $\mathsf{ALG}_S$ against the respective cases. In particular, $\mathsf{ALG}_L$ is the $(1/e)$-competitive algorithm [16] for the standard secretary problem and run with probability $e/(e+1)$; $\mathsf{ALG}_S$ is the $(1-o(1))$-competitive algorithm for $k$-secretary by Kleinberg [24] and run with probability $1/(e+1)$. The competitive ratio follows by a simple case distinction. A small subtlety that we need to take care of is that these subroutines require the number of items as input. To deal with this problem, we introduce dummy items. In the following, we make this idea formal.

*Proof (Algorithm).* The algorithm $\mathsf{ALG}_L$ treats all items as if they were large and then applies the standard secretary algorithm [13,26]. For the algorithm $\mathsf{ALG}_S$, whenever a large item arrives, we pretend that a small dummy item with value 0 arrives. These dummy items can be accepted and take up space in the capacity constraint, but they do not contribute to the solution value. On this adapted instance, we apply an optimal algorithm for the multiple-choice secretary problem, e.g. Kleinberg [24] or Kesselheim et al. [23]. Clearly, for both algorithms, any solution for the respective adapted instance can be translated back to a solution with equal value for the original instance. Also, both of these algorithms are ordinal.

For every input instance, our algorithm chooses $\mathsf{ALG}_L$ with probability $\frac{e}{e+1}$ and $\mathsf{ALG}_S$ otherwise. To analyze the competitive ratio, distinguish two cases. If $\mathsf{OPT} = \mathsf{OPT}_L$, we use that the algorithm chooses $\mathsf{ALG}_L$ with probability $e/(e+1)$ and conditioned on that achieves an expected value of $v(\mathsf{OPT}_L)/e$ [13,26], yielding an unconditional expected value of $v(\mathsf{OPT}_L)/(e+1)$. Otherwise, i.e., if $\mathsf{OPT} = \mathsf{OPT}_S$, we use that $\mathsf{ALG}_S$ is run with probability $1/(e+1)$ which achieves, as $B \to \infty$, an expected value of $(1-o(1)) \cdot v(\mathsf{OPT}_S)$, resulting in an unconditional expected value of $(1-o(1)) \cdot v(\mathsf{OPT}_S)/(e+1)$. $\square$

We now discuss the upper bound, i.e., the impossibility. In our construction, there are $B$ large and $B$ small items. All items have different values, and each large item is more valuable than each small item. The adversary chooses between two ways of setting the values: The first option is to make the solution consisting of *all* small items much more valuable than any single large item; the second option is to make a single large item much more valuable than any other solution.

Ideally, we would like to analyze algorithms in the following setting: In each of $n$ rounds, the algorithm is presented with both a uniformly random small and a uniformly random large item out of the items not presented thus far. Upon presentation of any such two items, the algorithm has to choose whether to select all small items from now on or to select the current large item. While the actual setting, in which all items arrive in uniformly random order, is clearly different, we show below that working with the other setting is only with a $(1 \pm o(1))$-factor loss in the impossibility by reductions between our problem and an auxiliary batched-arrival model.

Assuming the latter setting, we can write a linear program similar to that of Buchbinder et al. [8]. Like in that approach, each LP solution corresponds to an algorithm and vice versa. More specifically, our LP uses two variables (rather than one) for every time step, corresponding to the probabilities that the algorithm accepts a large item or the first small item, respectively. In addition, there is a variable representing the competitive ratio, and there are two upper bounds (rather than one) on that variable, representing the two instances the adversary can choose. A feasible dual solution then yields the desired impossibility. We formalize these ideas in the following.

*Proof (Impossibility)*. Consider the following two instances that are treated identically by ordinal algorithms. There are $n = 2B$ items where items $i \in \{1, \ldots, B\}$ are large and items $i \in \{B + 1, \ldots, 2B\}$ are small. In one instance, the item values are $v_i = 1 + (B - i) \cdot \varepsilon$ for $i \leq B$ and $v_i = 1 - i\varepsilon$ for $i > B$. In the other instance, the values are the same except for $v_1 = B^2$. So, for both instances, the rank of item $i$ is indeed $i$, for all $i \in \{1, \ldots, 2B\}$. The two optimal solutions are $\mathsf{OPT}_L = \{1\}$ and $\mathsf{OPT}_S = \{B + 1, B + 2, \ldots, 2B\}$. The adversary decides which of the two instances is the actual instance.

We consider the following batched-arrival setting parameterized with some constant $k$ and assume that $k$ divides $n$. The items still arrive in uniformly random order, but the algorithm does not always have to make a decision upon the arrival of an item. More specifically, for any $i \in \{1, \ldots, k\}$, upon the arrival of the $(i \cdot n/k)$-th item, the algorithm may make a decision about all items that have arrived in the current batch, i.e., after the $((i - 1) \cdot n/k)$-th item. Clearly, any upper bound on the competitive ratio achievable in this setting, is also an upper bound on the competitive ratio achievable in the original setting.

Note that the expected number of items of each type, i.e., small and large, in each batch is $n/(2k)$. Let $\delta > 0$ be some constant. As follows from a standard concentration (e.g., Chernoff) bound, when $n \to \infty$, the probability that the number of items from each type is between $(1 - \delta) \cdot n/(2k)$ and $(1 + \delta) \cdot n/(2k)$ approaches 1. From the union bound over all batches it then follows that also the probability that the number of items of each type *in each batch* is within the given range approaches 1. We may therefore assume that this is indeed the case at an arbitrarily small loss in our impossibility.

To analyze the algorithm in the batched-arrival setting, we write a linear program similar to that of Buchbinder et al. [8]. The LP encodes a probability distribution for the decisions that an algorithm $\mathsf{ALG}$ makes against the pair of instances. The variable $p_i$ represents the probability that the algorithm selects the best large item from the $i$-th batch. Similarly, the variable $q_i$ represents the probability that the algorithm selects all small items from both the $i$-th batch and forthcoming batches.

$$\max\ c \qquad\qquad\qquad \min\ \sum_{i=1}^{k}(x_i + y_i)$$

$$s.t.\quad c \le \frac{1}{k}\sum_{i=1}^{k}(i \cdot p_i) \qquad\qquad s.t.\qquad \alpha + \beta = 1$$

$$c \le \sum_{i=1}^{k}\left[\left(1 - \frac{i-1}{k}\right)\cdot q_i\right] \qquad i\cdot x_i + \sum_{j=i+1}^{k}(x_j + y_j) \ge \frac{i}{k}\cdot\alpha \qquad \forall\, i \in [k]$$

$$i\cdot p_i \le 1 - \sum_{j=1}^{i-1}(p_j + q_j) \quad \forall\, i \in [k] \qquad y_i + \sum_{j=i+1}^{k}(x_j + y_j) \ge \left(1 - \frac{i-1}{k}\right)\cdot\beta \quad \forall\, i \in [k]$$

$$q_i \le 1 - \sum_{j=1}^{i-1}(p_j + q_j) \quad \forall\, i \in [k] \qquad x_i, y_i \ge 0 \qquad\qquad \forall\, i \in [k]$$

$$p_i, q_i \ge 0 \qquad\qquad \forall\, i \in [k] \qquad \alpha, \beta \ge 0$$

**Fig. 2.** The primal and dual linear programs used in our proof of the upper bound in Theorem 2.

Note that the algorithm may make any such decision, i.e., selecting the best largest item or starting to select small items from a batch, for at most a single batch. Hence, we obtain $q_i \le 1 - \sum_{j=1}^{i-1} p_j + q_j$ as a constraint for our LP for all $i \in [k]$. Further, observe that we may assume that the algorithm only selects a large item when the best largest item so far is in the current batch. In batch $i$, the probability for this to happen is at most $(1+\delta)/((1+\delta) + (i-1)\cdot(1-\delta))$. As $\delta \to 0$, we obtain

$$p_i \le \left(1 - \sum_{j=1}^{i-1} p_j + q_j\right)\cdot\frac{1}{i}$$

for all $i \in [k]$, another constraint of the LP.

The objective function of the LP is $c$, an upper bound on the competitive ratio of the algorithm. For each of the two instances that the adversary could choose, we write an additional constraint upper bounding $c$. If the adversary chooses the first instance and the algorithm starts selecting small items at the end of the $i$-th batch, the fraction of $v(\mathsf{OPT}_S)$ the algorithm obtains is at most

$$\frac{(k-i+1)\cdot(1+\delta)}{(k-i+1)\cdot(1+\delta) + (i-1)\cdot(1-\delta)} \xrightarrow{\delta\to 0} 1 - \frac{i-1}{k}.$$

Hence, as $\delta \to 0$, we obtain the constraint $c \le \sum_{i=1}^{k}\left(1 - \frac{i-1}{k}\right) q_i$. Now consider the case that the adversary chooses the second instance. Suppose that the algorithm selects the best large item from the $i$-th batch, which is by assumption the best item that has already arrived. Since the order of large items is a uniformly random order, the probability that the chosen item is the globally best large item is the fraction of already observed large items within the whole instance, that is, at most

$$\frac{i\cdot(1+\delta)}{i\cdot(1+\delta) + (k-i)\cdot(1-\delta)} \xrightarrow{\delta\to 0} \frac{i}{k}.$$

Hence, as $\delta \to 0$, we get the constraint $c \leq \sum_{i=1}^{k}(p_i \cdot \frac{i}{k}) = \frac{1}{k}\sum_{i=1}^{k} i \cdot p_i$. We give both the resulting LP and its dual in Fig. 2.

We give a solution to the dual LP. Let $\tau$ be the integer number such that

$$\sum_{i=\tau}^{k-1} \frac{1}{i} < 1 \leq \sum_{i=\tau-1}^{k-1} \frac{1}{i}.$$

We set $y_i = 0$ for all $i < k$, $y_k = 1/((e+1) \cdot k)$, $x_i = 0$ for $i < \tau$, and

$$x_i = \frac{e}{(e+1) \cdot k} \cdot \left(1 - \sum_{j=i}^{k-1} \frac{1}{j}\right)$$

for $i \geq \tau$. Further, $\alpha = e/(e+1)$ and $\beta = 1/(e+1)$. Note that this choice of $x$ is analogous to the dual solution by Buchbinder et al. [8] but scaled by a factor of $\alpha$.

We argue that the solution is feasible when $x$ is scaled up by a $(1 + o(1))$ factor (where the Landau symbol is with respect to $k \to \infty$). Clearly, $\alpha + \beta = 1$. The inequality $i x_i + \sum_{j=i+1}^{k} x_j + y_j \geq \frac{i}{k} \cdot \alpha$ is the same as in the dual by Buchbinder et al., except for additional $y$ variables on the left-hand side and a scaling by $\alpha$ of the right-hand side. Therefore with our choice of $x$ (which is scaled up by $\alpha$ compared to Buchbinder et al.), the inequalities are identical and the previous proof of feasibility also holds, even without additional scaling of $x$. We consider the remaining (new) inequalities. For $i = 1$, we have

$$y_1 + \sum_{j=2}^{k} x_j + y_j \geq \frac{e}{(e+1) \cdot k} \cdot \sum_{j=\tau}^{k} \left(1 - \sum_{\ell=j}^{k-1} \frac{1}{\ell}\right)$$

$$= \frac{e}{(e+1) \cdot k} \cdot \left((k - \tau + 1) - \sum_{j=\tau}^{k} \sum_{\ell=j}^{k-1} \frac{1}{\ell}\right)$$

$$= \frac{e}{(e+1) \cdot k} \cdot \left((k - \tau + 1) - \sum_{j=\tau+1}^{k} \frac{j - \tau}{j}\right)$$

$$= \frac{e}{(e+1) \cdot k} \cdot \left(1 + \tau \sum_{j=\tau+1}^{k} \frac{1}{j}\right) \geq \frac{1}{1 + o(1)} \cdot \frac{1}{e+1}.$$

For $1 < i < \tau$ the corresponding inequality is weaker than the latter inequality. For $i \geq \tau$, we have

$$y_i + \sum_{j=i+1}^{k} x_j + y_j = y_k + \sum_{j=i+1}^{k} x_j.$$

Since $y_k = \beta/k$, we therefore have to show that, after scaling $x$ up by a $(1 + o(1))$-factor, $\sum_{j=i+1}^{k} x_j$ is at least as large as $(1 - i/k) \cdot \beta$. This is clear for $i = k$.

For $\tau \le i \le k - 1$,

$$\sum_{j=i+1}^{k} x_j = \frac{e}{(e+1) \cdot k} \cdot \sum_{j=i+1}^{k} \left( 1 - \sum_{\ell=j}^{k-1} \frac{1}{\ell} \right)$$

$$= \frac{e}{(e+1) \cdot k} \cdot \left( (k - i) - \sum_{j=i+1}^{k} \sum_{\ell=j}^{k-1} \frac{1}{\ell} \right)$$

$$= \frac{e}{(e+1) \cdot k} \cdot \left( (k - i) - \sum_{j=i+1}^{k-1} \frac{j - i}{j} \right)$$

$$= \frac{e}{(e+1) \cdot k} \cdot \left( 1 + i \sum_{j=i+1}^{k-1} \frac{1}{j} \right) \ge \frac{1}{1 + o(1)} \cdot \left( 1 - \frac{i}{k} \right) \cdot \frac{1}{e+1}.$$

Similar to the previous calculations, the objective-function value is

$$(1 + o(1)) \cdot \frac{e}{(e+1) \cdot k} \cdot \sum_{j=\tau}^{k} \left( 1 - \sum_{\ell=j}^{k-1} \frac{1}{\ell} \right)$$

$$= (1 + o(1)) \cdot \frac{e}{(e+1) \cdot k} \cdot \left( 1 + \tau \sum_{j=\tau+1}^{k} \frac{1}{j} \right)$$

$$\le (1 + o(1)) \cdot \frac{e}{(e+1) \cdot k} \cdot \tau$$

$$\le (1 + o(1)) \cdot \frac{1}{e+1},$$

as claimed.                                                                $\square$

## 5   Conclusion

In this paper, we have established that the 1-2-knapsack secretary problem is no harder than the classic secretary problem in a competitive-ratio sense. While we previously noticed that our technique cannot directly be extended to the general setting, we believe that our work is a first non-trivial step within the larger research plan of settling the achievable competitive ratio for general knapsack secretary.

It seems plausible that our result extends to the setting of arbitrary knapsack size $B$ and item sizes 1 or 2. One approach may be combining our techniques with simple $1/e$-competitive algorithms for $k$-secretary [4]. More general variants seem to require handling packings of items of various sizes. A variant that avoids considering such potentially complicated configurations and may still yield an impossibility of larger than $1/e$ is 1-$B$-knapsack.

# References

1. Albers, S., Khan, A., Ladewig, L.: Best fit bin packing with random order revisited. Algorithmica **83**(9), 2833–2858 (2021)
2. Albers, S., Khan, A., Ladewig, L.: Improved online algorithms for knapsack and GAP in the random order model. Algorithmica **83**(6), 1750–1785 (2021)
3. Albers, S., Ladewig, L.: New results for the k-secretary problem. Theor. Comput. Sci. **863**, 102–119 (2021)
4. Babaioff, M., Immorlica, N., Kempe, D., Kleinberg, R.: A knapsack secretary problem with applications. In: Charikar, M., Jansen, K., Reingold, O., Rolim, J.D.P. (eds.) APPROX/RANDOM - 2007. LNCS, vol. 4627, pp. 16–28. Springer, Heidelberg (2007). https://doi.org/10.1007/978-3-540-74208-1_2
5. Babaioff, M., Immorlica, N., Kempe, D., Kleinberg, R.: Matroid secretary problems. J. ACM **65**(6), 35:1–35:26 (2018)
6. Böckenhauer, H.-J., Komm, D., Královic, R., Rossmanith, P.: The online knapsack problem: advice and randomization. Theor. Comput. Sci. **527**, 61–72 (2014)
7. Boyar, J., Favrholdt, L.M., Larsen, K.S.: Online unit profit knapsack with untrusted predictions. In: Scandinavian Symposium and Workshops on Algorithm Theory (SWAT), pp. 20:1–20:17 (2022)
8. Buchbinder, N., Jain, K., Singh, M.: Secretary problems via linear programming. Math. Oper. Res. **39**(1), 190–206 (2014)
9. Chan, T.-H.H., Chen, F., Jiang, S.H.-C.: Revealing optimal thresholds for generalized secretary problem via continuous LP: impacts on online k-item auction and bipartite k-matching with random arrival order. In: ACM-SIAM Symposium on Discrete Algorithms (SODA), pp. 1169–1188 (2015)
10. Correa, J.R., Dütting, P., Fischer, F.A., Schewior, K.: Prophet inequalities for independent and identically distributed random variables from an unknown distribution. Math. Oper. Res. **47**(2), 1287–1309 (2022)
11. Correa, J.R., Dütting, P., Fischer, F.A., Schewior, K., Ziliotto, B.: Streaming algorithms for online selection problems. In: Innovations in Theoretical Computer Science (ITCS), p. 86:1 (2021)
12. Dinitz, M.: Recent advances on the matroid secretary problem. SIGACT News **44**(2), 126–142 (2013)
13. Dynkin, E.: The optimum choice of the instant for stopping a Markov process. Soviet Math. Dokl. **4**, 627–629 (1963)
14. Ezra, T., Feldman, M., Gravin, N., Tang, Z.G.: General graphs are easier than bipartite graphs: tight bounds for secretary matching. In: ACM Conference on Economics and Computation (EC), pp. 1148–1177 (2022)
15. Feldman, M., Svensson, O., Zenklusen, R.: A simple $O(\log \log(\text{rank}))$-competitive algorithm for the matroid secretary problem. Math. Oper. Res. **43**(2), 638–650 (2018)
16. Ferguson, T.S.: Who solved the secretary problem? Stat. Sci. **4**(3), 282–289 (1989)
17. Giliberti, J., Karrenbauer, A.: Improved online algorithm for fractional knapsack in the random order model. In: Koenemann, J., Peis, B. (eds.) WAOA 2021. LNCS, vol. 12982, pp. 188–205. Springer, Cham (2021). https://doi.org/10.1007/978-3-030-92702-8_12
18. Gupta, A., Singla, S.: Random-order models. In: Roughgarden, T. (ed.) Beyond the Worst-Case Analysis of Algorithms, pp. 234–258. Cambridge University Press (2021)

19. Hoefer, M., Kodric, B.: Combinatorial secretary problems with ordinal information. In: International Colloquium on Automata, Languages, and Programming (ICALP), pp. 133:1–133:14 (2017)
20. Kenyon, C.: Best-fit bin-packing with random order. In: ACM-SIAM Symposium on Discrete Algorithms (SODA), pp. 359–364 (1996)
21. Kesselheim, T., Molinaro, M.: Knapsack secretary with bursty adversary. In: International Colloquium on Automata, Languages, and Programming (ICALP), pp. 72:1–72:15 (2020)
22. Kesselheim, T., Radke, K., Tönnis, A., Vöcking, B.: An optimal online algorithm for weighted bipartite matching and extensions to combinatorial auctions. In: Bodlaender, H.L., Italiano, G.F. (eds.) ESA 2013. LNCS, vol. 8125, pp. 589–600. Springer, Heidelberg (2013). https://doi.org/10.1007/978-3-642-40450-4_50
23. Kesselheim, T., Radke, K., Tönnis, A., Vöcking, B.: Primal beats dual on online packing LPS in the random-order model. SIAM J. Comput. **47**(5), 1939–1964 (2018)
24. Kleinberg, R.D.: A multiple-choice secretary algorithm with applications to online auctions. In: ACM-SIAM Symposium on Discrete Algorithms (SODA), pp. 630–631 (2005)
25. Lachish, O.: $O(\log\log\operatorname{rank})$ competitive ratio for the matroid secretary problem. In: IEEE Symposium on Foundations of Computer Science (FOCS), pp. 326–335 (2014)
26. Lindley, D.: Dynamic programming and decision theory. Appl. Statist. **10**, 39–51 (1961)
27. Marchetti-Spaccamela, A., Vercellis, C.: Stochastic on-line knapsack problems. Math. Program. **68**, 73–104 (1995)
28. Naori, D., Raz, D.: Online multidimensional packing problems in the random-order model. In: International Symposium on Algorithms and Computation (ISAAC), pp. 10:1–10:15 (2019)
29. Soto, J.A., Turkieltaub, A., Verdugo, V.: Strong algorithms for the ordinal matroid secretary problem. Math. Oper. Res. **46**(2), 642–673 (2021)
30. Zhou, Y., Chakrabarty, D., Lukose, R.: Budget constrained bidding in keyword auctions and online knapsack problems. In: Papadimitriou, C., Zhang, S. (eds.) WINE 2008. LNCS, vol. 5385, pp. 566–576. Springer, Heidelberg (2008). https://doi.org/10.1007/978-3-540-92185-1_63

# Scheduling Appointments Online: The Power of Deferred Decision-Making

Devin Smedira$^{(\boxtimes)}$ and David Shmoys

Cornell University, Ithaca, NY 14850, USA
{dts88,david.shmoys}@cornell.edu

**Abstract.** The recently introduced online Minimum Peak Appointment Scheduling (MPAS) problem is a variant of the online bin-packing problem that allows for deferred decision making. Specifically, it allows for the problem to be split into an online phase where a stream of appointment requests arrive requiring a scheduled time, followed by an offline phase where those appointments are scheduled into rooms. Similar to the bin-packing problem, the aim is to use the minimum number of rooms in the final configuration. This model more accurately captures scheduling appointments than bin packing. For example, a dialysis patient needs to know what time to arrive for an appointment, but does not need to know the assigned station ahead of time.

Previous work developed a randomized algorithm for this problem which achieved an asymptotic competitive ratio of at most 1.5, proving that online MPAS was fundamentally different from the online bin-packing problem. Our main contribution is to develop a new randomized algorithm for the problem that achieves an asymptotic competitive ratio under 1.455, indicating the potential for further progress. This improvement is attained by modifying the process for scheduling appointments to increase the density of the packing in the worst case, along with utilizing the dual of the bin-packing linear programming relaxation to perform the analysis. We also present the first known lower bound of 1.2 on the asymptotic competitive ratio of both deterministic and randomized online MPAS algorithm. These results demonstrate how deferred decision-making can be leveraged to yield improved worst-case performance, a phenomenon which should be investigated in a broader class of settings.

**Keywords:** Bin packing · Online algorithm · Competitive analysis · Primal-dual algorithm

## 1 Introduction

The bin-packing problem is a classic and well-studied problem in algorithm design, with applications in a wide array of industries [8,9,12]. The problem

This research is supported in part by grants NSF/FDA SIR IIS-1935809, NSF CCF-1740822, NSF DMS-1839346, and NSF CCF-1522054.

P. Chalermsook and B. Laekhanukit (Eds.): WAOA 2022, LNCS 13538, pp. 82–115, 2022.
https://doi.org/10.1007/978-3-031-18367-6_5

requires packing items of given lengths into unit-length bins such that no two items in a bin overlap, with the goal of minimizing the total number of bins used. In particular, the "online" version of the problem is historically important and continues to be practically relevant. In this version, items are received one at a time, and each item must be placed in a bin before the next item is received. Existing approximation algorithms for it have become very well refined, with upper and lower bound results on the competitive ratio becoming very close in the last few years [1, 2, 7]. Further, the optimal absolute competitive ratio is known [3].

This paper studies a variant of the well-known bin-packing problem called the minimum peak appointment scheduling (MPAS) problem [10]. Recently proposed, an input to this problem is a set of jobs, each of a given length, that needs to be scheduled within a period of time. For example, one might be assigning each job to an interval within one 9AM–5PM work day. The objective is to minimize the peak utilization, that is, the maximum number of jobs being serviced simultaneously. In fact, Escribe et al. [10] highlight this objective, since they considered an application in which each job is also assigned to a facility, and the aim is to minimize the number of facilities needed. One can view each facility as a bin, and the position within the bin as the time interval to which a job is assigned. They observe that the minimum number of bins required (for a feasible schedule) is equal to the peak utilization, which follows directly from the fact that interval graphs are perfect (see, e.g., [11]). Therefore, in the offline setting, this scheduling problem is identical to the bin-packing problem. In the online setting, however, as requests for appointments arrive one after the other, only a time interval needs to be committed to, instead of a facility (as well as the time). This allows for added scheduling flexibility which can be exploited. As Escribe et al. [10] showed, this difference is sufficient to prove that the online MPAS problem is fundamentally different from bin packing. This is highlighted in Table 1, where the column denoted asymptotic ratio, referring to the asymptotic competitive ratio, gives a proven upper or lower bound on the ratio between objective function of the solution found to the optimal (taken in the limit over increasing input sizes).

**Table 1.** Comparison of previously known bin-packing and MPAS bounds

| Problem | Type of algorithm | Type of bound | Asymptotic ratio | Paper |
|---------|-------------------|---------------|------------------|-------|
| Bin-packing | Deterministic | Lower bound | 1.54278 | Balogh et al. (2021) [2] |
| Bin-packing | Deterministic | Upper bound | 1.57829 | Balogh et al. (2017) [1] |
| Bin-packing | Randomized | Lower bound | 1.536 | Chandra (1992) [7] |
| MPAS | Randomized | Upper bound | 1.5 | Escribe et al. (2021) [10] |

Many real-world decision-making problems can be solved by framing them as a bin-packing problem and applying existing algorithms. However, using the familiar but rigid framework of bin packing can create solutions that do not fully utilize the flexibility of the particular setting at hand. For example, an infusion center might need to determine how many chairs to configure each day. As appointments are made, the center may only need to commit to a particular time for the appointment - they do not necessarily have to commit to a particular chair for that appointment. An event host might need to determine how many rooms are necessary to host a particular event, but they may not need to tell the attendees which room they will be in until the day of the event. In both cases, traditional bin-packing algorithms can provide answers to these questions, but as this work will show, these solutions may result in an unnecessary waste of resources. In this sense, the MPAS framework is a natural generalization of the traditional bin-packing paradigm that is tailored for optimization problems that revolve around appointment scheduling. While this framework can be used to improve the solutions generated by approximation algorithms, the principles developed by the exploration of this framework can also be used to improve the efficiency of operations even when the scheduling is managed manually.

In additional to its practical use, this problem is also of theoretical interest because it is an effective vehicle for studying the power of delayed decision-making in online problems. This problem is partitioned into two fundamental steps, the first conducted in real time online and the second conducted later offline. One goal of this work is to highlight how this partition can be leveraged to improve on the solution to the normal online bin-packing problem, with the hope that similar delayed decision principles might lead to improvements in the solutions to other existing or future online problems. In particular, one could imagine other online bin-packing relaxations arising in different settings, the study of which would be aided by work on the online MPAS problem.

To underscore the connection of this work to bin-packing problems, this paper will use terminology commonly found in the bin-packing literature. Arriving jobs for the MPAS problem will be referred to as items, and their duration referred to as their size. Scheduling an item will be referred to as placing it into a position within a bin. While it is traditional to represent bin-packing solutions in which the bins are oriented vertically, in this paper, we will be view them as oriented horizontally, with position 0 on the left and position 1 on the right; this orientation provides a more intuitive connection to the time intervals the bins represent, analogous to a Gantt chart representation for a schedule. For example, scheduling an item at the left end of a bin is analogous to the first appointment of the interval, and scheduling an item to the right of another item corresponds to the item occurring later.

*An Overview of Known Algorithms.* In addition to defining the MPAS problem, Escribe et al. [10] contributed a novel randomized algorithm for the online MPAS problem called the Harmonic Rematching Algorithm and provided an analysis to prove the algorithm had an asymptotic competitive ratio of 1.5. The algorithm that we propose and analyze in this work is greatly inspired by the Harmonic Rematching algorithm, which itself was influenced by the harmonic algorithm for online bin-packing [13]. Like the problem, the algorithm is composed of two phases: an initial scheduling phase done in an online fashion and a subsequent rematching phase that places items into final bins. The number of bins used in the algorithm provides an upper bound on the peak appointments, from which an asymptotic competitive ratio for their algorithm can be calculated.

In the scheduling phase of the algorithm of [10], items are initially sorted into categories based on their size of the form $(\frac{1}{n+1}, \frac{1}{n}]$. Items are then scheduled according to unique rules for each category. Then, during the rematching phase of the algorithm, items of size under 0.5 are rematched into bins containing only other items of the same category and items of size over 0.5. The analysis of the algorithm then shows that an average bin density of $\frac{2}{3}$ is maintained so long as there are under 2 items of size over 0.5 for every 3 bins the algorithm outputs. The derivation of the asymptotic competitive ratio follows directly from this property.

*New Contributions.* The primary contribution of this work is to refute a conjecture [14] that no algorithm could achieve a lower asymptotic competitive ratio than 1.5 for the online MPAS problem. The new bound is achieved through modifications to the existing algorithm to increase the packing density and the use of a new analytic framework inspired by a primal-dual analysis of the bin-packing linear program. By refuting this conjecture, and providing a more sophisticated algorithm and analysis to do so, this work represents a step towards better understanding the degree to which deferred decision making can improve the performance of online algorithms more broadly.

A secondary contribution of this work is to present the first known lower bounds on the asymptotic competitive ratio of any algorithm for the online MPAS problem. Using a family of inputs consisting only of items with size $\frac{1}{3}$ and $\frac{2}{3}$, it is possible to prove no deterministic or randomized algorithm for the online MPAS problem can achieve an asymptotic competitive ratio below 1.2 (works building on an arXiv posting of this work improved the lower bound to 1.2691 [4]). For deterministic algorithms, this is relatively straightforward - an adversary can present $n$ items of size $\frac{1}{3}$ first, and depending on whether the algorithm leaves sufficiently many bins with only 1 of these, decides whether to add to the input $n$ items of size $\frac{2}{3}$. The lower bound for randomized algorithms extends this approach using Yao's framework for proving randomized lower bounds [6].

The majority of this paper will be dedicated to outlining the new algorithm for the online MPAS problem and providing an analysis to prove it achieves an asymptotic competitive ratio of $\frac{16}{11} \approx 1.455$. In particular, Sect. 2 will detail the behavior of the algorithm, whereas Sect. 3 will present an analysis of the desired performance guarantee. Section 4 will present the lower bound results, whereas Sect. 5 outlines the potential for further work.

## 2   Packing Algorithm

The following is a detailed description of the algorithm for solving the online MPAS problem. The algorithm will proceed in two main steps. First, there will be a section dedicated to scheduling the items as they arrive. Then, there will be a second separate section dedicated to rematching the scheduled items into final bins.

An online algorithm for the MPAS problem would be complete with only the first schedule creation phase of the algorithm, since the task is only to schedule appointments to minimize the peak. However, many practical applications would require some mechanism to take this appointments and actually group them together, say to schedule them to a room. For this reason the work includes the second algorithmic step, rematching. This step will also play a role in the analysis of the algorithm, with the number of bins produced by the rematching portion used in place of the number of appointments at the peak time, since the former is an upper bound on the latter.

## 2.1   Schedule Creation

The first phase of this online algorithm is the scheduling phase. During the scheduling phase, new items arrive in an online manner and each item is assigned to a position within a bin.

*Item Categories.* Every item will be given a type based on its size. How an item is initially scheduled, as well as in what manner it is rematched into bins, will depend on which type of item it is. These different item types are:

**Small Item** - Item of size $\leq 1/3$
**Third Item** - Item of size in $(1/3, 0.34375]$
**Medium Item** - Item of size in $(0.34375, 0.5]$
**Large Item** - Item of size in $(1/2, 0.6875]$
**Very Large Item** - Item of size $> 0.6875$.

Within the specified range of each item type, there will be one or more different categories. Each category will have its own specific procedure on how items from the category should be both initially scheduled and eventually rematched into bins. The full list of item categories appears in Appendix C, along with other relevant details. Two categories in particular will be important enough in the analysis to separately name. These categories are:

**Quarter Items** - A Small Item category for items of size in $(1/4, 0.27]$
**Half Items** - A Medium Item category for items of size in $(0.46, 1/2]$.

In particular, Quarter Items are the Small Item category that admits the least dense packing, thus requiring more special care in the analysis.

*One-Sided Bins.* The initial scheduling phase of the algorithm will consist of creating a set of **One-Sided Bins** that will later be rematched. A one-sided bin is a category-specific construct containing only items from one particular category. Items will be scheduled into these one-sided bins, and every item in the same one-sided bin will be scheduled in the same bin at the end of the algorithm (with a slight exception for Third Items detailed later). Specifically, each final bin returned by the algorithm will consist exclusively of items from one or two one-sided bins.

There will be three types of one-sided bins used by the algorithm. They are:

**Type 1 Bins** - A type 1 bin will be a one-sided bin that is over 0.6875 full. These bins will not be matched with other one-sided bins, so each type 1 bin will be assigned its own bin when the algorithm terminates.

**Type 2 Large Bins** - A type 2 large bin will be a one-sided bin that is over 0.3125 full but under 0.6875 full. These bins might or might not be matched during the course of the algorithm and so might be paired with another one-sided bin in the algorithm's output.

**Type 2 Small Bins** - A type 2 small bin will be a one-sided bin that is at most 0.3125 full. All but a constant number of these bins will be matched with another Type 2 one-sided bin, and share an output bin when the algorithm terminates.

Each Small Item category will have a specified number of items per type 1, type 2 large, and type 2 small bin listed in the appendix. Very Large Items will be packed exclusively into type 1 bins containing 1 item, while Large and Medium Items are packed exclusively into type 2 large bins containing 1 item. Third Items will be exclusively packed into type 2 large bins, although some will contain 1 item while others will contain 2. Further, the type 2 bins containing 2 items may be shuffled so that the items initially in the same one-sided bin are not packed in the same bin when the algorithm terminates. However, every Third Item will end in a bin with the same number of items as the one it was initially assigned to. The exact packing mechanism is detailed below.

Type 2 bins will have a further designation of being a left or right sided bin. Simply, a left sided type 2 bin will have the first item placed at position 0 with subsequent items placed closer to 1, while a right sided type 2 bin will have the first item placed at position 1 with subsequent items placed closer to 0. The scheduling process detailed below will ensure some amount of balance between the left and right sided bins.

*Complete Sets of One-Sided Bins.* Items from each particular category will be packed into **Complete Sets of One-Sided Bins**, referred to interchangeably as complete sets. For each item category, a complete set of one-sided bins will be a certain number of type 1, type 2 large, and type 2 small bins. Further, every complete set will contain an even number of type 2 small and type 2 large bins, with half packed on either side. For Very Large Items, a complete set will have 1 type 1 bin, while for Large and Medium Item categories a complete set will have 2 type 2 large bins. For Quarter Items, a complete set will have 1 type 1 bin containing 3 items, 2 type 2 large bins containing two items, and 4 type 2 small bins containing 1 item.

For Third Items, a complete set will have 4 type 2 large bins containing 1 item each, and 2 type 2 large bins containing 2 items each. These will be referred to as **Outer Bins** and **Inner Bins** respectively. This distinction will be important in later analysis. Further, unlike other one-sided bins with multiple items, it will be important for inner bins to be packed so that either item can be placed first. Therefore, the interior item of inner bins will be placed exactly 0.34375 from the left or right border, depending on the side of the bin, to accommodate another Third Item of any size. This will also allow any interior left item in an inner bin to be paired with any exterior left item in an inner bin of the same side.

During the scheduling phase, items arrive one at a time. Therefore, it is not always possible to maintain complete sets throughout the algorithm's execution. A **Partial Set of One-Sided Bins** will refer to a complete set of one-sided bins that has been allocated by the algorithm but has not yet had every item spot filled.

*Arrival Scheduling.* This section will detail the actual procedure used for the scheduling phase of the algorithm. When an item arrives, the first step to scheduling is to determine of what category it is a member. Then, the algorithm determines whether or not a partial set of matched bins exists for that category, creating a new one if one does not exist. Then,

For **Very Large** items, a partial complete set only contains 1 type 1 bin, so the item is placed in it and the set is marked as being completed.

For **Large and Medium** items, a complete set of bins only contains two 1 item bins. So if the partial set has one of the type 2 large bins filled, the other bin is filled with the item and the set is marked as completed. Otherwise, the side of the bin the item is placed on is chosen randomly with equal probability.

For **Third** items, the item scheduling process has several steps. Within the partial set, if there are more items in inner bins than outer bins then mark the new item for an outer pair and vice versa. If the item is marked for an outer bin, place it in a type 2 large bin on the side with fewer items, or randomly with equal probability if there are the same number of items on each side. If the item is marked for an inner bin, first choose whether to place it in the left or right side bin based on which has fewer items, or randomly if they are equally full. Finally, place it on the edge of the bin if there is an item in the interior already and vice versa. Again choose randomly if both spots are not filled. If there are no more open spots in the partial set, mark the set as completed. This complicated procedure will play an important role in future analysis.

For **Small** items, items can be placed arbitrarily among the open spots within a partial set of the correct category. If there are no more open spots in the partial set after the item is placed, mark the set as completed.

The above procedure will ensure that each item category has at most 1 partial set of one-sided bins at any time during execution. Thus, when the scheduling phase terminates, there will be a finite number of partial sets, allowing them to be ignored in the final analysis.

## 2.2   Rematching

The second phase of the algorithm will be an offline rematching phase. This step will take all of the items that are already committed to a position within the bin and assign them to final bins in a manner such that no two items within a bin overlap. For the sake of this overview and the analysis, it will be sufficient to simply say that items from some category were rematched with items from another category. The exact details of how the rematching puts items into bins is detailed in Appendices B and C.

*Large and Very Large Items.* Very Large Items are exclusively put into type 1 bins, and as such will not need to be rematched by the algorithm. Each Very Large Item will end up in its own bin when the algorithm finishes.

Large Items are exclusively put into type 2 large bins, and can not fit in a bin with another Large Item. They will be used to rematch with smaller items, detailed below. Though complete sets of Large Items contain two Large Items, the individual items may be rematched with different sets during the rematching process.

*Medium Items.* As is the case with Large Items, Medium Items may not be packed with the other item originally in the same complete set. Rather, each item will be used individually in the rematching stage. A Medium Item can be matched in a bin with another Medium Item of the same category or matched in a bin with a Large Item on the opposite side if the Large Item is small enough to fit with the Medium Item. At most 1 Medium Item per category will not be matched in one of these two ways. Depending on the distribution of items, five bins of Medium Item pairs may be grouped with 11 bins containing single Large Items for the algorithm's analysis.

Medium Items not in the Half Item category can also be rematched with Large Items and a complete set of Quarter Items.

*Third Items.* As is the case with Large and Medium Items, Third Items in the same complete set initially may not be placed in the same group of bins during the rematching process. In fact, Third Items in inner bins may end up placed in an inner bin with a different item than originally matched with. However, Third Items will always be matched in units of complete sets, unlike the Large Items which may not be.

Third Items can be rematched with Large Items if the Third Items in the outer bins and the Large Items can fit in a bin. In this case, the four outer bins will be matched with a Large Item in a bin, the two inner bins will be assigned their own bin, and a 5th Large Item will be assigned its own bin and grouped with the rematched set. Third Items may also be grouped with Large Items that do not fit in a bin with the outer bins. In this case, 11 bins containing a Large Item will be grouped for every 5 bins containing two Third Items. Complete sets of Third Items may also be rematched with complete sets of Small Items as detailed below.

Third Items can also be rematched with Large Items too large to share a bin with and a complete set of Quarter Items.

*Small Item Categories.* During the rematching process, every complete set of Small Items may or may not be rematched with other complete sets to assign one-sided bins to bins. Any complete set of Small Items not rematched with other items will have every one-sided bin assigned to bins only with other one-sided bins from the same set. This set of bins will be referred to as a **Complete Set of Matched Bins**.

All Small Item categories can have their complete sets grouped with other complete sets from different categories to form a **Complete Set of Rematched Bins**. A complete set of rematched bins will consist of bins containing some combinations of one-sided bins from a complete set of Small Items and one-sided bins from a different item category. The terms complete set of rematched bins and a complete set of bins rematched with, followed by whatever the bins are rematched with, will be used interchangeably.

For any small category besides the Quarter Item category, a complete set of bins can be rematched with Large Items, one or more complete sets of Third Items, Large Items and one or more complete sets of Quarter Items, or one or more complete sets of Third Items and one or more complete sets of Quarter Items.

In addition to all the previously listed cases, a complete set of Quarter Items can also be rematched with Large Items.

*Rematching Procedure.* The algorithm will use the following outlined procedure.

**Step 1** - Rematch Large Items with Third Items
Sort both the left and right Third Items in outer bins in increasing order, and do the same for all four item positions in the inner bins. Also sort the left and right Large Items in increasing order.
As long as there remains complete sets of Third Items and Large Items on both sides, check if the first two outer left Third Items fit in a bin pairwise with the first two right Large Items, and vice versa. If they do not, then move to the next step. If they do, place the outer one-sided bins in a final bin with the Large Item. Then form a right and left inner bin by taking the first item from each respective list and place both inner bins in separate bins. Finally, take the largest Large Item from whichever side has more items (arbitrarily if a tie) and place it in a bin by itself.
**Step 2** - Rematch Large Items with Medium Items
Sort both the left and right Medium Items in increasing order, and do the same for the left and right Large Items. As long as there remains Large and Medium Items on both sides, check if the first left Medium Item fits in a bin with the first right Large Item and vice versa. If they do not, move to step 3. Otherwise, place both pairs into a bin together and repeat.
**Step 3** - Rematch Large Items with Third and Quarter Items
Sort both the left and right Third Items in outer bins in increasing order, and do the same for all four item positions in the inner bins. Also sort the left and right Large Items in increasing order. As long as there remains complete sets of Third Items, complete sets of Quarter Items, and enough Large Items on both sides, do the following. Take 7 Large Items from each side, a complete set of Quarter Items, and a complete set of Third Items and rematch them together.

**Step 4** - Rematch Large Items with Medium and Quarter Items

Sort both the left and right Medium Items in increasing order, and do the same for the left and right Large Items. As long as there remains non-Half Medium Items on each side, complete sets of Quarter Items, and enough Large Items on both sides, do the following. Take 4 Large Items from each side, a complete set of Quarter Items, and one Medium Item from each side and rematch them together.

**Step 5** - Rematch Large Items with Complete Sets of Small Items

While there remains complete sets of Quarter Items and complete sets of small non-Quarter Items, rematch them together with the appropriate number of Large Items. Do this until there are not enough Large Items or complete quarter sets, or until all non-quarter complete sets have been used.

If complete sets of non-quarter Small Items and enough Large Items remain, continue rematching them until one runs out.

**Step 6** - Rematch Third Item sets with Complete Small Item Sets

While there remains complete sets of Quarter Items and complete sets of small non-Quarter Items, rematch them together with the appropriate number of complete Third Item sets. Do this until there are not enough complete Third Item sets or complete quarter sets, or until all non-quarter complete sets have been used.

If complete sets of non-quarter Small Items and enough Third Item sets remain, continue rematching them until one runs out.

**Step 7** - Rematch Large Items with Complete Sets of Quarter Items

Rematch Large Items with complete sets of Quarter Items until one runs out.

**Step 8** - Rematch Complete Sets of Third Items with Complete Sets of Quarter Items

Rematch complete sets of Third Items with complete sets of Quarter Items until one runs out.

**Step 9** - Group Large Items with Third Items and Medium Items.

While there remains Large Items and either Medium or Third Items that are not placed in a bin, group bins containing two Third/Medium Items with bins containing Large Items at a ratio of 5 third/medium bins to 11 large bins.

**Step 10** - Assign Remaining Large and Very Large Items to Their Own Bins.

## 2.3  Example Category Packing

In the interest of clarifying the algorithm the following section will be dedicated to describing the matching procedure for the category of items in the range $(0.215, 0.23]$.

*Matched Set.* A complete set of items from this category will consist of four type 2 large bins each containing two items and two type 2 small bins each containing 1 item. A complete set of matched bins will be formed from this category by placing both type 2 small one-sided bins in a bin together, and by putting the four type 2 large one sided bins into two bins. Figure 1 below demonstrates how this matching is done, with the blue representing items from the category and the grey representing empty space in the bin.

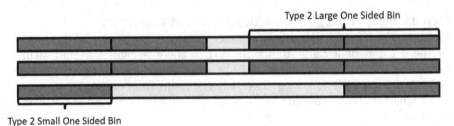

Type 2 Large One Sided Bin

Type 2 Small One Sided Bin

**Fig. 1.** A complete set of matched bins (not to scale)

*Rematched Set.* To fully specify the behavior of rematching with this item category, five different rematching alignments would need to be specified. These would be the complete set rematched with

1) Large Items of size under 0.54
2) Large Items of size over 0.54
3) A complete set of Third Items
4) Large Items and a complete set of Quarter Items
5) A complete set of Third Items and a complete set of Quarter Items.

Figure 2 demonstrates case 2 on the left and case 4 on the right, with blue representing items in the category, orange representing Quarter Items, red representing Large Items, and grey representing empty space in the bins. The presented matchings will only have a Large Item in 2 out of every 3 bins, instead of the later purported 0.6875 Large Items per bin. This can be rectified by assigning an additional Large Item for every 15 rematched bins, a process detailed in the appendix.

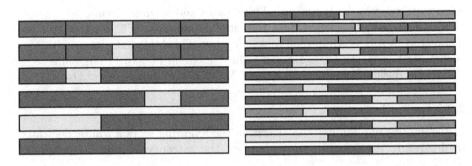

**Fig. 2.** Example complete sets of rematched bins (not to scale)

Different item categories will have different packing protocols and different cases than the 5 listed above. But, they will all follow this general pattern on small type 2 bins being matched with Large Items (or 2 Medium Items), while type 2 large bins are matched with small Large Items, each other, or Quarter Item type 2 large bins.

# 3  Analysis

In this section, we will upper bound the asymptotic competitive ratio of our algorithm. More formally, we define the asymptotic competitive ratio to be the infimum of the set of all $\alpha$ satisfying

$$\lim_{N \to \infty} \sup_{I:\text{OPT}(I) \geq N} \Pr\left[\frac{\text{ALG}(I)}{\text{OPT}(I)} \leq \alpha + \epsilon\right] = 1$$

where $I$ ranges over all possible inputs to the MPAS problem, $\text{OPT}(I)$ is the number of bins used in an optimal solution to the MPAS problem instance $I$ (equivalent to the number of appointments at the peak time), $\text{ALG}(I)$ is the number of bins used by the proposed algorithm for that instance, and $\epsilon$ is any real number greater than 0. The probability in the above statement is taken with respect to the random choices of the algorithm. Notice that the rematching phase of the algorithm is not guaranteed to use the fewest number of bins possible for the given schedule, so $\text{ALG}(I)$ might actually be an overestimate of the MPAS objective of the algorithm's scheduling.

**Theorem 1 (Asymptotic Competitive Ratio).** *The asymptotic competitive ratio of the algorithm presented in Sect. 2 is at most* $\frac{16}{11}$.

To prove the bound, the analysis that follows will show that for any execution of the algorithm on an instance of the online MPAS problem $I$, all of the bins used by the algorithm will fall into one of three categories. The first will be the additive constants bins, the number of which will be bounded above by a constant $K$ independent of the input $I$. The second will be the imbalanced bins, the number of which will follow some distribution $K(I)$ for each input instance $I$ satisfying, for any positive $\epsilon$:

$$\lim_{N \to \infty} \sup_{I:\text{OPT}(I) \geq N} \Pr\left[\frac{K(I)}{\text{OPT}(I)} \geq \epsilon\right] = 0$$

The third category will be the packed bins, which will be the primary focus of this section and denoted as $P(I)$. The analysis of the packed bins is based on the dual of a bin-packing linear programming relaxation (often referred to as the configuration LP for bin-packing). For each of five separate possible scenarios in which the rematching process might end, a feasible solution to the dual program will be created such that the dual objective $D$ will satisfy $\frac{11}{16}P(I) \leq D$, which will in turn imply that $\frac{11}{16}P(I) \leq \text{OPT}(I)$ by weak duality and the fact that the linear relaxation optimal solution is a lower bound on $\text{OPT}(I)$. This will be sufficient to prove the result.

## 3.1  Properties

In order to conduct the analysis of the algorithm's performance, a few core properties of the algorithm must be established. Each of the following properties have been verified for each of the relevant categories.

*Property 1 (Average Fullness).* For each Small Item category, a complete set of matched bins will have an average fullness of at least 0.6875 per bin.

*Property 2 (Rematched Fullness).* For each Small Item category excluding Quarter Items, a complete set of bins rematched with Large or Third Items will have an average fullness of at least 0.6875 per bin. Further, this packing will have at least 0.6875 Large Items for every bin or 1.375 Third Items for every bin on average.

*Property 3 (Quarter Rematched Fullness).* For each Small Item category excluding Quarter Items, a complete set of bins rematched with Quarter Items and Large or Third Items will have an average fullness of at least 0.6875. Further, this packing will have at least 0.6875 Large Items for every bin or 1.375 Third Items for every bin on average.

*Property 4 (Quarter Size Reallocation).* For each Small Item category excluding Quarter Items, a complete set of bins rematched with Quarter Items and Large or Third Items will satisfy $0.3125q + 0.34375t + 0.6875l \geq 0.6875b$, where $q$ is the number of items of size in $(1/4, 1/3]$; $t$ is the number of Third Items in the set; $l$ is the number of Large Items in the set; $b$ is the number of bins the set uses.

Complete sets of Quarter Items will satisfy this property whether or not they are rematched with something else.

Like the exact details for matching each item category, the complete verification of these four properties appears in the appendix. These properties will be used in the remainder of the work.

## 3.2   Configuration Bin-Packing LP

The analysis will rely on providing a bound on the minimal number of bins the items could be packed in using a bin-packing linear program relaxation. Define $C$ to be the set of all valid configurations of items from an instance to the MPAS problem, where a *configuration* is a set of items that can be feasibly packed together within one bin. Let $I$ be the set of all items in the MPAS instance, with each individual element receiving its own entry; in other words, even if there are two items of the same size, they will be viewed as distinct items. Further define $C(i)$ to be the set of configurations containing the element $i$. One possible linear program relaxation of the bin-packing problem is as follows:

$$\text{Minimize} \sum_{c \in C} y_c$$

$$\text{Subject To} \sum_{c \in C(i)} y_c \geq 1 \qquad \forall i \in I$$

$$y_c \geq 0 \qquad \forall c \in C$$

Intuitively, the $y_c$ variables in this program will represent the number of bins which are packed according to configuration $c$ in a solution. Since this is a relaxation, it is entirely possible that a particular $y_c$ is not an integer value. However, since each configuration contains specific items, no optimal solution will ever have $y_c > 1$. The dual of this program will be:

$$\text{Maximize} \sum_{i \in I} x_i$$

$$\text{Subject To} \sum_{i \in c} x_i \leq 1 \qquad\qquad \forall c \in C$$

$$x_i \geq 0 \qquad\qquad \forall i \in I$$

This program will be the one utilized to complete the analysis. Each dual variable $x_i$ intuitively represents assigning a size to a particular item. In this sense, it will be possible to give more or less weight to certain specific items to change the dual objective. The set of additive constraints essentially says that no valid bin configuration can have the weights of its items sum to more than 1. It will be convenient going forward to label the size of an item $i$ as $w_i$.

## 3.3   Additive Constants Bins

This section will identify the sources of the Additive Constants Bins, as well as justify why there must be at most a constant number of them without explicitly calculating the constant $K$. Throughout the algorithm's execution, there are several stages where items from a specific category will need to be used in some quantity, but that quantity will not be available. These shortages, detailed below, will not play a role in the final analysis.

*Small Category Imbalance of Size-Dependent Scheduling.* Some Small Item categories have analysis of fullness dependent on being rematched with Large Items either all larger or all smaller than a certain threshold. There may not be enough large items of the correct size to rematch with, but if that is the case then there must be at most a certain number of Large Items remaining of that size that have not already been rematched. Each such threshold is a boundary between different Large Item categories, so this problem cannot arise from an imbalance of item sizes on each side. Thus, if this problem arises, every remaining Large Item can be placed into its own bin and those bins can be incorporated into the additive constant for the analysis.

*Partial Sets.* Each category can only have at most 1 partial set of one-sided bins at any point during the scheduling phase. Thus there will be only a finite number of items in partial sets when the scheduling phase ends using only a finite number of bins. These bins can be grouped into the additive constant in the analysis.

*Insufficient Items to Rematch.* The rematching phase has several sections with instructions that simply state something along the lines of rematch this item group with some other specified item group. This may not be possible even if items still remain if, for example, there are only 2 Large Items left but 6 needed.

But, if such a problem arises, the category causing the problem will have at most only a certain number of items left to rematch. These leftover items can be placed in their own bins and grouped into the additive constant for the analysis.

## 3.4   Bounded Imbalanced Bins

In the following sections where the multiplicative bound is derived for the packed bins, the following property will be necessary in some cases of the analysis:

*Property 5 (Balanced Bins).* After the rematching phase of the algorithm is complete, all Large and Very Large Items not matched to a bin with a Medium or Third sized item can not fit in a bin with any Medium sized item not matched to a bin with a Large Item. Further, all Large and Very Large Items not matched to a bin with a Medium or Third sized item can not fit in a bin with any Third sized item in a complete set which does not have a Large and Third Item sharing a bin in its final matching.

It is relatively easy to see that this property does not necessarily hold in the described algorithm. The remainder of this section will be dedicated to proving that, for any positive $\epsilon$,

$$\lim_{N \to \infty} \sup_{I:\mathrm{OPT}(I) \geq N} \Pr \left[ \frac{\mathrm{IMB}(I)}{\mathrm{OPT}(I)} \geq \epsilon \right] = 0$$

where $\mathrm{IMB}(I)$ represents the number of items that do not satisfy the balanced bins property. With this proven, the analysis can be completed by assuming the balanced bins property and taking items that defy it and placing them into their own bins. These extra bins will be the imbalanced bins and counted in the $K(I)$ component of the competitive analysis.

*Large and Medium Items.* Let $L_c$ denote the number of Large Items in each large category $c$ and $M_c$ the Medium Items in each medium category $c$. For any large category $c$ and any $x$, define $n_c(x)$ to be the number of large items of size greater than or equal to $x$ in instance $I$. Further, define $L_c(x)$ to be the number of large items of size greater than or equal to $x$ placed on the right side minus the number of such items placed on the left in category $c$. Define $M_c(x)$ similarly for medium jobs.

Consider the items in the category $c$ smaller than $x$. Some of those items may arrive consecutively and be placed on opposite sides in a guaranteed manner, contributing a net of 0 to $L_c(x)$. The remainder will be randomly assigned a side to be placed on, independent of the placement for the other items by the algorithm's design. Thus, $L_c(x)$ can be represented as a sum of at most $n_c(x)$ independent random variables $S_i$, which take value 1 (if the item is placed on the right) or $-1$ (if the item is placed on the left) with equal probability.

Now, consider the behavior of $L_c(x)$ for some $x$ as the optimal solution of the inputs $I$ grows arbitrarily large. If $n_c(x)$ does not grow with $\mathrm{OPT}(I)$, then

$\frac{|L_c(x)|}{OPT(I)}$ must approach 0, since $L_c(x)$ is bounded above by $n_c(x)$. However, if $n_c(x)$ does grow with $OPT(I)$, then, in the worst case, $L_c(x)$ is the sum of $n_c(x)$ independent random variables $S_i$. Since each $S_i$ has a mean of 0, the law of large numbers [5] will imply

$$\lim_{n_c(x) \to \infty} \Pr\left[\frac{|L_c(x)|}{n_c(x)} \geq \epsilon\right] = 0 \text{ for any } \epsilon > 0.$$

Therefore, observing that for any input instance $I$, at least $n_c(x)$ bins must be used, the values $L_c(x)$ for each category $c$ and each cutoff $x$ must satisfy the desired property of IMB. Notice a similar argument can be made to show that the values of $M_c(x)$ also satisfy the property.

Now, consider the first time a Large and Medium Item on opposite sides cannot be matched together in the same bin. Assume without loss of generality this happens on the left for the Medium Items and right for the Large Item. Let $x_m$ be the size of the Medium Item which cannot be matched, let $c_m$ be the category of the Medium Item, let $x_l$ be the size of the large item which cannot be matched, and let $c_l$ be the category of the Large Item. At that point, any violations of the balanced bins property must be by items in the same category as either the unmatched large or the unmatched medium item. These violations must also happen because an item could fit in a bin with either a Medium Item on the right or a Large Item on the left respectively.

Removing the next $L_{c_l}(x_l)$ Large Items from both sides and the next $M_{c_m}(x_m)$ Medium Items from both sides will lead to the Balanced Bins Property being satisfied for Large and Medium Items. These items can be stored in $2L_{c_l}(x_l) + M_{c_m}(x_m)$ bins, which by the above reasoning is sufficiently small in the limit.

*Third Items.* The proof of the property for Third Items follows the same pattern as the above section, but applies the law of large numbers all four times there is a binary split on item placement. Then, the number of Third Items which do not satisfy the Balanced Bins property can be bounded above by the sum of the 4 imbalances. These imbalances will again be asymptotically small as the number of items increases, just as in the above section. Since this analysis offers no new insights from the previous one, it will be omitted.

## 3.5   Multiplicative Bound on Packed Bins

The following section will prove the multiplicative bound for the asymptotic competitive ratio result. Specifically, this section will bound the number of packed bins after the algorithm finishes in terms of the linear program established above. In what follows, all five properties above will be assumed, and the additive constants addressed above will be ignored. Thus, this section will provide an input to the dual linear program which produces an objective over $\frac{11}{16}$ times the number of bins used in complete sets of matched bins, complete sets of rematched bins, and additional excess Large, Medium, and Third Item bins.

*Case 1 - No Adjustment.* Suppose during the algorithms execution, every Large Item is either rematched with Medium/Third Items or rematched into a Small Item complete set. Further suppose that every complete set of Third Items is rematched in some manner, whether with Large Items, Large Items and Quarter Items, a complete set of Small Items, or Quarter Items and a complete set of Small Items.

Complete matched Quarter Item sets have an average fullness of at least 0.6875, as do any other matched complete sets. Further, every rematched complete set will have an average fullness of at least 0.6875, as will rematched sets containing Quarter Items. Medium Items matched with Large Items have a per bin fullness over 0.6875, as do bins containing two Medium Items. Finally, bins containing one Very Large Item are at least 0.6875 full. Thus, for each item $i$, one can set $x_i = w_i$ in the dual to get a dual objective of at least $\frac{11}{16}$ times the number of bins used.

*Case 2 - Excess of Third Items.* Suppose during the algorithms execution, every Large Item is either rematched with Medium Items or rematched into a Small Item complete set. Further suppose that not every complete set of Third Items is rematched. This will imply every complete set of Small Items (including potentially Quarter Item sets by themselves) are in complete sets of rematched bins.

In this case, it is possible to use the dual program to achieve the necessary bound. Set $x_i = 1/2$ for all Medium Items (including Third Items) and Large Items matched with Medium Items, and set $x_i = 1$ for unmatched Large Items and Very Large Items. Set $x_i = 0$ for the remaining items. This will be a valid input to the dual program by the balanced bins property above. By the fullness properties, every set will have 0.6875 Large Items or 1.375 Third Items per bin, so will contribute 0.6875 to the dual per bin. Every other bin will contain either 2 Medium Items, a Medium Item and Large Item matched, or 1 Large Item, and thus contribute 1 to the dual objective per bin. Therefore, every bin will on average contribute at least 0.6875 to the dual objective, so this objective will be at least $\frac{11}{16}$ times the number of bins the algorithm uses.

*Case 3 - Excess of Quarter Items.* Suppose there are complete Quarter Item sets directly rematched with Large Items or Third Items, but not every such set is rematched. This will imply that every Large and Third Item set is rematched with other items in this instance. Further, every non-quarter Small Item category will be in rematched complete sets of bins and every item size under $1/4$ will be rematched with Quarter Item sets.

In this case, it is possible to set $x_i = 0.3125$ for every item with size in $(1/4, 1/3]$, Third Items, and Medium Items matched with Large Items, $x_i = 0.34375$ for Medium Items not matched with Large Items, $x_i = 0.5$ for Large Items matched with a Medium Item, and $x_i = 0.6875$ for Large Items not matched with a Medium Item. The balanced bins property will ensure this is a valid input to the dual program.

With these dual variable values, the quarter size reallocation property will guarantee that every complete set of rematched bins contributes at least 0.6875 the dual objective per bin. Further, every bin containing two Medium Items will contribute 0.6875 to the objective, and every bin containing a Medium and Large Item will contribute well over this amount. Therefore, every bin will on average contribute at least 0.6875 to the dual objective, so this objective will be at least the $\frac{11}{16}$ times the number of bins the algorithm uses.

*Case 4 - Unmatchable Large Items.* Suppose that every Small Item set is rematched with Large Items, but not all such sets are rematched with Quarter Items. Further, suppose there are Medium Items or Third Item sets that cannot be matched with Large Items, and not enough Large Items to assign 11 Large Items for every 5 bins of unmatched Third and Medium Items.

Take the smallest unmatched Medium Item or smallest Third Item in a complete set not rematched with Large Items, and assume it has a weight of $1 - w$. This implies every Large Item not matched with a Medium Item will have size at least $w$ by the balanced bins property. Now, for every 5 bins containing two Medium or Third Items, there will need to be 11 bins containing 1 Large Item for every set to have at least 0.6875 Large Items per bin. Further, this set must contribute 0.6875 per bin to the dual value. Thus, each Large Item not paired with a Medium Item will have its dual variable set to some value $w + x$. Solving $(2(1 - w) \cdot 5 + 11(w + x))/16 = 0.6875$, it must be that $x = 1/11 - w/11$.

Each Large Item not paired with a Medium Item will have $x_i = 10w_i/11 + 1/11$. Medium Items not paired with a Large Item will have $x_i = w_i$, as will Large Items paired with Medium Items. For the remaining items to fit, it must be that $x_i = \frac{1 - 10w/11 - 1/11}{1 - w} w_i = 10w_i/11$. A bin containing a Medium and Large Item will contribute at least $1/2 + 0.34375 \cdot 10/11 > 0.6875$ to the dual objective. Further, a set of Third Items rematched with Large Items will have 6 Large Items matched with Third Items and 1 Large Item not, so will contribute at least $(0.3125 \cdot 8 \cdot 10/11 + 0.5 \cdot 4 + .5 \cdot 10/11 + 1/11)/7 > 0.6875$ per bin to the dual objective. For all other complete rematched sets, they will have an average fullness of at least 0.6875 by properties 2 and 3. Thus, for every 16 bins in complete rematched sets with Small Items, there will be at least 11 Large Items and a density of at least 0.6875 per bin, leading to at least $(10(0.6875 \cdot 16 - 11w)/11 + 11(10w/11 + 1/11))/16 = 0.6875$ being contributed to the dual objective per bin. Therefore, every bin will on average contribute at least 0.6875 to the dual objective, so this objective will be at least $\frac{11}{16}$ times the number of bins the algorithm uses.

*Case 5 - Excess Large Items.* Suppose there are enough Large Items to be rematched with every other item category and group with unmatched Medium or Third Items with some Large Items left over. If this is the case, simply setting $x_i = 1$ for each item of size above 0.5 and $x_i = 0$ for all other items will contribute over 0.6875 per bin to the dual objective by the rematched fullness property, the Quarter Item rematched fullness property, and the manner with which Large Items are grouped with unmatched Medium and Third Items. So this objective will be at least $\frac{11}{16}$ times the number of bins the algorithm uses.

### 3.6  Completing the Proof of Theorem 1

The work in Sect. 3.5 is sufficient to show that for any online MPAS instance $I$ and any execution of the algorithm, the number of packed bins $P(I)$ established satisfies

$$\frac{11}{16}P(I) = \frac{11}{16}(\text{ALG}(I) - K - K(I)) \leq D$$

with $K$ being the upper bound on the number of additive constants bins from Sect. 3.3, $K(I)$ being all of the bounded imbalance bins identified in Sect. 3.4, and $\text{ALG}(I)$ being the number of bins the algorithm uses after its rematching process. Therefore, it must be that for any $\epsilon > 0$,

$$\lim_{N \to \infty} \sup_{I : \text{OPT}(I) \geq N} \Pr\left[\frac{\text{ALG}(I)}{\text{OPT}(I)} \leq \alpha + \epsilon\right]$$

$$\geq \lim_{N \to \infty} \sup_{I : \text{OPT}(I) \geq N} \Pr\left[\frac{P(I)}{\text{OPT}(I)} \leq \alpha\right] - \Pr\left[\frac{K + K(I)}{\text{OPT}(I)} \geq \epsilon\right]$$

$$= 1 - 0 = 1$$

Thus, the asymptotic competitive ratio of the algorithm is at most $\frac{16}{11}$, proving the claim.

## 4  Lower Bounds

This section will prove a lower bound of 1.2 on the asymptotic competitive ratio for any deterministic online MPAS algorithms. Then, a similar technique will be used to prove no randomized algorithm can achieve an asymptotic competitive ratio below 1.2 in expectation.

### 4.1  Deterministic Lower Bound

The same family of inputs is used to construct both the deterministic and randomized lower bounds for the MPAS problem. Though the deterministic result is implied by the randomized result, a deterministic lower bound is easier to prove and conceptualize. Thus, this subsection will be dedicated to proving the theorem below, before the full result in the next subsection.

**Theorem 2 (Asymptotic Deterministic Lower Bound).** *Any deterministic algorithm for the online MPAS problem must have an asymptotic competitive ratio of at least $\frac{6}{5} = 1.2$.*

*Proof.* Consider the set of inputs to the MPAS problem restricted to items of size $\frac{1}{3}$ and $\frac{2}{3}$. For any such input, there must be an optimal solution that only places items of size $\frac{1}{3}$ on either edge or in the exact center and only places items of size $\frac{2}{3}$ on an edge. Moving an item of size $\frac{2}{3}$ from wherever it is placed to its closest edge can not increase the number of items scheduled at the same time, since each item is at least of size $\frac{1}{3}$. The same is true for items of size $\frac{1}{3}$ not placed exactly in the middle of a bin.

Now, take any deterministic algorithm for the online MPAS problem which only schedules $\frac{1}{3}$ items in the middle or on an edge, and $\frac{2}{3}$ sized items on an edge, and choose some integer $n$. Consider an input to the algorithm with $n$ items of size $\frac{1}{3}$. If the algorithm schedules at least $\frac{1}{5}$ of those items in the center, then subsequently send $n$ items of size $\frac{2}{3}$; otherwise the input consists only of the first $n$ items.

If under $\frac{1}{5}$ of the $\frac{1}{3}$ sized items were placed in the middle, then at least $\frac{2}{5}$ of the items would have been placed on either side. Thus, the algorithm will have at least $\frac{2n}{5}$ items at the peak appointment time, whereas the off-line optimum would have packed 3 items to a bin for a peak appointment of $\frac{n}{3}$. Thus, the algorithm must have achieved a competitive ratio of at least $\frac{6}{5}$ on this input.

Conversely, if over $\frac{1}{5}$ of the $\frac{1}{3}$-sized items were placed in the middle, then there must be at least $n + \frac{n}{5}$ items scheduled at the middle point in the day. But, an optimal solution would schedule each $\frac{2}{3}$ item with a $\frac{1}{3}$ item, for a peak appointment of $n$. Thus, the algorithm would have achieved a competitive ratio of at least $\frac{6}{5}$ on this input.

Since these properties hold for all $n$, taking $n$ to infinity and using this same pattern will create a sequence of inputs with an optimal solution tending toward infinity where every algorithm achieves a competitive ratio of at least $\frac{6}{5}$ on every input. Thus, the asymptotic competitive ratio of any deterministic algorithm must be at least $\frac{6}{5}$.

## 4.2   Randomized Lower Bound

A lower bound on the performance of any randomized algorithm for the MPAS problem will be developed in this section by first analyzing a simpler request-answer game (see, e.g., [6]). Consider a game where a requester can request an item be placed, either of size $\frac{1}{3}$ or $\frac{2}{3}$, and a responder needs to answer with a location within a bin to place the item. Further, define the cost function for this game to be the peak utilization within the bin, as in the MPAS problem. It is easy to see that this game is equivalent to the MPAS problem with item sizes restricted to $\frac{1}{3}$ and $\frac{2}{3}$, and thus any lower bounds on the game's asymptotic competitive ratio will bound the asymptotic competitive ratio of the MPAS problem.

**Theorem 3 (Request-Answer   Asymptotic   Randomized   Lower Bound).** *The above request-answer game has an asymptotic competitive ratio of at least $\frac{6}{5} = 1.2$.*

*Proof.* Let $\sigma_n$ be the family of request sequences consisting of $n$ consecutive $\frac{1}{3}$ requests, and let $\gamma_n$ be the family of request sequences consisting of $n$ consecutive $\frac{1}{3}$ requests followed by $n$ consecutive $\frac{2}{3}$ requests. Let $y_n$ be a family of probability distributions over request sequences, with $y_n(\sigma_n) = \frac{2}{5}$, $y_n(\gamma_n) = \frac{3}{5}$, and $y_n(j) = 0$ for all other $j$.

Choose some $n$ a multiple of 3. Now, take any deterministic algorithm ALG for the responder, which when presented with $n$ consecutive items of size $\frac{1}{3}$, places $c$ of them in the middle position of a bin. Since every bin with 3 items must have an item in the middle and every item in the middle of a bin will have to conflict with an item of size $\frac{2}{3}$, the following must be true:

$$E_{y \in y_n} \left[ \frac{\text{ALG}(y)}{\text{OPT}(y)} \right] = \frac{2}{5} \frac{\text{ALG}(\sigma_n)}{\text{OPT}(\sigma_n)} + \frac{3}{5} \frac{\text{ALG}(\gamma_n)}{\text{OPT}(\gamma_n)} \geq \frac{2}{5} \frac{c + (n - 3c)/2}{n/3} + \frac{3}{5} \frac{n + c}{n} = \frac{6}{5}$$

By Yao's principle, it must be that the competitive ratio of any randomized algorithm for this game is at most $\frac{6}{5}$ against an oblivious adversary [6]. Further, the above inequality will hold for any $n$ which is a multiple of 3, which allows for the creation of arbitrarily large hard inputs. Adding a lower bound to the game on the cost of valid inputs will not change the lower bound, since one can just take $n$ to be sufficiently large to satisfy the cost lower bound. Thus, the competitive ratio must remain bounded below by $\frac{6}{5}$ for any minimum optimal cost, implying the asymptotic competitive ratio for this problem is at least $\frac{6}{5}$ as well.

This result will directly imply the main theorem for this section, by the line of reasoning at the start of this section.

**Theorem 4 (Asymptotic Randomized Lower Bound).** *Any random algorithm for the online MPAS problem must have an asymptotic competitive ratio of at least $\frac{6}{5} = 1.2$.*

## 5    Conclusions and Future Work

The work done in this paper provides an improvement to the asymptotic competitive ratio of the MPAS problem compared to previous work. It further proves that 1.5 is not a lower bound for this problem, and invites even further improvements. Moving forward, proving a lower bound on the asymptotic competitive ratio would help future analysis of this problem be more focused and guided. It is also a hope to reimagine other known online problems and study how decisions can be delayed and what properties emerge from such analysis.

## A    Particularly Hard Input

It is possible to show that the asymptotic competitive ratio derived for the given algorithm is nearly the best possible. For any integer $n$, construct an input to the MPAS problem containing $2n$ items of size 0.3438, $n$ items of size 0.2501, and $n$ items of size 0.0623. Since this input has no Large or Third Items, the algorithm will pack this input in the same manner regardless of the order items arrive. The optimal way to pack these items is to put two of the items of size 0.3438 in a bin with one of each of the other items, requiring exactly $n$ bins. However, as $n$ goes to infinity, the algorithm will pack sets of two items of size 0.3438 to a bin, sets of 11 items of size 0.2501 to 4 bins, and sets of 54 items of size 0.0623 to 4 bins. This will result in the algorithm using roughly $n(1 + 4/11 + 4/54) = n\frac{427}{297}$ bins, for a competitive ratio of $\frac{427}{297} > 1.437$.

# B   Configuration Strategies

This section will detail a few important configuration strategies, which are used implicitly in the following section.

## B.1   Alternating Sides

Every complete set will have an even number of type 2 bins, and in the following section half of them will be on either side. Further, when rematching with Large Items, half of the Large Items will be on each side.

## B.2   Type 2 Small Bins

When rematching a Small Item category with Large or Third Items, any type 2 small bins within the complete set of bins will be placed in a bin with a Large Item or inner bin respectively.

## B.3   Rematching with Different Large Item Categories

Several of the Small Item categories will have different rematching guidelines depending on the size of the Large Items (when rematching with only Large Items). In every such case, the smaller size will have every type 2 large bin assigned to the same bin as a Large Item, while the larger case will have type 2 large bins assigned to a bin with another type 2 large bin.

## B.4   Rematching with Quarter Item Sets

When rematching Small Item sets with a Quarter Item set and some other items, any type 2 large bins in the Small Item set will be rematched with the type 2 large Quarter Item bins if possible. This will leave the Quarter Item type 2 small bins to be rematched with Large Items or inner bins.

## B.5   Rematching with Large Items

The properties the above analysis uses rely on there being 0.6875 Large Items for every bin after a rematching with a Small Item category, with an average fullness of at least 0.6875. In what follows, what will be shown is how to rematch so there are at least two Large Items per 3 bins, and an average density of at least 0.7. This is an equivalent statement, since for every 15 rematched bins an extra Large Item in its own bin can be assigned, leading to 11 Large Items for 16 bins with an average fullness of at least $11/16$.

# C   Category List and Matching Process Description

This section will be dedicated to going through each Small Item category, detailing the manner in which they are packed, and verifying all relevant properties used above for the category.

## C.1    Very Large Items

Very Large Items will only have one category, and do not get rematched.

## C.2    Large Items

The interior cutoffs for the Large Item categories will be:
$\frac{2}{3}, 0.642, 0.635, 0.625, 0.6, 0.588, 0.57, \frac{6}{11}, 0.54$
    Rematching with Large Items is detailed in the relevant item category.

## C.3    Medium Items

The interior cutoffs for the Medium Item categories will be:
$0.358, 0.365, 0.375, 0.4, 0.412, 0.43, \frac{5}{11}, 0.46$
    Medium Items rematch with Large Items by being placed in the same bin as a Large Item on the opposite side, if they both fit in a bin.
    Two Non-half Medium Items can rematch with a Quarter Item set and 8 Large Items. The two Medium Items are placed in a bin with the two Quarter Item type 2 large bins. The Quarter Item type 1 bin is placed in its own bin. The 4 Quarter Item type 2 small bins are placed in bins with Large Items, and the remaining 4 Large Items are placed in their own bin. This process will require 20 bins, and lead to an average fullness of at least $(0.34375 \cdot 2 + 0.25 \cdot 11 + 0.54 \cdot 8)/11 > 0.7$ over the bins. This process requires the Large Items to not fit in a bin with the relevant Medium Items remaining (otherwise this rematching would not happen in the algorithm's execution).

## C.4    Third Items

Third Items will only have one category.
    Third Item sets can be rematched with 6 Large Items. Four left and right Large Items will be placed in the same bins as right and left outer bins respectively. The remaining 2 Large Items will be placed alone in a bin, as will the two inner bins. This process will require 8 bins, and lead to an average fullness of at least $((1/3) \cdot 8 + .5 \cdot 6)/8 > 0.7$ over the bins. This process requires there to be Large Items which fit in a bin with the relevant Third Items remaining.
    Third Item sets can be rematched with 1 Quarter Item set and 14 Large Items. Two left and right Quarter Item type 2 large bins will be placed in the same bins as right and left outer bins respectively. The 4 type 2 small Quarter Item bins will be placed in a bin with a Large Item. The Quarter Item type 1 bins will remain in their own bin, as will the inner bins. The remaining two outer bins will be placed in their own bin. The remaining 10 Large Items will be placed alone in a bin. This process will require 20 bins, and lead to an average fullness of at least $((1/3) \cdot 8 + 0.25 \cdot 11 + (2/3) \cdot 14)/20 > 0.7$ over the bins. This process requires the Large Items to not fit in a bin with the relevant Third Items remaining (otherwise this rematching would not happen in the algorithm's execution).

## C.5    Small Items

**Sup-Category 3 (0.3125, 1/3].** A **Type 2 Large Bin** for this category will contain 1 item.

A **Type 1 Bin** for this category will contain 3 items.

A **Complete Set of Matched Bins** for this category will contain

- 2 Type 2 Large Bins Matched Pairwise
- 2 Type 1 Bins Unmatched

Filling 3 bins for an average fullness of at least $(0.3125 \cdot 8)/3 \geq 0.6875$.

A **Complete Set of Rematched Bins** can be obtained by rematching with
4 Large Items under size $2/3$ in 6 bins, for an average fullness of at least
$(0.5 \cdot 4 + 0.3125 \cdot 8)/6 \geq 0.7$
6 Large Items over size $2/3$ in 9 bins, for an average fullness of at least
$(\frac{2}{3} \cdot 6 + 0.3125 \cdot 8)/9 \geq 0.7$
2 Third Item sets (16 Third Items) in 11 bins, for an average fullness of at least
$(\frac{1}{3} \cdot 16 + 0.3125 \cdot 8)/11 \geq 0.6875$.

This category will rematch with 0 Quarter Items sets and satisfy the quarter size reallocation property.

**Category 3 (0.27, 0.3125].** A **Type 2 Small Bin** for this category will contain 1 item.

A **Type 1 Bin** for this category will contain 3 items.

A **Complete Set of Matched Bins** for this category will contain

- 4 Type 2 Small Bins Matched Pairwise
- 3 Type 1 Bins Unmatched

Filling 5 bins for an average fullness of at least $(0.27 \cdot 13)/5 \geq 0.6875$.

A **Complete Set of Rematched Bins** can be obtained by rematching with
6 Large Items in 9 bins, for an average fullness of at least $(0.5 \cdot 6 + 0.27 \cdot 13)/9 \geq 0.7$
2 Third Item sets (16 Third Items) in 11 bins, for an average fullness of at least
$(\frac{1}{3} \cdot 16 + 0.27 \cdot 13)/11 \geq 0.6875$.

This category will rematch with 0 Quarter Items sets and satisfy the quarter size reallocation property.

**Quarter Items (0.25, 0.27].** A **Type 2 Small Bin** for this category will contain 1 item.

A **Type 2 Large Bin** for this category will contain 2 items.

A **Type 1 Bin** for this category will contain 3 items.

A **Complete Set of Matched Bins** for this category will contain

- 2 Type 2 Small Bins Matched Pairwise
- 2 Type 2 Small Bins Matched with 2 Type 2 Large Bins Pairwise
- 1 Type 1 Bin Unmatched

Filling 4 bins for an average fullness of at least $(0.25 \cdot 11)/4 = 0.6875$.

This category rematches in a unique way with almost every other item type, explained where relevant. When matched directly with Large Items, Quarter Item and Large Item bins are grouped together, with no items being rematched into new bins.

**Sup-Category 4 (0.23, 0.25].** A **Type 2 Small Bin** for this category will contain 1 item.
A **Type 1 Bin** for this category will contain 4 items.
    A **Complete Set of Matched Bins** for this category will contain

- 4 Type 2 Small Bins Matched Pairwise
- 2 Type 1 Bins Unmatched

Filling 4 bins for an average fullness of at least $(0.23 \cdot 12)/4 \geq 0.6875$.

A **Complete Set of Rematched Bins** can be obtained by rematching with
4 Large Items in 6 bins, for an average fullness of at least $(0.5 \cdot 4 + 0.23 \cdot 12)/6 \geq 0.7$
2 Third Item sets (16 Third Items) in 10 bins, for an average fullness of at least
$(\frac{1}{3} \cdot 16 + 0.23 \cdot 12)/10 \geq 0.6875$
10 Large Items & 1 Quarter Item set in 15 bins, for an average fullness of at least $(0.5 \cdot 10 + 0.25 \cdot 11 + 0.23 \cdot 12)/15 \geq 0.7$
    and satisfying $(0.3125 \cdot 11 + 0.6875 \cdot 10)/15 \geq 0.6875$
3 Third Item sets (24 Third Items) & 1 Quarter Item set in 17 bins, for an average fullness of at least
    $(\frac{1}{3} \cdot 24 + 0.25 \cdot 11 + 0.23 \cdot 12)/17 \geq 0.6875$ and satisfying $(0.3125 \cdot 11 + 0.34375 \cdot 24)/17 \geq 0.6875$.

**Category 4 (0.215, 0.23].** A **Type 2 Small Bin** for this category will contain 1 item.
A **Type 2 Large Bin** for this category will contain 2 items.
    A **Complete Set of Matched Bins** for this category will contain

- 2 Type 2 Small Bins Matched Pairwise
- 4 Type 2 Large Bins Matched Pairwise

Filling 3 bins for an average fullness of at least $(0.215 \cdot 10)/3 \geq 0.6875$.

A **Complete Set of Rematched Bins** can be obtained by rematching with
6 Large Items under size 0.54 in 6 bins, for an average fullness of at least $(0.5 \cdot 6 + 0.215 \cdot 10)/6 \geq 0.7$
4 Large Items over size 0.54 in 6 bins, for an average fullness of at least $(0.54 \cdot 4 + 0.215 \cdot 10)/6 \geq 0.7$
2 Third Item sets (16 Third Items) in 10 bins, for an average fullness of at least
$(\frac{1}{3} \cdot 16 + 0.215 \cdot 10)/10 \geq 0.6875$
8 Large Items & 1 Quarter Item set in 12 bins, for an average fullness of at least $(0.5 \cdot 8 + 0.25 \cdot 11 + 0.215 \cdot 10)/12 \geq 0.7$
    and satisfying $(0.3125 \cdot 11 + 0.6875 \cdot 8)/12 \geq 0.6875$

3 Third Item sets (24 Third Items) & 1 Quarter Item set in 16 bins, for an average fullness of at least
$$(\tfrac{1}{3} \cdot 24 + 0.25 \cdot 11 + 0.215 \cdot 10)/16 \geq 0.6875 \text{ and satisfying } (0.3125 \cdot 11 + 0.34375 \cdot 24)/16 \geq 0.6875.$$

**Sub-category 4 (0.206, 0.215].** A **Type 2 Small Bin** for this category will contain 1 item.
A **Type 2 Large Bin** for this category will contain 2 items.
   A **Complete Set of Matched Bins** for this category will contain

- 2 Type 2 Small Bins Matched Pairwise
- 6 Type 2 Large Bins Matched Pairwise

Filling 4 bins for an average fullness of at least $(0.206 \cdot 14)/4 \geq 0.6875$.
   A **Complete Set of Rematched Bins** can be obtained by rematching with
8 Large Items under size 0.57 in 8 bins, for an average fullness of at least $(0.5 \cdot 8 + 0.206 \cdot 14)/8 \geq 0.7$
6 Large Items over size 0.57 in 9 bins, for an average fullness of at least $(0.57 \cdot 6 + 0.206 \cdot 14)/9 \geq 0.7$
2 Third Item sets (16 Third Items) in 11 bins, for an average fullness of at least $(\tfrac{1}{3} \cdot 16 + 0.206 \cdot 14)/11 \geq 0.6875$
14 Large Items & 2 Quarter Item sets in 21 bins, for an average fullness of at least $(0.5 \cdot 14 + 0.25 \cdot 22 + 0.206 \cdot 14)/21 \geq 0.7$
      and satisfying $(0.3125 \cdot 22 + 0.6875 \cdot 14)/21 \geq 0.6875$
3 Third Item sets (24 Third Items) & 1 Quarter Item set in 17 bins, for an average fullness of at least
$$(\tfrac{1}{3} \cdot 24 + 0.25 \cdot 11 + 0.206 \cdot 14)/17 \geq 0.6875 \text{ and satisfying } (0.3125 \cdot 11 + 0.34375 \cdot 24)/17 \geq 0.6875.$$

**Sub-Sub-Category 4 (0.2, 0.206].** A **Type 2 Small Bin** for this category will contain 1 item.
A **Type 2 Large Bin** for this category will contain 2 items.
   A **Complete Set of Matched Bins** for this category will contain

- 2 Type 2 Small Bins Matched Pairwise
- 6 Type 2 Large Bins Matched Pairwise

Filling 4 bins for an average fullness of at least $(0.2 \cdot 14)/4 \geq 0.6875$.
   A **Complete Set of Rematched Bins** can be obtained by rematching with
8 Large Items under size 0.588 in 8 bins, for an average fullness of at least $(0.5 \cdot 8 + 0.2 \cdot 14)/8 \geq 0.7$
6 Large Items over size 0.588 in 9 bins, for an average fullness of at least $(0.588 \cdot 6 + 0.2 \cdot 14)/9 \geq 0.7$
2 Third Item sets (16 Third Items) in 11 bins, for an average fullness of at least $(\tfrac{1}{3} \cdot 16 + 0.2 \cdot 14)/11 \geq 0.6875$

14 Large Items & 2 Quarter Item sets in 21 bins, for an average fullness of at least $(0.5 \cdot 14 + 0.25 \cdot 22 + 0.2 \cdot 14)/21 \geq 0.7$

and satisfying $(0.3125 \cdot 22 + 0.6875 \cdot 14)/21 \geq 0.6875$

3 Third Item sets (24 Third Items) & 1 Quarter Item set in 17 bins, for an average fullness of at least

$(\frac{1}{3} \cdot 24 + 0.25 \cdot 11 + 0.2 \cdot 14)/17 \geq 0.6875$ and satisfying $(0.3125 \cdot 11 + 0.34375 \cdot 24)/17 \geq 0.6875$.

**Sup-Category 5 (0.1825, 0.2].** A **Type 2 Small Bin** for this category will contain 1 item.

A **Type 2 Large Bin** for this category will contain 2 items.

A **Type 1 Bin** for this category will contain 5 items.

A **Complete Set of Matched Bins** for this category will contain

- 2 Type 2 Small Bins Matched Pairwise
- 4 Type 2 Large Bins Matched Pairwise
- 2 Type 1 Bins Unmatched

Filling 5 bins for an average fullness of at least $(0.1825 \cdot 20)/5 \geq 0.6875$.

A **Complete Set of Rematched Bins** can be obtained by rematching with

6 Large Items under size 0.6 in 8 bins, for an average fullness of at least $(0.5 \cdot 6 + 0.1825 \cdot 20)/8 \geq 0.7$

8 Large Items over size 0.6 in 12 bins, for an average fullness of at least $(0.6 \cdot 8 + 0.1825 \cdot 20)/12 \geq 0.7$

3 Third Item sets (24 Third Items) in 16 bins, for an average fullness of at least $(\frac{1}{3} \cdot 24 + 0.1825 \cdot 20)/16 \geq 0.6875$

16 Large Items & 2 Quarter Item sets in 24 bins, for an average fullness of at least $(0.5 \cdot 16 + 0.25 \cdot 22 + 0.1825 \cdot 20)/24 \geq 0.7$

and satisfying $(0.3125 \cdot 22 + 0.6875 \cdot 16)/24 \geq 0.6875$

5 Third Item sets (40 Third Items) & 2 Quarter Item sets in 28 bins, for an average fullness of at least

$(\frac{1}{3} \cdot 40 + 0.25 \cdot 22 + 0.1825 \cdot 20)/28 \geq 0.6875$ and satisfying $(0.3125 \cdot 22 + 0.34375 \cdot 40)/28 \geq 0.6875$.

**Category 5 (0.179, 0.1825].** A **Type 2 Small Bin** for this category will contain 1 item.

A **Type 2 Large Bin** for this category will contain 2 items.

A **Type 1 Bin** for this category will contain 5 items.

A **Complete Set of Matched Bins** for this category will contain

- 4 Type 2 Small Bins Matched Pairwise
- 10 Type 2 Large Bins Matched Pairwise
- 6 Type 1 Bins Unmatched

Filling 13 bins for an average fullness of at least $(0.179 \cdot 54)/13 \geq 0.6875$.

**A Complete Set of Rematched Bins** can be obtained by rematching with

14 Large Items under size 0.635 in 20 bins, for an average fullness of at least $(0.5 \cdot 14 + 0.179 \cdot 54)/20 \geq 0.7$

22 Large Items over size 0.635 in 33 bins, for an average fullness of at least $(0.635 \cdot 22 + 0.179 \cdot 54)/33 \geq 0.7$

7 Third Item sets (56 Third Items) in 39 bins, for an average fullness of at least $(\frac{1}{3} \cdot 56 + 0.179 \cdot 54)/39 \geq 0.6875$

42 Large Items & 5 Quarter Item sets in 63 bins, for an average fullness of at least $(0.5 \cdot 42 + 0.25 \cdot 55 + 0.179 \cdot 54)/63 \geq 0.7$

and satisfying $(0.3125 \cdot 55 + 0.6875 \cdot 42)/63 \geq 0.6875$

12 Third Item sets (96 Third Items) & 5 Quarter Item sets in 69 bins, for an average fullness of at least

$(\frac{1}{3} \cdot 96 + 0.25 \cdot 55 + 0.179 \cdot 54)/69 \geq 0.6875$ and satisfying $(0.3125 \cdot 55 + 0.34375 \cdot 96)/69 \geq 0.6875$.

**Sub-category 5 (1/6, 0.179].** A **Type 2 Small Bin** for this category will contain 1 item.

A **Type 2 Large Bin** for this category will contain 2 items.

A **Type 1 Bin** for this category will contain 5 items.

A **Complete Set of Matched Bins** for this category will contain

– 4 Type 2 Small Bins Matched Pairwise
– 10 Type 2 Large Bins Matched Pairwise
– 6 Type 1 Bins Unmatched

Filling 13 bins for an average fullness of at least $((1/6) \cdot 54)/13 \geq 0.6875$.

A **Complete Set of Rematched Bins** can be obtained by rematching with

14 Large Items under size 0.642 in 20 bins, for an average fullness of at least $(0.5 \cdot 14 + (1/6) \cdot 54)/20 \geq 0.7$

22 Large Items over size 0.642 in 33 bins, for an average fullness of at least $(0.642 \cdot 22 + (1/6) \cdot 54)/33 \geq 0.7$

7 Third Item sets (56 Third Items) in 39 bins, for an average fullness of at least $(\frac{1}{3} \cdot 56 + (1/6) \cdot 54)/39 \geq 0.6875$

36 Large Items under size 0.642 & 4 Quarter Item sets in 54 bins, for an average fullness of at least

$(0.5 \cdot 36 + 0.25 \cdot 44 + (1/6) \cdot 54)/54 \geq 0.7$ and satisfying $(0.3125 \cdot 44 + 0.6875 \cdot 36)/54 \geq 0.6875$

42 Large Items over size 0.642 & 5 Quarter Item sets in 63 bins, for an average fullness of at least

$(0.642 \cdot 42 + 0.25 \cdot 55 + (1/6) \cdot 54)/63 \geq 0.7$ and satisfying $(0.3125 \cdot 55 + 0.6875 \cdot 42)/63 \geq 0.6875$

12 Third Item sets (96 Third Items) & 5 Quarter Item sets in 69 bins, for an average fullness of at least

$(\frac{1}{3} \cdot 96 + 0.25 \cdot 55 + (1/6) \cdot 54)/69 \geq 0.6875$ and satisfying $(0.3125 \cdot 55 + 0.34375 \cdot 96)/69 \geq 0.6875$.

**Sup-Category 6 (0.15625, 1/6].** A **Type 2 Small Bin** for this category will contain 1 item.
A **Type 2 Large Bin** for this category will contain 2 items.
A **Type 1 Bin** for this category will contain 6 items.
  A **Complete Set of Matched Bins** for this category will contain

- 2 Type 2 Small Bins Matched Pairwise
- 2 Type 2 Large Bins Matched Pairwise
- 2 Type 1 Bins Unmatched

Filling 4 bins for an average fullness of at least $(0.15625 \cdot 18)/4 \geq 0.6875$.
  A **Complete Set of Rematched Bins** can be obtained by rematching with
4 Large Items under size 2/3 in 6 bins, for an average fullness of at least $(0.5 \cdot 4 + 0.15625 \cdot 18)/6 \geq 0.7$
6 Large Items over size 2/3 in 9 bins, for an average fullness of at least $((2/3) \cdot 6 + 0.15625 \cdot 18)/9 \geq 0.7$
2 Third Item sets (16 Third Items) in 11 bins, for an average fullness of at least $(\frac{1}{3} \cdot 16 + 0.15625 \cdot 18)/11 \geq 0.6875$
10 Large Items & 1 Quarter Item set in 15 bins, for an average fullness of at least $(0.5 \cdot 10 + 0.25 \cdot 11 + 0.15625 \cdot 18)/15 \geq 0.7$
  and satisfying $(0.3125 \cdot 11 + 0.6875 \cdot 10)/15 \geq 0.6875$
3 Third Item sets (24 Third Items) & 1 Quarter Item set in 17 bins, for an average fullness of at least
  $(\frac{1}{3} \cdot 24 + 0.25 \cdot 11 + 0.15625 \cdot 18)/17 \geq 0.6875$ and satisfying $(0.3125 \cdot 11 + 0.34375 \cdot 24)/17 \geq 0.6875$.


**Category 6 (1/7, 0.15625].** A **Type 2 Small Bin** for this category will contain 2 items.
A **Type 1 Bin** for this category will contain 6 items.
  A **Complete Set of Matched Bins** for this category will contain

- 4 Type 2 Small Bins Matched Pairwise
- 2 Type 1 Bins Unmatched

Filling 4 bins for an average fullness of at least $((1/7) \cdot 20)/4 \geq 0.6875$.
  A **Complete Set of Rematched Bins** can be obtained by rematching with
4 Large Items in 6 bins, for an average fullness of at least $(0.5 \cdot 4 + (1/7) \cdot 20)/6 \geq 0.7$
2 Third Item sets (16 Third Items) in 10 bins, for an average fullness of at least $(\frac{1}{3} \cdot 16 + (1/7) \cdot 20)/10 \geq 0.6875$
10 Large Items & 1 Quarter Item set in 15 bins, for an average fullness of at least $(0.5 \cdot 10 + 0.25 \cdot 11 + (1/7) \cdot 20)/15 \geq 0.7$
  and satisfying $(0.3125 \cdot 11 + 0.6875 \cdot 10)/15 \geq 0.6875$
3 Third Item sets (24 Third Items) & 1 Quarter Item set in 17 bins, for an average fullness of at least
  $(\frac{1}{3} \cdot 24 + 0.25 \cdot 11 + (1/7) \cdot 20)/17 \geq 0.6875$ and satisfying $(0.3125 \cdot 11 + 0.34375 \cdot 24)/17 \geq 0.6875$.

**Category 7 (1/8, 1/7].** A **Type 2 Small Bin** for this category will contain 2 item.

A **Type 2 Large Bin** for this category will contain 3 items.

A **Type 1 Bin** for this category will contain 7 items.

A **Complete Set of Matched Bins** for this category will contain

- 4 Type 2 Small Bins Matched Pairwise
- 2 Type 2 Large Bins Matched Pairwise
- 2 Type 1 Bins Unmatched

Filling 5 bins for an average fullness of at least $((1/8) \cdot 28)/5 \geq 0.6875$.

A **Complete Set of Rematched Bins** can be obtained by rematching with 6 Large Items in 9 bins, for an average fullness of at least $(0.5 \cdot 6 + (1/8) \cdot 28)/9 \geq 0.7$

2 Third Item sets (16 Third Items) in 11 bins, for an average fullness of at least $(\frac{1}{3} \cdot 16 + (1/8) \cdot 28)/11 \geq 0.6875$

10 Large Items & 1 Quarter Item set in 15 bins, for an average fullness of at least $(0.5 \cdot 10 + 0.25 \cdot 11 + (1/8) \cdot 28)/15 \geq 0.7$

and satisfying $(0.3125 \cdot 11 + 0.6875 \cdot 10)/15 \geq 0.6875$

4 Third Item sets (32 Third Items) & 1 Quarter Item set in 21 bins, for an average fullness of at least

$(\frac{1}{3} \cdot 32 + 0.25 \cdot 11 + (1/8) \cdot 28)/21 \geq 0.6875$ and satisfying $(0.3125 \cdot 11 + 0.34375 \cdot 32)/21 \geq 0.6875$.

**Category 8 (1/9, 1/8].** A **Type 2 Small Bin** for this category will contain 2 item.

A **Type 2 Large Bin** for this category will contain 3 items.

A **Type 1 Bin** for this category will contain 8 items.

A **Complete Set of Matched Bins** for this category will contain

- 2 Type 2 Small Bins Matched Pairwise
- 2 Type 2 Large Bins Matched Pairwise
- 2 Type 1 Bins Unmatched

Filling 4 bins for an average fullness of at least $((1/9) \cdot 26)/4 \geq 0.6875$.

A **Complete Set of Rematched Bins** can be obtained by rematching with 4 Large Items under size 5/8 in 6 bins, for an average fullness of at least $(0.5 \cdot 4 + (1/9) \cdot 26)/6 \geq 0.7$

6 Large Items over size 5/8 in 9 bins, for an average fullness of at least $((5/8) \cdot 6 + (1/9) \cdot 26)/9 \geq 0.7$

2 Third Item sets (16 Third Items) in 11 bins, for an average fullness of at least $(\frac{1}{3} \cdot 16 + (1/9) \cdot 26)/11 \geq 0.6875$

10 Large Items & 1 Quarter Item set in 15 bins, for an average fullness of at least $(0.5 \cdot 10 + 0.25 \cdot 11 + (1/9) \cdot 26)/15 \geq 0.7$

and satisfying $(0.3125 \cdot 11 + 0.6875 \cdot 10)/15 \geq 0.6875$

4 Third Item sets (32 Third Items) & 1 Quarter Item set in 21 bins, for an average fullness of at least

$(\frac{1}{3} \cdot 32 + 0.25 \cdot 11 + (1/9) \cdot 26)/21 \geq 0.6875$ and satisfying $(0.3125 \cdot 11 + 0.34375 \cdot 32)/21 \geq 0.6875$.

**Category 9 (1/10, 1/9].** A **Type 2 Small Bin** for this category will contain 2 item.
A **Type 2 Large Bin** for this category will contain 3 items.
A **Type 1 Bin** for this category will contain 9 items.
   A **Complete Set of Matched Bins** for this category will contain

- 2 Type 2 Small Bins Matched Pairwise
- 2 Type 2 Large Bins Matched Pairwise
- 2 Type 1 Bins Unmatched

Filling 4 bins for an average fullness of at least $((1/10) \cdot 28)/4 \geq 0.6875$.
   A **Complete Set of Rematched Bins** can be obtained by rematching with
4 Large Items under size 2/3 in 6 bins, for an average fullness of at least $(0.5 \cdot 4 + (1/10) \cdot 28)/6 \geq 0.7$
6 Large Items over size 2/3 in 9 bins, for an average fullness of at least $((2/3) \cdot 6 + (1/10) \cdot 28)/9 \geq 0.7$
2 Third Item sets (16 Third Items) in 11 bins, for an average fullness of at least $(\frac{1}{3} \cdot 16 + (1/10) \cdot 28)/11 \geq 0.6875$
10 Large Items & 1 Quarter Item set in 15 bins, for an average fullness of at least $(0.5 \cdot 10 + 0.25 \cdot 11 + (1/10) \cdot 28)/15 \geq 0.7$
   and satisfying $(0.3125 \cdot 11 + 0.6875 \cdot 10)/15 \geq 0.6875$
4 Third Item sets (32 Third Items) & 1 Quarter Item set in 21 bins, for an average fullness of at least
   $(\frac{1}{3} \cdot 32 + 0.25 \cdot 11 + (1/10) \cdot 28)/21 \geq 0.6875$ and satisfying $(0.3125 \cdot 11 + 0.34375 \cdot 32)/21 \geq 0.6875$.

**Category 10 (1/11, 1/10].** A **Type 2 Small Bin** for this category will contain 3 item.
A **Type 2 Large Bin** for this category will contain 4 items.
A **Type 1 Bin** for this category will contain 10 items.
   A **Complete Set of Matched Bins** for this category will contain

- 2 Type 2 Small Bins Matched Pairwise
- 2 Type 2 Large Bins Matched Pairwise
- 2 Type 1 Bins Unmatched

Filling 4 bins for an average fullness of at least $((1/11) \cdot 34)/4 \geq 0.6875$.
   A **Complete Set of Rematched Bins** can be obtained by rematching with
4 Large Items under size 0.6 in 6 bins, for an average fullness of at least $(0.5 \cdot 4 + (1/11) \cdot 34)/6 \geq 0.7$
6 Large Items over size 0.6 in 9 bins, for an average fullness of at least $(0.6 \cdot 6 + (1/11) \cdot 34)/9 \geq 0.7$
2 Third Item sets (16 Third Items) in 11 bins, for an average fullness of at least $(\frac{1}{3} \cdot 16 + (1/11) \cdot 34)/11 \geq 0.6875$

10 Large Items & 1 Quarter Item set in 15 bins, for an average fullness of at least $(0.5 \cdot 10 + 0.25 \cdot 11 + (1/11) \cdot 34)/15 \geq 0.7$

and satisfying $(0.3125 \cdot 11 + 0.6875 \cdot 10)/15 \geq 0.6875$

4 Third Item sets (32 Third Items) & 1 Quarter Item set in 21 bins, for an average fullness of at least

$(\frac{1}{3} \cdot 32 + 0.25 \cdot 11 + (1/11) \cdot 34)/21 \geq 0.6875$ and satisfying $(0.3125 \cdot 11 + 0.34375 \cdot 32)/21 \geq 0.6875$.

**Category 11 (1/12, 1/11].** A **Type 2 Small Bin** for this category will contain 3 item.

A **Type 2 Large Bin** for this category will contain 5 items.

A **Type 1 Bin** for this category will contain 11 items.

A **Complete Set of Matched Bins** for this category will contain

- 2 Type 2 Small Bins Matched Pairwise
- 2 Type 2 Large Bins Matched Pairwise
- 2 Type 1 Bins Unmatched

Filling 4 bins for an average fullness of at least $((1/12) \cdot 38)/4 \geq 0.6875$.

A **Complete Set of Rematched Bins** can be obtained by rematching with

4 Large Items under size 6/11 in 6 bins, for an average fullness of at least $(0.5 \cdot 4 + (1/12) \cdot 38)/6 \geq 0.7$

6 Large Items over size 6/11 in 9 bins, for an average fullness of at least $((6/11) \cdot 6 + (1/12) \cdot 38)/9 \geq 0.7$

2 Third Item sets (16 Third Items) in 11 bins, for an average fullness of at least $(\frac{1}{3} \cdot 16 + (1/12) \cdot 38)/11 \geq 0.6875$

10 Large Items & 1 Quarter Item set in 15 bins, for an average fullness of at least $(0.5 \cdot 10 + 0.25 \cdot 11 + (1/12) \cdot 38)/15 \geq 0.7$

and satisfying $(0.3125 \cdot 11 + 0.6875 \cdot 10)/15 \geq 0.6875$

4 Third Item sets (32 Third Items) & 1 Quarter Item set in 21 bins, for an average fullness of at least

$(\frac{1}{3} \cdot 32 + 0.25 \cdot 11 + (1/12) \cdot 38)/21 \geq 0.6875$ and satisfying $(0.3125 \cdot 11 + 0.34375 \cdot 32)/21 \geq 0.6875$.

**Category 12 [0, 1/12].** This category will be slightly different, since it includes all items of size at most 1/12. So each bin type will not specify a number of items, but rather a cutoff a which point items will stop being added. For example, a Type 1 bin will be until 11/12 full, which means that it will not be "full" and able to accept more items until it is 11/12ths full, at which point it may not be able to accept another item without going over the limit and is thus at capacity.

A **Type 2 Small Bin** for this category will be filled until 1/4 full.

A **Type 2 Large Bin** for this category will be filled until 113/300 full.

A **Type 1 Bin** for this category will be filled until 11/12 full.

A **Complete Set of Matched Bins** for this category will contain

– 2 Type 2 Small Bins Matched Pairwise
– 2 Type 2 Large Bins Matched Pairwise
– 2 Type 1 Bins Unmatched

Filling 4 bins for an average fullness of at least $(((1/4) + (113/300) + (11/12)) \cdot 2)/4 \geq 0.6875$.

A **Complete Set of Rematched Bins** can be obtained by rematching with
4 Large Items under size 0.54 in 6 bins, for an average fullness of at least $(0.5 \cdot 4 + ((1/4) + (113/300) + (11/12)) \cdot 2)/6 \geq 0.7$
6 Large Items over size 0.54 in 9 bins, for an average fullness of at least $(0.54 \cdot 6 + ((1/4) + (113/300) + (11/12)) \cdot 2)/9 \geq 0.7$
2 Third Item sets (16 Third Items) in 11 bins, for an average fullness of at least
$(\frac{1}{3} \cdot 16 + ((1/4) + (113/300) + (11/12)) \cdot 2)/11 \geq 0.6875$
10 Large Items & 1 Quarter Item set in 15 bins, for an average fullness of at least
$(0.5 \cdot 10 + 0.25 \cdot 11 + ((1/4) + (113/300) + (11/12)) \cdot 2)/15 \geq 0.7$ and satisfying $(0.3125 \cdot 11 + 0.6875 \cdot 10)/15 \geq 0.6875$
4 Third Item sets (32 Third Items) & 1 Quarter Item set in 21 bins, for an average fullness of at least
$(\frac{1}{3} \cdot 32 + 0.25 \cdot 11 + ((1/4) + (113/300) + (11/12)) \cdot 2)/21 \geq 0.6875$ and satisfying $(0.3125 \cdot 11 + 0.34375 \cdot 32)/21 \geq 0.6875$.

# References

1. Balogh, J., Békési, J., Dósa, G., Epstein, L., Levin, A.: A new and improved algorithm for online bin packing. arXiv preprint arXiv:1707.01728 (2017)
2. Balogh, J., Békési, J., Dósa, G., Epstein, L., Levin, A.: A new lower bound for classic online bin packing. Algorithmica **83**, 1–16 (2021)
3. Balogh, J., Békési, J., Dósa, G., Sgall, J., van Stee, R.: The optimal absolute ratio for online bin packing. J. Comput. Syst. Sci. **102**, 1–17 (2019)
4. Balogh, J., Dósa, G., Epstein, L., Jeż, L: Lower bounds on the performance of online algorithms for relaxed packing problems. In: Bazgan, C., Fernau, H. (eds.) IWOCA 2022. LNCS, vol. 13270, pp. 101–113. Springer, Cham (2022). https://doi.org/10.1007/978-3-031-06678-8_8
5. Billingsley, P.: Probability and Measure. Wiley, Hoboken (2008)
6. Borodin, A., El-Yaniv, R.: Online Computation and Competitive Analysis. Cambridge University Press, Cambridge (2005)
7. Chandra, B.: Does randomization help in on-line bin packing? Inf. Process. Lett. **43**(1), 15–19 (1992)
8. Coffman, E.G., Csirik, J., Galambos, G., Martello, S., Vigo, D.: Bin packing approximation algorithms: survey and classification. In: Pardalos, P.M., Du, D.-Z., Graham, R.L. (eds.) Handbook of Combinatorial Optimization, pp. 455–531. Springer, New York (2013). https://doi.org/10.1007/978-1-4419-7997-1_35
9. Csirik, J., Woeginger, G.J.: On-line packing and covering problems. Online Algorithms 147–177 (1998)
10. Escribe, C., Hu, M., Levi, R.: Competitive algorithms for the online minimum peak appointment scheduling. Available at SSRN 3787306 (2021)

11. Golumbic, M.C.: Algorithmic Graph Theory and Perfect Graphs. Elsevier, Amsterdam (2004)
12. Johnson, D.S.: Fast algorithms for bin packing. J. Comput. Syst. Sci. **8**(3), 272–314 (1974)
13. Lee, C.C., Lee, D.T.: A simple on-line bin-packing algorithm. J. ACM (JACM) **32**(3), 562–572 (1985)
14. Levi, R.: Private communication (2021)

# Canadian Traveller Problem
# with Predictions

Evripidis Bampis[1], Bruno Escoffier[1,2], and Michalis Xefteris[1(✉)]

[1] Sorbonne Université, CNRS, LIP6, 75005 Paris, France
{evripidis.bampis,Bruno.Escoffier,michail.xefteris}@lip6.fr
[2] Institut Universitaire de France, Paris, France

**Abstract.** In this work, we consider the $k$-Canadian Traveller Problem ($k$-CTP) under the learning-augmented framework proposed by Lykouris & Vassilvitskii [23]. $k$-CTP is a generalization of the shortest path problem, and involves a traveller who knows the entire graph in advance and wishes to find the shortest route from a source vertex $s$ to a destination vertex $t$, but discovers online that some edges (up to $k$) are blocked once reaching them. A potentially imperfect predictor gives us the number and the locations of the blocked edges.

We present a deterministic and a randomized online algorithm for the learning-augmented $k$-CTP that achieve a tradeoff between consistency (quality of the solution when the prediction is correct) and robustness (quality of the solution when there are errors in the prediction). Moreover, we prove a matching lower bound for the deterministic case establishing that the tradeoff between consistency and robustness is optimal, and show a lower bound for the randomized algorithm. Finally, we prove several deterministic and randomized lower bounds on the competitive ratio of $k$-CTP depending on the prediction error, and complement them, in most cases, with matching upper bounds.

**Keywords:** Canadian Traveller Problem · Online algorithm · Learning augmented algorithm

## 1   Introduction

Motivated by various applications including online route planning in road networks, or message routing in communication networks, the Canadian Traveller problem (CTP), introduced in 1991 by Papadimitriou and Yannakakis [29], is a generalization of one of the most prominent problems in Computer Science, the Shortest Path Problem [20, 28]. In CTP the underlying graph is given in advance, but it is unreliable, i.e. some edges may become unavailable (e.g. because of snowfall, or link failure) in an online manner. The blockage of an edge becomes known to the algorithm only when it arrives at one of its extremities. The objective is to devise an efficient adaptive strategy minimizing the ratio between the length of the path found and the optimum (where the blocked edges are removed). Papadimitriou and Yannakakis [29] proved that the problem of devising an algorithm that guarantees a given competitive ratio is PSPACE-complete

P. Chalermsook and B. Laekhanukit (Eds.): WAOA 2022, LNCS 13538, pp. 116–133, 2022.
https://doi.org/10.1007/978-3-031-18367-6_6

if the number of blocked edges is not fixed. Given the intractability of CTP, Bar-Noy and Schieber [7] focused on $k$-CTP, a special case of CTP where the number of blocked edges is bounded by $k$. Here, we consider $k$-CTP in the framework of *learning-augmented* online algorithms [23,26]. It is natural to consider that in applications, like route planning, or message routing in communication networks, predictions may be provided on the input data. Our aim is to study the impact of the quality of such predictions on the performance of online algorithms for $k$-CTP.

Formally, in $k$-CTP we consider a connected undirected graph $\mathcal{G} = (V, E)$ with a source node $s$, a destination node $t$ and a non-negative cost function $c : E \to \mathbb{R}^+$ representing the cost to traverse each edge. An agent seeks to find a shortest path from $s$ to $t$. However, one or more edges (up to $k$) might be blocked, and thus cannot be traversed. An agent only learns that an edge is blocked when reaching one of its endpoints.

In classical competitive analysis, a deterministic online algorithm $ALG$ for $k$-CTP is $c$-competitive if the total length $ALG(\sigma)$ traversed by $ALG$ for input $\sigma$ is at most $c \cdot OPT(\sigma)$, where $OPT(\sigma)$ is the length of a shortest $s - t$ path in $G$ without the blocked edges [33]. A randomized algorithm is $c$-competitive against an oblivious adversary if the expected cost $\mathbb{E}[ALG(\sigma)]$ is at most $c \cdot OPT(\sigma)$ [10].

Bar-Noy and Schieber [7] considered $k$-CTP and they proposed a polynomial time algorithm that minimizes the maximum travel length. Westphal in [38] gave a simple online deterministic algorithm for $k$-CTP which is $(2k+1)$-competitive. He also proved that no deterministic online algorithm with a better competitive ratio exists. Furthermore, he showed a lower bound for any randomized algorithm of $k + 1$, even if all $s - t$ paths are node disjoint. Xu et al. [39] proposed a deterministic algorithm that is also $(2k + 1)$-competitive for $k$-CTP. They also proved that a natural greedy strategy based on the available blockage information is exponential in $k$. A $(k + 1)$-competitive randomized online algorithm for $k$-CTP is known on graphs where all $s - t$ paths are node-disjoint [8,32]. Demaine et al. [12] proposed a polynomial time randomized algorithm that improves the deterministic lower bound of $2k + 1$ by an $o(1)$ factor for arbitrary graphs. They also showed that the competitive ratio is even better if the randomized algorithm runs in pseudo-polynomial time. More recently, Bergé et al. [9] proved that the competitive ratio of any randomized memoryless strategy (agent's strategy does not depend on his/her anterior moves) cannot be better than $2k + O(1)$. Several other variants of the problem have been studied in the recent years [16], [27].

Given the widespread of Machine Learning technology, in the last years, predictions from ML are used in order to improve the worst case analysis of online algorithms [11,13,24,34]. The formal framework for these learning-augmented algorithms has been presented by Lykouris and Vassilvitskii in their seminal paper [23], where they studied the caching problem. In this framework, no assumption is made concerning the quality of the predictor and the challenge is to design a learning-augmented online algorithm that finds a good tradeoff between the two extreme alternatives, i.e. following blindly the predictions, or simply ignore them. Ideally, the objective is to produce algorithms using predictions that are *consistent*, i.e. whose performance is close to the best offline algorithm when the prediction is accurate, and *robust*, i.e. whose performance is close to the online algorithm without predictions when the prediction is bad.

Antoniadis et al. [1], Rohatgi [31] and Wei [36] subsequently gave simpler and improved algorithms for the caching problem. Kumar et al. in [30] applied the learning-augmented setting to ski rental and online scheduling. For the same problems, Wei and Zhang, [37], provided a set of non-trivial lower bounds for competitive analysis of learning-augmented online algorithms. Many other papers have been published in this direction for ski rental [5,6,15,35], scheduling [3,4,17,18,25], the online $k$-server problem [21], $k$-means clustering [14] and others [2,19,22].

## 1.1   Our Contribution

In this work, we study the $k$-Canadian Traveller Problem through the lens of online algorithms with predictions. We present both deterministic and randomized upper and lower bounds for the problem. Following previous works we focus on path-disjoint graphs for the randomized case[1]. We use a simple model where we are given predictions on the locations of the blocked edges. For example, consider a situation wherein you need to follow the shortest route to a destination. You usually open the Maps app on your phone to find you the best route. Maps app does that using predictions about the weather condition, the traffic jam etc. These predictions capture additional side information about the route we should follow. In our model, the error of the prediction is just the total number of false predictions we get. The parameter $k$ upper bounds the number of real blocked edges (denoted by $\kappa$, usually unknown when the algorithm starts) and the number of predicted blocked edges (denoted by $k_p$) in the graph, meaning that we want to design algorithms that hedge against all situations where up to $k$ edges can be blocked (and up to $k$ predicted to be blocked).

In Sect. 3, we give the main results of this paper which are algorithms with predictions for $k$-CTP (deterministic and randomized ones) that are as consistent and robust as possible. More precisely, we say that an algorithm is $(a, b)$-competitive ($[a, b]$-competitive), when it achieves a competitive ratio smaller than (no more than) $a$ when the prediction is correct and no more than $b$ otherwise. Our aim is to answer the following question: if we want an algorithm which is $(1 + \epsilon)$-competitive if the prediction is correct (consistency $1 + \epsilon$), what is the best competitive ratio we can get when the prediction is not correct (robustness)? The parameter $\epsilon > 0$ is user defined, possibly adjusted depending on her/his level of trust in the predictions. The results are presented in Table 1. We give a deterministic $\left(1+\epsilon, 2k-1+\frac{4k}{\epsilon}\right)$-lower bound and a matching upper bound. For the randomized case, we give a randomized $\left[1 + \epsilon, k + \frac{k}{\epsilon}\right]$-lower bound and an algorithm that achieves a tradeoff of $\left[1 + \epsilon, k + \frac{4k}{\epsilon}\right]$ on path-disjoint graphs, when $k$ is considered as known. We note that the above lower bounds are also valid when the parameter is $\kappa$ (the real number of blocked edges). In most real

---

[1] As mentioned in earlier, while a $(2k + 1)$-competitive (matching the lower bound) deterministic algorithm is known for general graph, a $(k+1)$-competitive randomized algorithm (matching the lower bound) is only known for path-disjoint graphs.

world problems such as the ones described earlier, the number of blocked edges is usually small and we can get interesting tradeoffs between consistency and robustness.

**Table 1.** Our bounds on the tradeoffs between consistency and robustness for our learning-augmented model ($0 < \epsilon \leq 2k$ for the deterministic case, and $0 < \epsilon \leq k$ for the randomized one).

| Deterministic algorithms | Lower bound | $\left(1 + \epsilon, 2k - 1 + \frac{4k}{\epsilon}\right]$ Theorem 1 |
|---|---|---|
| | Upper bound | $\left(1 + \epsilon, 2k - 1 + \frac{4k}{\epsilon}\right]$ Theorem 2 |
| Randomized algorithms | Lower bound | $\left[1 + \epsilon, k + \frac{k}{\epsilon}\right]$ Theorem 3 |
| | Upper bound | $\left[1 + \epsilon, k + \frac{4k}{\epsilon}\right]$ Theorem 4 |

In Sect. 4, we explore the competitive ratios of $k$-CTP that can be achieved depending on the error of the predictor. Besides consistency and robustness, most works in this area classically study smooth error dependencies for the competitive ratio [5,31]. In this paper this is not the case, since the error is highly non-continuous. Our analysis contains both deterministic and randomized lower bounds complemented with matching upper bounds in almost all cases. These results are presented in Table 2 and justify the model of consistency-robustness tradeoff we chose in the previous section. All lower bounds are also valid if the parameter is $\kappa$. Note that for all upper bounds (except for $c^* = 2k + 1$) the parameter $k$ is considered to be known in advance. For the randomized case, the upper bound is given for path-disjoint graphs.

## 2    Preliminaries

We introduce two algorithms of the literature that are useful for our work and some notation we use in the rest of the technical sections.

### 2.1    Deterministic and Randomized Algorithms

As mentioned in the introduction, Westphal in [38] gave an optimal deterministic algorithm BACKTRACK for $k$-CTP, which is $(2k + 1)$-competitive (note that the algorithm does not need to know $k$).

BACKTRACK: An agent begins at source $s$ and follows the cheapest $s - t$ path on the graph. When the agent learns about a blocked edge on the path to $t$, he/she returns to $s$ and takes the cheapest $s - t$ path without the blocked edge discovered. The agent repeats this strategy until he/she arrives at $t$. Observe that he/she backtracks at most $k$ times, since there are no more than $k$ edges blocked, and thus BACKTRACK is $(2k + 1)$-competitive.

**Table 2.** Our bounds on the competitive ratio $c^*$ with respect to *error*, $k$. The upper bounds $2k+1$ and $k+1$ are also valid for *error* $\leq t$, for any $t \geq 2$.

| | | $k = 1$ | $k = 2$ | $k \geq 3$ |
|---|---|---|---|---|
| Deterministic algorithms | *error* $\leq 1$ | $c^* = 3$ <br> Theorem 10 | $c^* = \frac{3+\sqrt{17}}{2} \simeq 3.56$ <br> Theorems 11 and 12 | $c^* = 2k - 1$ <br> Theorems 8 and 9 |
| | *error* $\leq 2$ | $c^* = 2k + 1$ <br> Theorem 7 | | |
| Randomized algorithms | *error* $\leq 1$ | $c^* \geq k$ <br> Theorem 14 | | |
| | *error* $\leq 2$ | $c^* = k + 1$ <br> Theorem 13 | | |

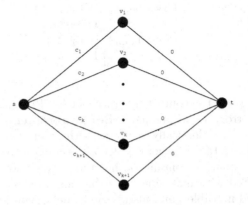

**Fig. 1.** The graph $\mathcal{G}^*$ for the proof of lower bounds.

Concerning randomized algorithm, the proof of the lower bound of $(k+1)$ [38], holding even if all $s - t$ paths are node-disjoint (besides $s$ and $t$), uses the graph in Fig. 1 with $c_1 = c_2 = \cdots = c_{k+1}$ (note that the lower bound holds when restricting to positive length by replacing 0 with $\epsilon > 0$). The matching upper bound of $(k + 1)$ [8,32], which is known to hold only when the paths are node-disjoint, is based on a randomized algorithm that we will call RANDBACKTRACK in the sequel. The very general idea of this algorithm, which can be seen as a randomized version of BACKTRACK, is the following.

RANDBACKTRACK: Consider the $k + 1$ shortest $s - t$ paths in the graph. The algorithm defines an appropriate probability distribution and then chooses a path according to this distribution that the agent tries to traverse. If this path is feasible, the algorithm terminates. If it is blocked, the agent returns to $s$ and repeats the procedure for a smaller set of paths.

## 2.2   Notations

For every edge $e \in E$, we get a prediction on whether $e$ is blocked or not. We define the error of the predictions as the total number of false predictions we have compared to the real instance. Formally every edge has prediction error $\text{ER}(e) \in \{0, 1\}$.

The total prediction error is given by:

$$error = \sum_{e \in E} \text{ER}(e)$$

For proving lower bounds, we will refer several times to the graph $\mathcal{G}^*$, which means the graph in Fig. 1. When we say that $P_i$ is blocked, we mean that the edge with cost 0 of $P_i$ is blocked. We omit these details in the proofs for ease of explanation. Moreover, when we refer to a path we always mean an $s - t$ path in the rest of the paper. Note that the lower bounds we get occur with strictly positive values on the edges, simply by replacing 0 with $\epsilon > 0$ in $\mathcal{G}^*$.

We denote by $OPT$ the optimal offline cost of the $k$-CTP instance, $ALG$ the value of an algorithm under study, and by $r$ the competitive ratio of an online algorithm to avoid any confusion with the cost $c$ of an edge on the graph. More specifically, $r = \frac{ALG}{OPT}$ in the deterministic case and $r = \frac{\mathbb{E}[ALG]}{OPT}$ in the randomized one.

# 3   Tradeoffs Between Consistency and Robustness

In this section, we study the tradeoffs between consistency and robustness. As explained before, we express tradeoffs by answering the following question: if we want our algorithm to be $(1 + \epsilon)$-competitive if the prediction is correct (consistency $1 + \epsilon$), what is the best competitive ratio we can get when the prediction is not correct (robustness)?

We deal with the deterministic case in Sect. 3.1 and the randomized one in Sect. 3.2.

## 3.1   Tradeoffs for Deterministic Algorithms

**Theorem 1.** *Any deterministic algorithm that achieves competitive ratio smaller than $1 + \epsilon$ when the prediction is correct, achieves a ratio of at least $2k - 1 + \frac{4k}{\epsilon}$ when the prediction is not correct, even when the error is at most 2 and the graph is path-disjoint.*

*Proof.* Consider a graph $\mathcal{G}^*$ with $k + 1$ paths $P_1$, $P_2$, ..., $P_k$, $P_{k+1}$, which are node-disjoint. The paths $P_1$, $P_2$, ..., $P_{k-1}$, $P_k$ have costs equal to 1 ($c_1 = c_2 = ... = c_k = 1$) and path $P_{k+1}$ has cost $c_{k+1} = \frac{2k}{\epsilon}$. $P_1$, $P_2$,...,$P_k$ are predicted to be blocked ($k$ predicted blocked edges) and $P_{k+1}$ is feasible.

If there is no error, the predicted instance is also the real one, $P_{k+1}$ is optimal. To get a competitive ratio smaller than $1 + \epsilon$ (consistency bound), a deterministic online algorithm cannot follow all paths $P_1, P_2, \ldots, P_k$ before exploring $P_{k+1}$ as the ratio would be $r = \frac{2c_1}{c_{k+1}} + \ldots + \frac{2c_k}{c_{k+1}} + 1 = 1 + \epsilon$. Therefore, $P_{k+1}$ is visited before at least one path $P_1, P_2, \ldots, P_k$.

When an adversary blocks $P_{k+1}$ and all the other paths visited by the algorithm except for the last one ($k$ blocks in total), it creates a new instance with $error = 2$. The optimal cost is 1 and the algorithm now has competitive ratio:

$$r = \frac{2(k-1) + 2c_{k+1} + 1}{1} = 2k - 1 + \frac{4k}{\epsilon}$$

Consequently, we have a Pareto lower bound $(1 + \epsilon, 2k - 1 + \frac{4k}{\epsilon}]$.    □

We now give an algorithm that matches the previous lower bound.

E-BACKTRACK is formally described in Algorithm 1. It basically executes BACKTRACK, but interrupts at some point its execution in order to explore the shortest unblocked-predicted path. The interruption point (determined by Eq. (1) in the description of the algorithm) is chosen sufficiently early to ensure good consistency and not too early to ensure good robustness.

---

**Algorithm 1: E-BACKTRACK**

    **Input**   : An instance of CTP with prediction with parameter $k$, $\epsilon > 0$
    **Output:** An $s - t$ path

1  $P_{pred}, c_{pred} \leftarrow$ shortest path and its cost after removing all predicted blocked edges
2  Execute BACKTRACK and explore paths $P_1, \ldots, P_j$, of cost $c_1, \ldots, c_j$ until one of the following cases occurs:
3      (a) $t$ is reached
4      (b) the next path $P_{j+1}$ to explore is such that:

$$2c_1 + 2c_2 + \ldots + 2c_j + 2c_{j+1} \geq \epsilon \cdot c_{pred} \qquad (1)$$

5  **if** *(a) occurs* **then** Return the found path;
6  **else**
7     |  Explore $P_{pred}$ (if not yet known to be blocked)
8     |  **if** $P_{pred}$ *is not blocked* **then** output it;
9     |  **else** Resume the execution of BACKTRACK;
10 **end**

---

**Theorem 2.** *For $0 < \epsilon \leq 2k$, E-BACKTRACK is a deterministic $(1 + \epsilon, 2k - 1 + \frac{4k}{\epsilon}]$-competitive algorithm.*

*Proof.* We denote by $ALG$ the cost of algorithm E-BACKTRACK.

Suppose first that case $(a)$ occurs. Then $j \leq k + 1$ as there are at most $k$ blocked edges, and $2c_1 + \cdots + 2c_j < \epsilon \cdot c_{pred}$ (otherwise case $(b)$ would have occurred earlier). In the case where the prediction is correct, $OPT = c_{pred}$ and, using $\epsilon \leq 2k$:

$$ALG \leq 2c_1 + \cdots + 2c_{j-1} + c_j < \epsilon \cdot c_{pred} < (1 + \epsilon)c_{pred}$$

If the prediction is not correct, then $OPT = c_j$ and:

$$ALG \leq 2c_1 + \cdots + 2c_{j-1} + c_j \leq (2k + 1)OPT \leq (2k - 1 + 4k/\epsilon)OPT$$

Suppose now that case $(b)$ occurs. As explained earlier $2c_1 + \cdots + 2c_j < \epsilon \cdot c_{pred}$. In the case where the prediction is correct, $OPT = c_{pred}$ and:

$$ALG \leq 2c_1 + \cdots + 2c_{j-1} + 2c_j + c_{pred} < (1 + \epsilon)c_{pred}$$

In the case where the prediction is not correct, if $P_{pred}$ were already known to be blocked, then we directly get a ratio $2k + 1 \leq 2k - 1 + 4k/\epsilon$. Otherwise, let $P_1, \ldots, P_j, P_{pred}, P_{j+1}, \ldots, P_t$ be the paths explored by E-BACKTRACK. As there are at most $k$ blocked edges, $t \leq k$ (note that the exploration of $P_{pred}$ does give a previously unknown blocked edge). Moreover, $c_1 \leq c_2 \leq \cdots \leq c_t = OPT$. We get:

$$ALG = 2\sum_{i=1}^{t-1} c_i + 2c_{pred} + c_t \leq (2k - 1)OPT + 2c_{pred} \tag{2}$$

Using (1) we know that $c_{pred} \leq 2\sum_{i=1}^{j+1} c_i/\epsilon \leq 2\sum_{i=1}^{t} c_i/\epsilon \leq 2kc_t/\epsilon$. Then Eq. (2) gives:

$$ALG \leq \left(2k - 1 + \frac{4k}{\epsilon}\right)OPT$$

$\square$

## 3.2   Randomized Bounds and Algorithms

We now consider the randomized case. As explained in the introduction, we restrict ourselves to the path-disjoint graphs for randomized algorithms.

**Theorem 3.** *Any randomized algorithm that achieves competitive ratio at most $1 + \epsilon$ when the prediction is correct, achieves a ratio of at least $k + \frac{k}{\epsilon}$ when the prediction is not correct, even when the error is at most 2 and the graph is path-disjoint.*

*Proof.* Consider a graph $\mathcal{G}^*$ with $k + 1$ paths $P_1, P_2, \ldots, P_k, P_{k+1}$, which are node-disjoint. The paths $P_1, P_2, \ldots, P_{k-1}, P_k$ have costs equal to 1 ($c_1 = c_2 = \ldots = c_k = 1$) and path $P_{k+1}$ has cost $c_{k+1} = \frac{k}{\epsilon}$. $P_1, P_2, \ldots, P_k$ are predicted to be blocked ($k$ predicted blocked edges) and $P_{k+1}$ is feasible.

   If there is no error, the predicted instance is also the real one, $P_{k+1}$ is optimal ($OPT = c_{k+1} = \frac{k}{\epsilon}$) and the competitive ratio must be at most $1 + \epsilon$. In the above instance, any deterministic algorithm can achieve one of the following competitive ratios:

- $r = 1$, when choosing only $P_{k+1}$.
- $r = \frac{2\epsilon}{k} + 1$, when choosing paths $P_i$, $P_{k+1}$ with $i \neq k + 1$.
- $r = \frac{4\epsilon}{k} + 1$, when choosing paths $P_i$, $P_j$, $P_{k+1}$ with $i, j \neq k + 1$ and $i \neq j$.
- $\ldots$
- $r = \frac{2k \cdot \epsilon}{k} + 1$, when choosing all $k$ paths $P_1, \ldots, P_k$ and then $P_{k+1}$.

A randomized algorithm can be viewed as a probability distribution over all deterministic algorithms. We assume that an arbitrary randomized algorithm chooses with cumulative probability $p_1$ the deterministic algorithms that achieve a ratio of 1 (here there is only one algorithm), with cumulative probability $p_2$ the deterministic algorithms that achieve a ratio of $\frac{2\epsilon}{k} + 1$, and so on. We also have that:

$$\sum_{i=1}^{k+1} p_i = 1 \tag{3}$$

Hence, the competitive ratio of an arbitrary randomized algorithm is:

$$r = p_1 \cdot 1 + p_2 \left( \frac{2\epsilon}{k} + 1 \right) + p_3 \left( \frac{4\epsilon}{k} + 1 \right) + \ldots + p_{k+1} \left( \frac{2k \cdot \epsilon}{k} + 1 \right)$$

Since $r \leq 1 + \epsilon$, we have that:

$$p_1 + p_2 \left( \frac{2\epsilon}{k} + 1 \right) + p_3 \left( \frac{4\epsilon}{k} + 1 \right) + \ldots + p_{k+1} \left( \frac{2k \cdot \epsilon}{k} + 1 \right) \leq 1 + \epsilon$$

$$\Rightarrow \sum_{i=1}^{k+1} p_i + p_2 \cdot \frac{2\epsilon}{k} + p_3 \cdot \frac{4\epsilon}{k} + \ldots + p_{k+1} \cdot \frac{2k \cdot \epsilon}{k} \leq 1 + \epsilon$$

From (3) it follows that:

$$p_2 \cdot \frac{2\epsilon}{k} + p_3 \cdot \frac{4\epsilon}{k} + \ldots + p_{k+1} \cdot \frac{2k \cdot \epsilon}{k} \leq \epsilon$$

$$\Rightarrow p_2 + 2p_3 + \ldots + k \cdot p_{k+1} \leq \frac{k}{2} \tag{4}$$

We now look at the case that the prediction is wrong. Consider the (randomized) set of instances where the path $P_i$ is unblocked, where $i$ is chosen uniformly at random in $\{1, \ldots, k\}$, and path $P_{k+1}$ is blocked. Note that these instances have $error = 2$. So, only path $P_i$ is feasible with $cost = 1$ and $OPT = 1$. Consider a deterministic algorithm which explores (until it finds an unblocked path) $\ell \geq 0$ paths among $P_1, \ldots, P_k$, then $P_{k+1}$, then the remaining paths among the first $k$. On the previously given randomized set of instances, it will explore $P_{k+1}$ with probability $(1-\ell/k)$, and will find the unblocked path after exactly $t$ explorations with probability $1/k$ (for any $t$). Thus, the expected cost of such an algorithm on the considered randomized set of instances is:

$$\mathbb{E}_\ell = \left(1 - \frac{\ell}{k}\right) 2c_{k+1} + \frac{(1 + 3 + \cdots + (2k - 1))}{k} = \left(1 - \frac{\ell}{k}\right) 2c_{k+1} + k$$

Then, the expected cost of the randomized algorithm (which chooses such an algorithm with probability $p_{\ell+1}$) on the given randomized set of instances verifies:

$$\mathbb{E}[ALG] \geq \sum_{\ell=0}^{k} p_{\ell+1} \mathbb{E}_\ell = \sum_{\ell=0}^{k} p_{\ell+1} \left(\left(1 - \frac{\ell}{k}\right) 2c_{k+1} + k\right)$$

$$= 2c_{k+1}\left(1 - \frac{\sum_{\ell=0}^{k} p_{\ell+1}\ell}{k}\right) + k$$

Equation (4) gives $\sum_{\ell=0}^{k} p_{\ell+1}\ell \leq k/2$, so we have $\mathbb{E}[ALG] \geq c_{k+1} + k = k + \frac{k}{\epsilon}$. $\qquad \square$

We now give a randomized algorithm that is $[1 + \epsilon, k + \frac{4k}{\epsilon}]$-competitive. Similarly as E-BACKTRACK, it executes RANDBACKTRACK but interrupts at some point (determined by Eq. (5)) its execution in order to explore the shortest unblocked-predicted path.

**Theorem 4.** *For $0 < \epsilon \leq k$, E-RANDBACKTRACK is a randomized $[1 + \epsilon, k + \frac{4k}{\epsilon}]$-competitive algorithm.*

*Proof.* We denote by $ALG$ the cost of algorithm E-RANDBACKTRACK and by $A_{k-1}$ the cost of RANDBACKTRACK (both $ALG$ and $A_{k-1}$ are random variables).

The proof is based on the following observations.

*Observation 1. At the time of the algorithm when (a), (b) or (c) occurs, $TVL \leq \epsilon \cdot c_{pred}$. In particular, if (a) or (b) occurs, $A_{k-1} \leq \epsilon \cdot c_{pred}$.*

Indeed, otherwise case $(c)$ would have occurred earlier.

*Observation 2. If $P_1, \ldots, P_k$ are blocked, then $P_{pred}$ is unblocked and optimal. In particular, if (b) occurs then $P_{pred}$ is unblocked and optimal.*

Indeed, if $P_1, \ldots, P_k$ are blocked, there is no other blocked path (as $k$ upper bounds the number of blocked paths) so $P_{pred}$ is unblocked. If the set of $k$ blocked edges are exactly the predicted ones, then $P_{pred}$ is by definition optimal. Otherwise, one path $P_i$ is not predicted to be blocked (as $k$ upper bounds the number of predicted blocked paths), hence $c_{pred} \leq c_i \leq c_k$. But by definition of paths $P_1, \ldots, P_k$, if they are all blocked then $OPT \geq c_k$, so again $OPT = c_{pred}$.

---

**Algorithm 2:** E-RANDBACKTRACK

---

    **Input**   : An instance of CTP with prediction with parameter $k$, $\epsilon > 0$
    **Output:** An $s - t$ path

1   $P_{pred}, c_{pred} \leftarrow$ shortest path and its cost after removing all predicted
     blocked edges

2   $P_1, \ldots, P_k$ of cost $c_1, \ldots, c_k \leftarrow k$ shortest paths except for $P_{pred}$ [2]

3   TVL $\leftarrow$ total visited length of RANDBACKTRACK before exploring the
     next path

4   Execute RANDBACKTRACK on paths $P_1, \ldots, P_k$ with parameter $k - 1$
     until one of the following cases occurs:

5       (a) $t$ is reached

6       (b) $t$ is not reached and RANDBACKTRACK terminates

7       (c) the next path $P_{new}$ of cost $c_{new}$ to explore is such that:

$$\text{TVL} + 2c_{new} > \epsilon \cdot c_{pred} \tag{5}$$

8   **if** *(a) occurs* **then** Return the found path;

9   **else if** *(b) occurs* **then**

10     |   Explore $P_{pred}$ and output it

11 **end**

12 **else**

13     |   Explore $P_{pred}$

14     |   **if** $P_{pred}$ *is not blocked* **then** output it;

15     |   **else** Resume the execution of RANDBACKTRACK;

16 **end**

---

*Observation 3. If (c) occurs, then $A_{k-1} > \frac{\epsilon \cdot c_{pred}}{2}$.*

Indeed, when (c) occurs RANDBACKTRACK has cost at least $TVL + c_{new} > \epsilon \cdot c_{pred}/2$.

Then, suppose first that $P_{pred}$ is unblocked and optimal. Following observation 1, if (a) or (b) occurs we have $ALG \leq A_{k-1} + c_{pred} \leq (1 + \epsilon)c_{pred}$, and in case (c) also $ALG \leq (1 + \epsilon)c_{pred}$. So anyway $ALG \leq (1 + \epsilon)OPT$, and in particular $\mathbb{E}[ALG] \leq (1 + \epsilon)OPT$. So E-RANDBACKTRACK is $(1 + \epsilon)$-competitive when there is no error. As $(1 + \epsilon) \leq k + 4k/\epsilon$ ($\epsilon \leq k$), the robustness bound is also verified in this case.

Now, suppose that we are in the other case, i.e., $P_{pred}$ is either blocked, or unblocked but not optimal. Note that the prediction is not correct here. Following Observation 2, (b) cannot occur, and one path in $P_1 \ldots, P_k$ is unblocked (so RANDBACKTRACK does find a path before terminating). Then in case (a) $ALG \leq A_{k-1}$, and in case (c), anyway, $ALG \leq A_{k-1} + 2c_{pred}$. So we get:

$$\mathbb{E}[ALG] \leq \mathbb{E}[A_{k-1}] + 2c_{pred} \cdot Pr(c) \tag{6}$$

---

[2] If the graph contains less than $k$ disjoint paths, then choose the maximum number
of paths $l < k$ and run RANDBACKTRACK with parameter $l - 1$. The analysis remains
the same.

where $Pr(c)$ denotes the probability that case $c$ occurs. Following Observation 3, if $(c)$ occurs $A_{k-1} > \epsilon \cdot c_{pred}/2$. Using Markov Inequality, we have:

$$Pr(c) \leq Pr\left(A_{k-1} > \frac{\epsilon \cdot c_{pred}}{2}\right) \leq \frac{2\mathbb{E}[A_{k-1}]}{\epsilon \cdot c_{pred}} \tag{7}$$

Using Eqs. (6) and (7) we get $\mathbb{E}[ALG] \leq \mathbb{E}[A_{k-1}] + \frac{4\mathbb{E}[A_{k-1}]}{\epsilon}$. As (at least) one path is unblocked in $P_1, \ldots, P_k$, $\mathbb{E}[A_{k-1}] \leq k \cdot OPT$, and the result follows. □

The guarantee provided by E-RANDBACKTRACK (Theorem 4) does not exactly match the lower bound of Theorem 3. While closing this gap is left as an open question, we conjecture that the exact (optimal) tradeoff corresponds to the lower bound $[1 + \epsilon, k + \frac{k}{\epsilon}]$ (for path-disjoint graphs). Towards this conjecture, we present two cases where we have an upper bound that matches the $[1 + \epsilon, k + \frac{k}{\epsilon}]$-lower bound. The two special cases (proofs omitted) are $k = 1$ (Theorem 5) and the case of uniform costs (Theorem 6).

For the case $k = 1$, the algorithm (full description omitted) is a modified version of RANDBACKTRACK with different probabilities, specifically settled exploiting the fact that there is (at most) one blocked edge and one predicted blocked edge.

For the case of uniform cost, the algorithm (full description omitted) explores at first a path with an appropriate probability. If the path is blocked, it then executes RANDBACKTRACK on the remaining graph.

**Theorem 5.** *For $0 < \epsilon \leq 1$, there exists a randomized $[1 + \epsilon, 1 + \frac{1}{\epsilon}]$-competitive algorithm when the graph is path-disjoint and $k = 1$.*

**Theorem 6.** *For $0 < \epsilon \leq k$, there exists a randomized $[1 + \epsilon, k + \frac{k}{\epsilon}]$-competitive algorithm when the graph is path-disjoint and the costs are uniform.*

## 4   Robustness Analysis

This section is devoted to the analysis of competitive ratios that can be achieved depending on the error made in the prediction. Section 4.1 deals with deterministic bounds, and Sect. 4.2 with randomized ones.

### 4.1   Deterministic Bounds

As a first result, we show that the lower bound of $2k + 1$ on (deterministic) competitive ratios still holds in our model with prediction even if the prediction error is (at most) 2 (proof omitted).

**Theorem 7.** *There is no deterministic algorithm that achieves competitive ratio smaller than $2k + 1$, even when the prediction has error at most 2 and the graph is path-disjoint.*

We can easily achieve a matching $2k + 1$ upper bound using the optimal deterministic algorithm BACKTRACK ignoring the predictions completely. Hence, the lower bound is tight.

We now consider the remaining case, when the error is (at most) 1. In this case we show that an improvement can be achieved with respect to the $2k + 1$ bound. More precisely, we first show in Theorem 8 that for any $k \geq 3$ a ratio $2k - 1$ can be achieved, and in Theorem 12 that a ratio $\frac{3+\sqrt{17}}{2} \simeq 3.56$ can be achieved for $k = 2$. We show in Theorems 9 and 11 respectively that these bounds are tight. Theorem 10 settles the case $k = 1$.

The two main ingredients of the claimed algorithm (Theorem 8) are (1) a careful comparison of the lengths of the shortest path and the shortest path without predicted blocked edges, to decide which one to explore, and (2) the fact that when the error is at most 1, if a blocked edge is discovered and was not predicted to be so, then we know exactly the set of blocked edges (i.e., the predicted ones and the new one) and thus we can determine directly the optimal solution without further testing.

**Theorem 8.** *There is a $(2k-1)$-competitive algorithm when the prediction error is at most 1 and $k \geq 3$ is known.*

We now show that the upper bound of $2k - 1$ (for $k \geq 3$) is tight (proof omitted).

**Theorem 9.** *There is no deterministic algorithm that achieves competitive ratio smaller than $2k - 1$, even when the prediction has error at most 1 and the graph is path-disjoint.*

We next examine separately the cases for $k = 1$ (Theorem 10, proof omitted) and $k = 2$.

**Theorem 10.** *When $k = 1$, there is no deterministic algorithm that achieves competitive ratio smaller than 3, even when the prediction has error at most 1 and the graph is path-disjoint.*

The previous 3-lower bound is clearly tight using the optimal deterministic algorithm BACKTRACK ($2 \cdot 1 + 1 = 3$). For $k = 2$, we have matching lower and upper bounds of $\frac{3+\sqrt{17}}{2} \simeq 3.56$ (the proof of the lower bound (Theorem 11) is omitted). The upper bound (Theorem 12) follows from an adaptation of ERR1-BACKTRACK for the case $k = 2$ (the description of Algorithm ERR1-BACKTRACK2 and the proof of the theorem are omitted).

**Theorem 11.** *When $k = 2$, there is no deterministic algorithm that achieves competitive ratio smaller than $\frac{3+\sqrt{17}}{2}$, even when the prediction has error at most 1 and the graph is path-disjoint.*

**Theorem 12.** *There exists a $\frac{3+\sqrt{17}}{2}$-competitive algorithm when the prediction error is at most 1 and $k = 2$.*

## 4.2   Randomized Bounds

We now consider randomized algorithms, and as explained in introduction we focus on graphs with node disjoint paths. Similarly as for the deterministic case, we first show that the lower bound of $k+1$ on (randomized) competitive ratios still holds in our model with prediction even if the prediction error is (at most) 2.

**Theorem 13.** *There is no randomized algorithm that achieves competitive ratio smaller than $k+1$ against an oblivious adversary, even when the prediction has error at most 2 and the graph is path-disjoint.*

*Proof.* In what follows we provide a randomized set of instances on which the expected cost of any deterministic algorithm is at least $k+1$ times the optimal cost. It follows from Yao's Principle [40] that the competitive ratio of any randomized algorithm is at least $k+1$.

Consider a graph $\mathcal{G}^*$ with $k+1$ paths $P_1, P_2, ..., P_k, P_{k+1}$, which are node-disjoint. All the paths have costs equal to 1, meaning that $c_1 = c_2 = ... = c_k = c_{k+1} = 1$.

Paths $P_1, P_2, ... , P_k$ are predicted to be blocked. We choose $i \in \{1, ..., k+1\}$ uniformly at random and block all paths $P_j$ with $j \neq i$ ($k$ blocked edges as all paths are node-disjoint). The prediction has an error of at most 2.

So, only the path $P_i$ is feasible at cost 1 and the optimal offline cost is 1. Furthermore, an arbitrary deterministic online algorithm finds path $P_i$ on the $l$th trial for $l = 1, ..., k+1$ with probability $\frac{1}{k+1}$.

If the algorithm is successful on its $l$th try, it incurs a cost of $2l - 1$, and thus it has an expected cost of at least $\frac{1}{k+1} \sum_{l=1}^{k+1} (2l - 1) = \frac{1}{k+1} \cdot (k+1)^2 = k+1$. $\square$

If $k$ is known and the graph is path-disjoint, then the above $k+1$ lower bound is tight using the optimal randomized online algorithm RANDBACKTRACK without predictions.

We finally consider the case of error (at most) 1. We show in Theorem 14 (proof omitted) a lower bound of $k$. We leave as an open question closing this gap between $k$ and $k+1$ when the error is at most 1 and $k \geq 2$. For the special case of $k = 1$, it is easy to show a matching lower bound of 2.

**Theorem 14.** *There is no randomized algorithm that achieves competitive ratio smaller than $k$ against an oblivious adversary, even when the prediction has error at most 1 and the graph is path-disjoint.*

**Acknowledgements.** This work was partially funded by the grant ANR-19-CE48-0016 from the French National Research Agency (ANR).

# References

1. Antoniadis, A., Coester, C., Eliás, M., Polak, A., Simon, B.: Online metric algorithms with untrusted predictions. In: Proceedings of the 37th International Conference on Machine Learning, ICML 2020, 13–18 July 2020, Virtual Event. Proceedings of Machine Learning Research, vol. 119, pp. 345–355. PMLR (2020). http://proceedings.mlr.press/v119/antoniadis20a.html

2. Antoniadis, A., Gouleakis, T., Kleer, P., Kolev, P.: Secretary and online matching problems with machine learned advice. In: Larochelle, H., Ranzato, M., Hadsell, R., Balcan, M., Lin, H. (eds.) Advances in Neural Information Processing Systems 33: Annual Conference on Neural Information Processing Systems 2020, NeurIPS 2020, 6–12 December 2020, Virtual (2020). https://proceedings.neurips.cc/paper/2020/hash/5a378f8490c8d6af8647a753812f6e31-Abstract.html

3. Azar, Y., Leonardi, S., Touitou, N.: Flow time scheduling with uncertain processing time. In: Khuller, S., Williams, V.V. (eds.) STOC 2021: 53rd Annual ACM SIGACT Symposium on Theory of Computing, Virtual Event, Italy, 21–25 June 2021, pp. 1070–1080. ACM (2021). https://doi.org/10.1145/3406325.3451023

4. Bamas, É., Maggiori, A., Rohwedder, L., Svensson, O.: Learning augmented energy minimization via speed scaling. In: Larochelle, H., Ranzato, M., Hadsell, R., Balcan, M., Lin, H. (eds.) Advances in Neural Information Processing Systems 33: Annual Conference on Neural Information Processing Systems 2020, NeurIPS 2020, 6–12 December 2020, Virtual (2020). https://proceedings.neurips.cc/paper/2020/hash/af94ed0d6f5acc95f97170e3685f16c0-Abstract.html

5. Bamas, É., Maggiori, A., Svensson, O.: The primal-dual method for learning augmented algorithms. In: Larochelle, H., Ranzato, M., Hadsell, R., Balcan, M., Lin, H. (eds.) Advances in Neural Information Processing Systems 33: Annual Conference on Neural Information Processing Systems 2020, NeurIPS 2020, 6–12 December 2020, Virtual (2020). https://proceedings.neurips.cc/paper/2020/hash/e834cb114d33f729dbc9c7fb0c6bb607-Abstract.html

6. Banerjee, S.: Improving online rent-or-buy algorithms with sequential decision making and ML predictions. In: Larochelle, H., Ranzato, M., Hadsell, R., Balcan, M., Lin, H. (eds.) Advances in Neural Information Processing Systems 33: Annual Conference on Neural Information Processing Systems 2020, NeurIPS 2020, 6–12 December 2020, Virtual (2020). https://proceedings.neurips.cc/paper/2020/hash/f12a6a7477077af66212ef0813bcf332-Abstract.html

7. Bar-Noy, A., Schieber, B.: The Canadian Traveller problem. In: Aggarwal, A. (ed.) Proceedings of the Second Annual ACM/SIGACT-SIAM Symposium on Discrete Algorithms, 28–30 January 1991, San Francisco, California, USA, pp. 261–270. ACM/SIAM (1991). http://dl.acm.org/citation.cfm?id=127787.127835

8. Bender, M., Westphal, S.: An optimal randomized online algorithm for the $k$-Canadian Traveller Problem on node-disjoint paths. J. Comb. Optim. **30**(1), 87–96 (2013). https://doi.org/10.1007/s10878-013-9634-8

9. Bergé, P., Hemery, J., Rimmel, A., Tomasik, J.: On the competitiveness of memoryless strategies for the $k$-Canadian Traveller Problem. In: International Conference Combinatorial Optimization and Applications (COCOA), Atlanta, United States, December 2018. https://hal.archives-ouvertes.fr/hal-02022459

10. Borodin, A., El-Yaniv, R.: Online Computation and Competitive Analysis. Cambridge University Press, Cambridge (1998)

11. Cole, R., Roughgarden, T.: The sample complexity of revenue maximization. In: Shmoys, D.B. (ed.) Symposium on Theory of Computing, STOC 2014, New York, NY, USA, 31 May–03 June 2014, pp. 243–252. ACM (2014). https://doi.org/10.1145/2591796.2591867

12. Demaine, E.D., Huang, Y., Liao, C.-S., Sadakane, K.: Approximating the Canadian Traveller problem with online randomization. Algorithmica **83**(5), 1524–1543 (2021). https://doi.org/10.1007/s00453-020-00792-6

13. Devanur, N.R., Hayes, T.P.: The adwords problem: online keyword matching with budgeted bidders under random permutations. In: Chuang, J., Fortnow, L., Pu, P. (eds.) Proceedings 10th ACM Conference on Electronic Commerce (EC-2009), Stanford, California, USA, 6–10 July 2009, pp. 71–78. ACM (2009). https://doi.org/10.1145/1566374.1566384

14. Ergun, J.C., Feng, Z., Silwal, S., Woodruff, D.P., Zhou, S.: Learning-augmented $k$-means clustering (2021). https://doi.org/10.48550/ARXIV.2110.14094. https://arxiv.org/abs/2110.14094

15. Gollapudi, S., Panigrahi, D.: Online algorithms for rent-or-buy with expert advice. In: Chaudhuri, K., Salakhutdinov, R. (eds.) Proceedings of the 36th International Conference on Machine Learning, ICML 2019, 9–15 June 2019, Long Beach, California, USA. Proceedings of Machine Learning Research, vol. 97, pp. 2319–2327. PMLR (2019). http://proceedings.mlr.press/v97/gollapudi19a.html

16. Huang, Y., Liao, C.-S.: The Canadian Traveller problem revisited. In: Chao, K.-M., Hsu, T., Lee, D.-T. (eds.) ISAAC 2012. LNCS, vol. 7676, pp. 352–361. Springer, Heidelberg (2012). https://doi.org/10.1007/978-3-642-35261-4_38

17. Im, S., Kumar, R., Qaem, M.M., Purohit, M.: Non-clairvoyant scheduling with predictions. In: Agrawal, K., Azar, Y. (eds.) SPAA 2021: 33rd ACM Symposium on Parallelism in Algorithms and Architectures, Virtual Event, USA, 6–8 July 2021, pp. 285–294. ACM (2021). https://doi.org/10.1145/3409964.3461790

18. Lattanzi, S., Lavastida, T., Moseley, B., Vassilvitskii, S.: Online scheduling via learned weights. In: Chawla, S. (ed.) Proceedings of the 2020 ACM-SIAM Symposium on Discrete Algorithms, SODA 2020, Salt Lake City, UT, USA, 5–8 January 2020, pp. 1859–1877. SIAM (2020). https://doi.org/10.1137/1.9781611975994.114

19. Lavastida, T., Moseley, B., Ravi, R., Xu, C.: Learnable and instance-robust predictions for online matching, flows and load balancing. In: Mutzel, P., Pagh, R., Herman, G. (eds.) 29th Annual European Symposium on Algorithms, ESA 2021, 6–8 September 2021, Lisbon, Portugal (Virtual Conference). LIPIcs, vol. 204, pp. 59:1–59:17. Schloss Dagstuhl - Leibniz-Zentrum für Informatik (2021). https://doi.org/10.4230/LIPIcs.ESA.2021.59

20. Lawler, E.: Combinatorial Optimization: Networks and Matroids. Holt Rinehart and Winston, New York (1977)

21. Lindermayr, A., Megow, N., Simon, B.: Double coverage with machine-learned advice. In: Braverman, M. (ed.) 13th Innovations in Theoretical Computer Science Conference, ITCS 2022, 31 January–3 February 2022, Berkeley, CA, USA. LIPIcs, vol. 215, pp. 99:1–99:18. Schloss Dagstuhl - Leibniz-Zentrum für Informatik (2022). https://doi.org/10.4230/LIPIcs.ITCS.2022.99

22. Lu, P., Ren, X., Sun, E., Zhang, Y.: Generalized sorting with predictions. In: Le, H.V., King, V. (eds.) 4th Symposium on Simplicity in Algorithms, SOSA 2021, Virtual Conference, 11–12 January 2021, pp. 111–117. SIAM (2021). https://doi.org/10.1137/1.9781611976496.13

23. Lykouris, T., Vassilvitskii, S.: Competitive caching with machine learned advice. In: Dy, J.G., Krause, A. (eds.) Proceedings of the 35th International Conference on Machine Learning, ICML 2018, Stockholmsmässan, Stockholm, Sweden, 10–15 July 2018. Proceedings of Machine Learning Research, vol. 80, pp. 3302–3311. PMLR (2018). http://proceedings.mlr.press/v80/lykouris18a.html

24. Medina, A.M., Vassilvitskii, S.: Revenue optimization with approximate bid predictions. CoRR abs/1706.04732 (2017). http://arxiv.org/abs/1706.04732

25. Mitzenmacher, M.: Scheduling with predictions and the price of misprediction. In: Vidick, T. (ed.) 11th Innovations in Theoretical Computer Science Conference, ITCS 2020, 12–14 January 2020, Seattle, Washington, USA. LIPIcs, vol. 151, pp. 14:1–14:18. Schloss Dagstuhl - Leibniz-Zentrum für Informatik (2020). https://doi.org/10.4230/LIPIcs.ITCS.2020.14

26. Mitzenmacher, M., Vassilvitskii, S.: Algorithms with predictions. In: Roughgarden, T. (ed.) Beyond the Worst-Case Analysis of Algorithms, pp. 646–662. Cambridge University Press, Cambridge (2020). https://doi.org/10.1017/9781108637435.037

27. Nikolova, E., Karger, D.R.: Route planning under uncertainty: the Canadian Traveller problem. In: Fox, D., Gomes, C.P. (eds.) Proceedings of the Twenty-Third AAAI Conference on Artificial Intelligence, AAAI 2008, Chicago, Illinois, USA, 13–17 July 2008, pp. 969–974. AAAI Press (2008). http://www.aaai.org/Library/AAAI/2008/aaai08-154.php

28. Papadimitriou, C.H., Steiglitz, K.: Combinatorial Optimization: Algorithms and Complexity. Prentice-Hall, Hoboken (1982)

29. Papadimitriou, C.H., Yannakakis, M.: Shortest paths without a map. Theor. Comput. Sci. **84**(1), 127–150 (1991). https://doi.org/10.1016/0304-3975(91)90263-2

30. Purohit, M., Svitkina, Z., Kumar, R.: Improving online algorithms via ML predictions. In: NeurIPS, Montréal, Canada, pp. 9684–9693 (2018). https://proceedings.neurips.cc/paper/2018/hash/73a427badebe0e32caa2e1fc7530b7f3-Abstract.html

31. Rohatgi, D.: Near-optimal bounds for online caching with machine learned advice. In: Chawla, S. (ed.) Proceedings of the 2020 ACM-SIAM Symposium on Discrete Algorithms, SODA 2020, Salt Lake City, UT, USA, 5–8 January 2020, pp. 1834–1845. SIAM (2020). https://doi.org/10.1137/1.9781611975994.112

32. Shiri, D., Salman, F.S.: On the randomized online strategies for the k-Canadian traveler problem. J. Comb. Optim. **38**, 254–267 (2019)

33. Sleator, D.D., Tarjan, R.E.: Amortized efficiency of list update and paging rules. Commun. ACM **28**(2), 202–208 (1985). https://doi.org/10.1145/2786.2793

34. Vee, E., Vassilvitskii, S., Shanmugasundaram, J.: Optimal online assignment with forecasts. In: Parkes, D.C., Dellarocas, C., Tennenholtz, M. (eds.) Proceedings 11th ACM Conference on Electronic Commerce (EC-2010), Cambridge, Massachusetts, USA, 7–11 June 2010, pp. 109–118. ACM (2010). https://doi.org/10.1145/1807342.1807360

35. Wang, S., Li, J., Wang, S.: Online algorithms for multi-shop ski rental with machine learned advice. In: Larochelle, H., Ranzato, M., Hadsell, R., Balcan, M., Lin, H. (eds.) Advances in Neural Information Processing Systems 33: Annual Conference on Neural Information Processing Systems 2020, NeurIPS 2020, 6–12 December 2020, Virtual (2020). https://proceedings.neurips.cc/paper/2020/hash/5cc4bb753030a3d804351b2dfec0d8b5-Abstract.html

36. Wei, A.: Better and simpler learning-augmented online caching. In: Byrka, J., Meka, R. (eds.) Approximation, Randomization, and Combinatorial Optimization. Algorithms and Techniques (APPROX/RANDOM 2020). Leibniz International Proceedings in Informatics (LIPIcs), vol. 176, pp. 60:1–60:17. Schloss Dagstuhl-Leibniz-Zentrum für Informatik, Dagstuhl, Germany (2020). https://doi.org/10.4230/LIPIcs.APPROX/RANDOM.2020.60. https://drops.dagstuhl.de/opus/volltexte/2020/12663

37. Wei, A., Zhang, F.: Optimal robustness-consistency trade-offs for learning-augmented online algorithms. In: Larochelle, H., Ranzato, M., Hadsell, R., Balcan, M., Lin, H. (eds.) Advances in Neural Information Processing Systems 33: Annual Conference on Neural Information Processing Systems 2020, NeurIPS 2020, 6–12 December 2020, Virtual (2020). https://proceedings.neurips.cc/paper/2020/hash/5bd844f11fa520d54fa5edec06ea2507-Abstract.html
38. Westphal, S.: A note on the k-Canadian Traveller problem. Inf. Process. Lett. **106**(3), 87–89 (2008). https://www.sciencedirect.com/science/article/pii/S0020019007002876
39. Xu, Y., Hu, M., Su, B., Zhu, B., Zhu, Z.: The Canadian Traveller problem and its competitive analysis. J. Comb. Optim. **18**(2), 195–205 (2009). https://doi.org/10.1007/s10878-008-9156-y
40. Yao, A.C.C.: Probabilistic computations: toward a unified measure of complexity. In: 18th Annual Symposium on Foundations of Computer Science (SFCS 1977), pp. 222–227 (1977)

# The Power of Amortized Recourse for Online Graph Problems

Alison Hsiang-Hsuan Liu$^{(\boxtimes)}$ (ID) and Jonathan Toole-Charignon (ID)

Department of Information and computing sciences, Utrecht University,
Utrecht, The Netherlands
{h.h.liu,j.c.f.toole-charignon}@uu.nl

**Abstract.** In this work, we study online graph problems with monotone-sum objectives, where the vertices or edges of the graph are revealed one by one and need to be assigned to a value such that certain properties of the solution hold. We propose a general two-fold greedy algorithm that augments its current solution greedily and references yardstick algorithms. The algorithm maintains competitiveness by strategically aligning to the yardstick solution and incurring recourse. We show that our general algorithm achieves $t$-competitiveness while incurring at most $\frac{w_{\max} \cdot (t+1)}{t-1}$ amortized recourse for any monotone-sum problems with integral solution, where $w_{\max}$ is the largest value that can be assigned to a vertex or an edge. For fractional monotone-sum problems where each of the assigned values is between $[0,1]$, our general algorithm incurs at most $\frac{t+1}{w_{\min} \cdot (t-1)}$ amortized recourse, where $w_{\min}$ is the smallest non-negative value that can be assigned. We further show that the general algorithm can be improved for three classical graph problems. For INDEPENDENT SET, we refine the analysis of our general algorithm and show that $t$-competitiveness can be achieved with $\frac{t}{t-1}$ amortized recourse. For MAXIMUM CARDINALITY MATCHING, we limit our algorithm's greed to show that $t$-competitiveness can be achieved with $\frac{(2-t^*)}{(t^*-1)(3-t^*)} + \frac{t^*-1}{3-t^*}$ amortized recourse, where $t^*$ is the largest number such that $t^* = 1 + \frac{1}{j} \leq t$ for some integer $j$. For VERTEX COVER, we show that our algorithm guarantees a competitive ratio strictly smaller than 2 for any finite instance in polynomial time while incurring at most 3.33 amortized recourse. We beat the almost unbreakable 2-approximation in polynomial time by using the optimal solution as the reference without computing it. We remark that this online result can be used as an offline approximation result (without violating the unique games conjecture [20]) to partially improve upon the constructive algorithm of Monien and Speckenmeyer [23].

## 1 Introduction

Graph optimization problems serve as stems for various practical problems. A solution for such a problem can be described as an assignment from the elements of the problem (e.g. vertices of a graph) to non-negative real numbers such that the constraints between the elements are satisfied. In the online setting, the

© The Author(s), under exclusive license to Springer Nature Switzerland AG 2022
P. Chalermsook and B. Laekhanukit (Eds.): WAOA 2022, LNCS 13538, pp. 134–153, 2022.
https://doi.org/10.1007/978-3-031-18367-6_7

most considered models are the *vertex-arrival* and *edge-arrival* models. That is, the graph is revealed vertex-by-vertex or edge-by-edge, and once an element arrives, the *online algorithm* has to immediately make an irrevocable decision on the new element. The performance of an online algorithm is measured by *competitive ratio* against the optimal offline solution. Many graph optimization problems are non-competitive: the larger the input size, the larger the competitive ratio of any deterministic online algorithm. In other words, a non-competitive problem has no constant-competitive online algorithm.

The pure online model is pessimistic, in that altering decisions may be possible (albeit expensive) or limited knowledge about the future may be available in the real world. In this work, we investigate online graph optimization problems in the *recourse* model. That is, decisions made by the online algorithm can be revoked. In particular, we aim at finding out the amount of amortized recourse that is sufficient and/or necessary for attaining a desirable competitive ratio for a given problem.

**Uncertainty and Amortized Recourse.** The competitive ratio can be seen as quantification of how far the quality of an online algorithm's solution is from that of a conceptual optimal offline algorithm that has complete knowledge of the input and unlimited computational power. Therefore, the non-competitiveness of graph optimization problems suggests that uncertainty of the input is critical to these problems. However, the online algorithm may perform better when the irrevocability constraint is relaxed or knowledge about future inputs is available. It is intriguing to investigate to what extent these problems remain non-competitive under these conditions, in particular to determine how much revocability or knowledge the online algorithm needs in order to attain a desirable competitive ratio.

Beyond the practical motivation of relaxing irrevocability of online algorithms' decisions, amortized recourse also provides insight on how a given online problem is affected by uncertainty. In particular, it captures how rapidly the structure of the offline optimal solution can change: the fewer elements required to do so, the larger the amortized recourse. Furthermore, the impact of uncertainty is directly correlated with this idea: the faster the optimal solution can change, the more impact uncertainty on future inputs will have. Different problems may attain constant competitive ratios using different amounts of (amortized) recourse, which implies variability in the impact of uncertainty. For example, to attain a constant competitive ratio, one needs exactly $O(\log n)$ recourse per edge for min-cost bipartite matching [21], while one only needs a constant amount of recourse per element for maximum independent set and minimum vertex cover [10].

**Online Monotone-Sum Problems.** We study *online* graph problems in the vertex-arrival or the edge-arrival models. Along with the newly-revealed element, which can be a vertex or an edge according to the arrival model, there may be constraints imposed upon some subset of the currently-revealed elements that a feasible solution should satisfy. An algorithm aims at finding a feasible solution that maximizes (or minimizes) the objective. A problem is a *sum* problem if the objective is a sum of the values assigned to each element. If the value of the optimal solution of an instance is always greater than or equal to that of a subset of the instance, then the problem is a *monotone* problem.[1]

An online algorithm makes decisions upon arrival of each element. In the *recourse* model, the online algorithm can also revoke an earlier decision that it made and pay for the revocation. We aim to reduce the competitive ratio with as little total recourse (i.e. as few revocations) as possible.

**Our Contribution.** We propose a general online algorithm Target-and-Switch ($\mathsf{TaS}_t$), which is parameterized by a target competitive ratio $t$ and uses a yardstick algorithm as a reference. The yardstick algorithm can be ant exact optimal algorithm or an incremental (defined later) approximation algorithm if one aims at polynomial time online algorithms. Throughout the process, the $\mathsf{TaS}_t$ algorithm compares itself to the yardstick algorithm's solution and strategically switches its solution to that of the yardstick algorithm. Overall, the $\mathsf{TaS}_t$ algorithm provides a trade-off between amortized recourse and competitive ratio for arbitrary monotone-sum graph problems. In particular, we consider two measurements of recourse cost: number of reassigned elements, or the amount of change in the reassigned values. Our result works for both unweighted and weighted problems, and it even works for fractional optimization problems, where the smallest non-zero value assigned to a single element can be a real number between 0 and 1. The following is the main result of our work, where the bound of amortized recourse works for both measurements of recourse cost (Theorem 1 and Corollary 1).

*Main result (informal). Using an optimal algorithm (resp. an incremental $\alpha$-approximation algorithm, defined formally in Sect. 2) as the yardstick, $\mathsf{TaS}_t$ is $t$-competitive (resp. $(t \cdot \alpha)$-competitive) and incurs at most $\frac{w_{max} \cdot (t+1)}{\min\{1, w_{min}\} \cdot (t-1)}$ amortized recourse for any monotone-sum graph problem where $w_{max}$ and $w_{min}$ are the maximum and minimum non-zero values that can be assigned to an element.[2]*

$\mathsf{TaS}_t$ is two-fold greedy. First, it assigns the value *greedily* once an element arrives. Second, the algorithm aligns its solution to the yardstick solution *completely* and incurs recourse when the current solution fails to be $t$-competitive against the yardstick solution.

In general, the $\mathsf{TaS}_t$ algorithm works for any optimization problem. The challenge is to bound the amortized recourse that it incurs, as the complete alignment may require a vast amount of recourse. By looking closer at a specific problem, we can show a tighter bound on the amount of recourse needed. We use a sophisticated analysis for the INDEPENDENT SET problem and improve the recourse bound (Theorem 2).

---

[1] The DOMINATING SET and MATCHING WITH DELAYS problems are sum problems but not monotone. The COLORING PROBLEM is monotone but not a sum problem.

[2] The bound of amortized recourse $\frac{w_{max} \cdot (t+1)}{w_{min} \cdot (t-1)}$ is larger when the elements can be assigned minimum non-zero values smaller than 1. For example, the fractional VERTEX COVER problem in [24].

The two-fold greedy algorithm may perform better when the greediness is relaxed. Moreover, by choosing different yardstick algorithms and tuning the alignment to the yardstick carefully, the amortized recourse can be further reduced. We show that for the MAXIMUM CARDINALITY MATCHING problem, partially aligning to the yardstick solution is more recourse-efficient (Theorem 5).

For the VERTEX COVER problem, we show that a special version of $\text{TaS}_t$ with $t = 2 - \frac{2}{\text{OPT}}$ incurs a very small amount of amortized recourse (Theorem 8) and is $(2 - \frac{2}{\text{OPT}})$-competitive, where OPT is the size of the optimal vertex cover[3] (Theorem 7). Our algorithm uses an optimal solution as a yardstick. The key to the polynomial time complexity is that instead of explicitly finding the yardstick assignment, we show that the yardstick cannot be too "far" from our solution at any moment if the target competitive ratio $2 - \frac{2}{\text{OPT}}$ is not already achieved. More specifically, by restricting the range of greedy choice, we can show that the yardstick solution can be aligned partially within a constant amount of amortized recourse. Thus, our result breaks the almost unbreakable 2-approximation for the VERTEX COVER problem and improves upon that of Monien and Specken-meyer [23] for a subset of the graphs containing odd cycles of length no less than $2k+3$ (for which $2 - \frac{2}{\text{OPT}} < 2 - \frac{1}{k+1}$), using an algorithm that is also constructive.

Our results are summarized in Table 1, which illustrates the power of amortization.

**Table 1.** Summary of our results. Note that $t$ can be any real number larger than 1. For Maximum Matching, $t^*$ is the largest number such that $t^* \leq t$ and $t^* = 1 + \frac{1}{j}$ for some integer $j$. The note P means that the algorithm is a polynomial-time online algorithm.

| | (Competitive ratio, worst case recourse) | (Competitive ratio, amortized recourse) |
|---|---|---|
| Monotone-sum problems | | $(t\alpha, \frac{w_{\max} \cdot (t+1)}{\min\{w_{\min}, 1\} \cdot (t-1)})$ (Theorem 1, Corollary 1, P with incremental $\alpha$-approximation algorithms) |
| Maximum independent set | $(2.598, 2)$ [10] | $(t, \frac{t}{t-1})$ (Theorem 2) $(2.598, 1.626)$ (Theorem 2) |
| Maximum matching | $(k, O(\frac{\log k}{k}) + 1))$ [2] $(1.5, 2)$ [10] | $(t, \frac{(2-t^*)}{(t^*-1)(3-t^*)} + \frac{t^*-1}{3-t^*})$ (Theorem 5, P) $((1.5, 1)$ with $t^* = 1.5)$ |
| Minimum vertex cover | $(2, 1)$ [10] | $(2 - \frac{2}{\text{OPT}}, \frac{10}{3})$ (Theorem 8, P) |

**Related Work.** Typically, online graph problems such as maximum independent set, maximum cardinality matching, minimum vertex cover, minimum dominating set, can be modeled as inclusion/exclusion problems. In these problems, any individual element (vertex or edge as appropriate) is either included in or

---

[3] Note that over all instances, OPT can be arbitrarily large. Thus, there is no $\varepsilon > 0$ for which $2 - \frac{2}{\text{OPT}} \leq 2 - \varepsilon$ over all instances. Therefore, our result does not violate the unique games conjecture [20].

excluded from the solution. In our terminology, the inclusion and exclusion of an element is assigning values 1 and 0 to it, respectively. Recourse is incurred each time the inclusion/exclusion status of an element is changed. The min-cost matching, Steiner tree, facility location, and routing problems are also inclusion/exclusion problems, but the cost of an edge/vertices inclusion is weighted. That is, the edges/vertices are assigned values 0 or 1, and the cost incurred by a value-1 edge/vertex is its weight. The scheme also works for non-monotone sum problems. For example, in coloring and bin-packing problems, the assignment of vertices/items to colors/bins can be modeled as assigning the vertices/items non-negative integral numbers, which are the indices of the colors/bins.

The closest previous result is the work by Boyar et al. [10]. The authors investigated the INDEPENDENT SET, MAXIMUM CARDINALITY MATCHING, VERTEX COVER, and MINIMUM SPANNING FOREST problems, which are all non-competitive in the pure online model. The authors showed that the competitive ratio of these problems can be massively reduced to a constant by incurring at most 2 recourse for any single element. Note that the bounds of the worst case recourse are upper bounds of the amortized recourse. Moreover, the algorithms in [10] incur at least 1.5 amortized recourse for the MAXIMUM CARDINALITY MATCHING problem and at least 0.5 amortized recourse for the VERTEX COVER problem.

There is a line of research on online matching problems with recourse. Angelopoulos et al. [2] studied a more general setting for MAXIMAL CARDINALITY MATCHING and showed that given that no element incurs more than $k$ recourse, there exists an algorithm that attains a competitive ratio of $1 + O(1/\sqrt{k})$. Megow and Nölke [21] showed that for the MIN-COST BIPARTITE MATCHING problem, constant competitiveness is achievable with amortized recourse $O(\log n)$, where $n$ is the number of requests. Bernstein et al. [6] showed that there exists an algorithm that achieves 1-competitiveness with $O(\log^2 n)$ amortized recourse for the BIPARTITE MATCHING problem, where $n$ is the number of vertices inserted. The result also shows that to achieve 1-competitiveness for VERTEX COVER, any online algorithm needs at least $\Omega(n)$ amortized recourse per vertex.

In addition, there has been extensive work on online algorithms in the recourse model for a variety of different problems. For amortized recourse, studied problems include online bipartite matching [6], graph coloring [9], minimum spanning tree and traveling salesperson [22], Steiner tree [12], online facility location [11], bin packing [13], submodular covering [14], and constrained optimization [3].

Graph problems model various real-world issues whose performance guarantees are often abysmal, as they are notoriously non-competitive in the pure online model. Prior work has shown curiosity about the conditions under which these problems become competitive, and these problems have been investigated under different models out of both practical and theoretical interests. Other than the recourse model, considered models include paying for a delay in the timing of decision making to achieve a better solution [5,7]. Another model for delayed decision making is the reordering buffer model [1], where the online algorithm can delay up to $k$ decisions by storing the elements in a size-$k$ buffer.

The impact of extra knowledge about the input has also been studied. For example, once a vertex arrives, the neighborhood is known to the algorithm [16]. In the lookahead model, an online algorithm is capable of foreseeing the next events [1]. Predictions provided by machine learning are also considered for graph problems [4]. Finally, there are also works where the integral assignment restrictions are relaxed for vertex cover and matching problems [24].

Another major area of related work for practically any problem considered in the online model is polynomial-time approximation algorithms for the equivalent problem in the offline setting. The link between the two is particularly salient when considering a polynomial-time online algorithm, as this online algorithm can also be run in polynomial time in the offline setting by processing the graph as if it were revealed in an online manner.

In the case of minimum vertex cover, assuming the unique games conjecture, it is not possible to obtain an approximation factor of $(2 - \varepsilon)$ for fixed $\varepsilon > 0$ [20]. However, results have been obtained for parameterized $\varepsilon$. In particular, Halperin [15] showed an approximation factor of $2 - (1 - o(1))\frac{2\ln\ln\Delta}{\ln\Delta}$ on graphs with maximum degree $\Delta$, and Karakostas [19] showed an approximation factor of $2 - \theta(\frac{1}{\sqrt{\log n}})$. Both of these results use semidefinite relaxations of the problem, whereas Monien and Speckenmeyer [23] had previously used a constructive approach to show an approximation factor of $2 - \frac{1}{k+1}$ for graphs without odd cycles of length at most $2k + 1$.

**Paper Organization.** Section 2 defines monotone-sum graph problems and the amortized recourse model. We propose a general algorithm $\mathsf{TaS}_t$ for finding the trade-off between the desired competitive ratio and the amortized recourse needed. Section 3 provides a refined analysis on the $\mathsf{TaS}_t$ algorithm on the INDEPENDENT SET problem. Section 4 discusses an existing algorithm [2], which is a variant of $\mathsf{TaS}_t$ algorithm, for the MAXIMUM CARDINALITY MATCHING problem that is less greedy in aligning its solution and obtains a better trade-off. Section 5 introduces a polynomial-time version of $\mathsf{TaS}_t$ algorithm for the VERTEX COVER problem that limits both greedy aspects. This algorithm can also be used as a novel offline approximation algorithm for certain graph classes. Due to space constraints, we only provide proof ideas for the theorem. The detailed proof for all lemmas and theorems can be found in the full version of this paper.

## 2   Monotone-Sum Graph Problems and a General Algorithm

For an *online graph problem* $Q$ on a graph $G = (V, E)$, which is unknown a priori, we consider either the *vertex-arrival* model or the *edge-arrival* model. In the vertex-arrival model (resp. edge-arrival model), the elements in $V$ (resp. elements in $E$) arrive one at a time, and an algorithm has to assign each element a non-negative value in $[0, w_{\max}]$ such that the assignment satisfies certain properties associated with $Q$. Formally, the *assignment* is defined as $\mathcal{A} : \mathcal{X} \to \mathbb{R}^+$, where

$\mathcal{X}$ is $V$ or $E$, such that $\mathcal{A}(\mathcal{X})$ satisfies a set of properties $\mathcal{P}_Q$. The *value* of a feasible assignment $\mathcal{A}$ is defined as a function *value* : $\mathcal{X} \times \mathcal{A}(\mathcal{X}) \to \mathbb{R}^+$, which should be minimized or maximized as appropriate. In this work, we focus on the problems with *sum* objectives, that is, $value(\mathcal{X}, \mathcal{A}(\mathcal{X})) = \sum_{x \in \mathcal{X}} \mathcal{A}(x)$. Moreover, we concern ourselves about the impact of lacking information on the optimality of the solution. Therefore, we consider monotone sum graph problems where given a feasible assignment and a newly-arrived element, there is always a value in $[0, w_{\max}]$ that can be assigned to the new element such that the new assignment is feasible.[4]

We denote the assignment on input $\mathcal{X}$ returned by the algorithm ALG by ALG($\mathcal{X}$). We abuse the notation $\mathcal{X}$ to denote the graph revealed by the input $\mathcal{X}$. We further abuse notation and denote the total value of the assignment by ALG($\mathcal{X}$) as well. That is, ALG($\mathcal{X}$) $= \sum_{x_i \in \mathcal{X}}$ ALG($x_i$). When the context is clear, the parameter $\mathcal{X}$ is dropped.

We study the family of monotone-sum graph problems, which is defined as follows. Similarly, we define the family of incremental algorithms. Note that a monotone-sum problem can be a maximization or a minimization problem.

**Definition 1.** *The* projection *of an assignment $\mathcal{A}(G)$ on an induced subgraph $H$ of $G$ assigns to each element in $H$ the same value that $\mathcal{A}(G)$ does in $G$.*

**Definition 2. Monotone-sum graph problems.** *A* sum problem *is* monotone *if for any graph $G$ and any induced subgraph $H$ of $G$, 1) the projection of any feasible assignment $\mathcal{A}(G)$ on $H$ is also feasible, and 2) OPT$(H) \leq$ OPT$(G)$, where OPT is an optimal solution.*

**Definition 3. Incremental algorithms.** *An algorithm ALG is incremental if for any graph $G$ corresponding to the instance $\mathcal{X}$ and any induced subgraph $H$ of $G$, ALG$(H) \leq$ ALG$(G)$. Furthermore, the projection of ALG$(\mathcal{X})$ on a prefix $\mathcal{X}'$ of instance $\mathcal{X}$ does not have a better objective value than the assignment ALG$(\mathcal{X}')$.*[5]

In this work, the performance of an online algorithm is measured by the *competitive ratio*. An online algorithm ALG attains a competitive ratio of $t$ if $\max\{\frac{\text{ALG}(\mathcal{X})}{\text{OPT}(\mathcal{X})}, \frac{\text{OPT}(\mathcal{X})}{\text{ALG}(\mathcal{X})}\} \leq t$ for any instance $\mathcal{X}$, where OPT is the optimal offline algorithm that knows all information necessary for solving the problem. In the recourse model, the online algorithm can revoke its decisions and incurs *recourse cost*. There are two types of recourse cost considered in this paper:

- **Type-1:** The recourse cost is defined as the *number* of elements which assignment values are changed. Formally, $\sum_{x_i \in \mathcal{X}} \mathbb{1}[A_1(x_i) \neq A_2(x_i)]$ when an assignment on instance $\mathcal{X}$ is changed from $A_1(\mathcal{X})$ to $A_2(\mathcal{X})$.
- **Type-2:** The recourse cost is defined as the *amount of change* of the assignment value. Formally, $\sum_{x_i \in \mathcal{X}} |A_1(x_i) - A_2(x_i)|$ when an assignment on instance $\mathcal{X}$ is changed from $A_1(\mathcal{X})$ to $A_2(\mathcal{X})$.

---

[4] Classical graph problems such as INDEPENDENT SET, MAXIMUM CARDINALITY MATCHING, and VERTEX COVER all satisfy this property.

[5] For example, the Ramsey algorithm in [8] is an incremental algorithm. Also note that any online algorithm is an incremental algorithm.

We study the trade-off between the competitive ratio and the *amortized recourse*. That is, the total incurred recourse cost divided by the number of elements that should be assigned a value in the final instance. We define a family of algorithms for monotone-sum problems.

**Target-and-Switch ($\mathtt{TaS}_t$) Algorithm.** The $\mathtt{TaS}_t$ algorithm uses a yardstick algorithm REF as a reference, where the yardstick can be the optimal algorithm or an incremental $\alpha$-approximation algorithm. Throughout the process, $\mathtt{TaS}_t$ keeps track of the yardstick solution value. Once a new element arrives, $\mathtt{TaS}_t$ greedily assigns a feasible value[6] to the newly-revealed element if this assignment remains $t$-competitive relative to the yardstick algorithm's solution. Otherwise, $\mathtt{TaS}_t$ *switches* its assignment to the one by the yardstick algorithm and incurs recourse. (See Algorithm 1).

---

**Algorithm 1.** $\mathtt{TaS}_t$ algorithm for monotone-sum graph problems

---

ALG $\leftarrow$ 0
**while** new element $v$ arrives **do**
    $g \leftarrow$ the best value from $[0, w_{\max}]$ such that no feasibility constraint is violated
    **if** the new assignment will fail to be $t$-competitive **then** $\triangleright \max\{\frac{\text{ALG}+g}{\text{OPT}}, \frac{\text{OPT}}{\text{ALG}+g}\} > t$
        SWITCH(OPT)
    **else**
        incorporate the greedy assignment
    **end if**
    ALG $\leftarrow$ the value of $\mathtt{TaS}_t$'s current assignment
**end while**

**function** SWITCH(assignment $A$)
    **for** every element $x$ **do**
        **if** $\mathtt{TaS}_t(x) \neq A(x)$ **then**
            change the assignment of element $x$ into $A(x)$
        **end if**
    **end for**
**end function**

---

Now, we show that the $\mathtt{TaS}_t$ algorithm achieves the desired competitive ratio $t$ with at most polynomial of $t$ amortized recourse. In our analysis, we use the following observation heavily (including for Theorem 1).

**Observation 1.** *For all $x_i \geq 0$ and $y_i > 0$, $\frac{\sum_i x_i}{\sum_i y_i} \leq \max_i \frac{x_i}{y_i}$.*

**Theorem 1.** *Using an optimal algorithm (resp. incremental $\alpha$-approximation algorithm) as the yardstick, $\mathtt{TaS}_t$ is $t$-competitive (resp. $(t \cdot \alpha)$-competitive) and incurs at most $\frac{w_{max} \cdot (t+1)}{t-1}$ **Type-2** amortized recourse for any monotone-sum graph problem where $w_{max}$ is the maximum value that can be assigned to an element. The bound also works for **Type-1** amortized recourse.*

---

[6] Note that there always exists a value such that the new assignment is feasible since the problem is monotone.

*Proof* (**Ideas.**) We can show that any optimal solution satisfies the incremental property (see the full version) and thus can be seen as an incremental 1-approximation algorithm.

Since recourse is incurred only at the moments when a switch happens in the $\mathtt{TaS}_t$ algorithm, we partition the process of the algorithm into *phases* according to the switches. Phase $i$ consists all the events after the $(i-1)$-th switch until the $i$-th switch. By Observation 1, the amortized recourse for the whole instance is bounded by the maximum amortized recourse incurred within a phase. Therefore, we consider the amortized recourse incurred by the $(i+1)$-th switch for arbitrary $i \geq 0$.

Let $\mathtt{REF}_i$ and $\mathtt{TaS}_i$ denote the value of the yardstick algorithm's solution and the $\mathtt{TaS}_t$ algorithm's solution right *after* the $i$-th switch, respectively. By construction, $\mathtt{TaS}_i = \mathtt{REF}_i$. Let $\mathtt{ALG}$ denote the value of $\mathtt{TaS}_t$'s solution right *before* the arrival of $x$, which triggers the $(i+1)$-th switch. The total **Type-2** recourse cost is at most $\mathtt{ALG} + \mathtt{REF}_{i+1}$ (where the $\mathtt{TaS}_t$ algorithm changes the value on every element to zero and then changes it to the $\mathtt{REF}$ assignment).

The main ingredients for the proof are:

- **Property 1:** Monotonicity of the problem and the incremental nature of $\mathtt{REF}$ implies that $\mathtt{REF}_i \leq \mathtt{REF}_{i+1}$.
- **Property 2:** The incremental nature of $\mathtt{TaS}_t$ during a phase implies that $\mathtt{ALG} \geq \mathtt{REF}_i$.
- **Property 3:** By the switching condition of $\mathtt{TaS}_t$, $\mathtt{ALG} < \mathtt{REF}_{i+1}/t$ for maximization problems, and $\mathtt{ALG} + w_{\max} > t \cdot \mathtt{REF}_{i+1}$ for minimization problems.

**Maximization Problems.** By **Property** 1 and the fact that the assigned values are at most $w_{\max}$, we can show that the adversary needs to release at least $\frac{\mathtt{REF}_{i+1} - \mathtt{REF}_i}{w_{\max}}$ elements such that the yardstick assignment value increases enough to trigger the switch. By **Property** 2 and **Property** 3, $\frac{\mathtt{REF}_{i+1} - \mathtt{REF}_i}{w_{\max}} \geq \frac{(1-1/t) \cdot \mathtt{REF}_{i+1}}{w_{\max}}$. By **Property** 3, the total recourse incurred by the $(i+1)$-th switch is at most $\mathtt{ALG} + \mathtt{REF}_{i+1} < (1+1/t) \cdot \mathtt{REF}_{i+1}$. Hence, the **Type-2** amortized recourse incurred in phase $i+1$ is bounded by $\frac{w_{\max} \cdot (1+1/t) \cdot \mathtt{REF}_{i+1}}{(1-1/t) \cdot \mathtt{REF}_{i+1}} = \frac{w_{\max} \cdot (t+1)}{t-1}$.

**Minimization Problems.** In minimization problems, the $(i+1)$-th switch may be triggered by shifting the $\mathtt{REF}$ assignment completely but without changing its value. In this case, a massive amount of recourse is incurred by a single input. However, we can show by **Property** 3 that in this case, the $\mathtt{ALG}$ value must be large enough to trigger the switch. Thus, we can bound the number of elements released during phase $i+1$ by the change of $\mathtt{TaS}_t$ assignment's total value. That is, it is at least $\frac{\mathtt{ALG} - \mathtt{TaS}_i}{w_{\max}} + 1 = \frac{\mathtt{ALG} - \mathtt{REF}_i}{w_{\max}} + 1$, where the 1 is the element which triggers the switching. By **Property** 1 and **Property** 2, the number is at least $\frac{(1-1/t) \cdot (\mathtt{ALG} + w_{\max})}{w_{\max}}$. By **Property** 3, the total recourse incurred by the $(i+1)$-th switch is at most $\mathtt{ALG} + \mathtt{REF}_{i+1} < (1+1/t) \cdot \mathtt{ALG} + w_{\max}/t$. Therefore, the **Type-2** amortized recourse incurred in phase $i+1$ is bounded by $\frac{w_{\max} \cdot ((1+1/t) \cdot \mathtt{ALG} + w_{\max}/t)}{(1-1/t) \cdot (\mathtt{ALG} + w_{\max})} \leq \frac{w_{\max} \cdot (t+1)}{t-1}$.     □

The yardstick algorithm can be the optimal offline algorithm. Since the problem is monotone, our algorithm can be $t$-competitive for arbitrary $t > 1$. Furthermore, if we apply a polynomial-time incremental $\alpha$-approximation algorithm as the yardstick, then our algorithm also runs in polynomial time.

The results work for weighted versions of problems, and it also work for fractional assignment problems, where the value assigned to any element is in $[0, 1]$ (for example, the fractional VERTEX COVER problem in [24]). In this case, the **Type-2** amortized recourse is bounded above by the **Type-1** amortized recourse:

**Corollary 1.** *For a fractional monotone-sum problem, $TaS_t$ is $(t\cdot\alpha)$-competitive and incurs at most $\frac{t+1}{w_{min}\cdot(t-1)}$ **Type-1** amortized recourse using an incremental $\alpha$-approximation algorithm as the yardstick. The bound also works for **Type-2** amortized recourse.*

The monotone-sum problem property captures many classical graph optimization problems such as INDEPENDENT SET, MAXIMUM CARDINALITY MATCHING, and VERTEX COVER. The three problems can be interpreted as a special case of general monotone-sum problems as follows.

**Independent Set Problem in Vertex-Arrival Model.** Vertices arrive one at a time and should be assigned a value 0 or 1. Once a vertex is revealed, the edges between it and its previously-revealed neighbors are known. The goal is to find a maximum value assignment such that for any edge, the sum of values assigned to the two endpoints is at most 1.

**Maximum Cardinality Matching Problem in Vertex-Arrival or Edge-Arrival Model.** Edges or vertices arrive one at a time and each of the edges should be assigned a value 0 or 1. The goal is to find a maximum value assignment such that for any vertex, the sum of values assigned to its incident edges is at most 1.

**Vertex Cover Problem in Vertex-Arrival Model.** Vertices arrive one at a time and should be assigned a value 0 or 1. Once a vertex is revealed, the edges between it and its previously-revealed neighbors are known. The goal is to find a minimum value assignment such that for any edge, the sum of values assigned to its two endpoints is at least 1.

Since the available value for each element is either 0 or 1 in these three problems, we say that an element is *accepted* if it is assigned a value 1. Similarly, an element is *rejected* if it is assigned a value 0. An element is *late-accepted* if its value is changed from 0 to 1 after its arrival, and *late-rejected* if its value is changed from 1 to 0 after its arrival. Furthermore, since the value for any element only changes between 0 and 1, the **Type-1** recourse cost and **Type-2** recourse cost are equivalent in these three problems. Therefore, we have the following corollary.

**Corollary 2.** *The $TaS_t$ algorithm attains competitive ratio $t > 1$ while incurring at most $\frac{t+1}{t-1}$ (**Type-1** or **Type-2**) amortized recourse for* INDEPENDENT SET, MAXIMUM CARDINALITY MATCHING, *and* VERTEX COVER *problems.*

# 3    Maximum Independent Set

For the maximum independent set problem in the vertex-arrival model, the algorithm proposed by Boyar et al. incurs at most 2 amortized recourse while maintaining a competitive ratio of 2.598 [10]. By Theorem 1, the general $TaS_t$ algorithm incurs at most $\frac{t+1}{t-1}$ amortized recourse and guarantees a competitive ratio of $t$. In this section, we show that the amortized recourse incurred by $TaS_t$ is even smaller by a more sophisticated analysis.

**Lemma 1 (Instance reduction).** *For any instance $(G, \sigma)$ of the maximum independent set problem, there exists an instance $(G', \sigma')$ for which any newly revealed vertex is either accepted by $TaS_t$ or is part of the optimal offline solution when $TaS_t$ incurs its next switch, but not both, such that the amortized recourse for $(G', \sigma')$ is at least that for $(G, \sigma)$.*

By Lemma 1, given any input $(G, \sigma)$, its amortized recourse incurred by $TaS_t$ is bounded above by that of its reduced instance $(G', \sigma')$. In Theorem 2, we provide an upper bound of the amortized recourse incurred by $TaS_t$ against any reduced instance, and thus that this upper bound holds for any instance.

**Theorem 2.** *For the maximum independent set problem, given a target competitive ratio $t > 1$, $TaS_t$ is $t$-competitive while incurring at most $\frac{t}{t-1}$ amortized recourse.*

*Proof* (**Ideas**). We prove this theorem by using the same phase partition argument in the proof of Theorem 1 on any reduced instance $(G', \sigma')$. Assume that every vertex is released with some "budget" $B$, which is a function of $t$, for later recourse. Our attempt is to find a sufficient $B$ such that the total recourse incurred by $TaS_t$ in one phase is no more than the total recourse budget carried by the vertices released in this phase. That is, the total recourse incurred by one switch can be "paid" by the recourse budget from the newly revealed vertices. By Lemma 1 and Observation 1, $TaS_t$ incurs at most $B$ amortized recourse.

By Lemma 1, in the reduced instance, any newly revealed vertex is either accepted by $TaS_t$ or is part of $OPT_{i+1}$. Therefore, we can show that the number of vertices revealed in phase $i + 1$ that are part of $OPT_{i+1}$ is bounded above by $OPT_{i+1} - OPT_{i-1}$, which implies that it is sufficient for the budget to satisfy $B \geq \frac{ALG + OPT_{i+1}}{OPT_{i+1} - OPT_{i-1}}$. Furthermore, we incorporate both the number of vertices in phase $i + 1$ that are accepted by $TaS_t$ and the number of vertices revealed in phase $i$ that are accepted by $TaS_t$ into our analysis and show that the lower bound on the required budget is largest when there are no such vertices.

We conclude that it is sufficient for each newly-revealed vertex to carry budget $B = \frac{t}{t-1}$. Thus, $TaS_t$ is $t$-competitive while incurring at most $\frac{t}{t-1}$ amortized recourse. □

**Theorem 3.** *For any $1 < t \le 2$, $\varepsilon > 0$, and $t$-competitive deterministic online algorithm, there exists an instance for which the algorithm incurs at least $\frac{1}{t-1} - \varepsilon$ amortized recourse.*

*Proof* **(Ideas).** Consider any $t$-competitive online algorithm against an adversary that constructs a complete bipartite graph and only reveals new vertices in the partition which does not contain the algorithm's current solution. This means that the maximum number of vertices that the algorithm's solution can contain will only increase if the algorithm moves its solution from one partition to the other. Furthermore, in doing so, the algorithm is forced to late-reject all vertices in its old solution in order to late-accept all vertices in its new solution.

By the structure of this adversary, we can show that 1) each partition-changing switch will incur at least $\frac{1}{t}$ recourse amortized over the size of the revealed graph when the switch occurs, and 2) there are at most $t$ times more vertices at switch $i + 1$ than at switch $i$. Therefore, the recourse incurred by switches up to switch $i$ amortized over the size of the revealed graph $f(i)$ satisfied the recurrence relation $f(i) = \frac{1}{t} + \frac{1}{t} \cdot f(i - 1)$, where $f(1) = 1$. Solving this recurrence relation, we conclude that for any $1 < t \le 2$, $\varepsilon > 0$, and $t$-competitive deterministic online algorithm, there exists an instance for which the algorithm incurs at least $\frac{1}{t-1} - \varepsilon$ amortized recourse. □

## 4   Maximum Cardinality Matching

The $\mathtt{TaS}_t$ algorithm greedily aligns with the yardstick solution completely and incurs a lot of recourse. However, for some of the elements whose value is changed, the alignment may not contribute to the improvement of the competitive ratio as much as the alignment of other elements. This observation suggests that it may be possible to reduce the amount of amortized recourse while maintaining $t$-competitiveness by switching the solution only *partially* into the yardstick. In this section, we show that the $L$-Greedy algorithm by Angelopoulos et al. [2], which is in fact a $\mathtt{TaS}_t$ algorithm that uses an optimal solution as the yardstick without aligning to it fully, incurs less amortized recourse for the MAXIMUM CARDINALITY MATCHING problem.

**$L$-Greedy Algorithm [2].** The algorithm is associated with a parameter $L$. Throughout the process, the $L$-Greedy algorithm partially switches its solution to the optimal once by eliminating all augmenting paths with length at most $2L + 1$. That is, it late rejects all the edges selected by itself and late accepts all the edges in the optimal solution on the path.

After applying late operations on all augmenting paths with at most $2L + 1$ edges, every remaining augmenting path has length at least $2L + 3$, and the ratio of the OPT solution value to the $L$-Greedy solution value is $\frac{\mathrm{OPT}(P)}{L\text{-}\mathrm{Greedy}(P)} \le \frac{L+2}{L+1}$ on the component $P$. Since the MAXIMUM CARDINALITY MATCHING problem can be solved in $O(n^{2.5})$ time, the following theorem holds by selecting $L = \lceil \frac{1}{t-1} \rceil - 1$.

**Theorem 4.** *The $L$-Greedy algorithm returns a valid matching with competitive ratio $\frac{L+2}{L+1}$ in $O(n^{3.5})$ time, where $n$ is the number of vertices in the final graph.*

*Proof* (**Ideas**). After applying the late operations on all the augmenting path with at most $2L + 1$ edges, every remaining augmenting path $P$ has length at least $2L + 3 = (L + 2) + (L + 1)$, and the ratio of the OPT size to the $L$-Greedy size $\frac{\text{OPT}(P)}{L\text{-Greedy}(P)} \leq \frac{L+2}{L+1}$ on the component $P$. By Observation 1, the ratio of an optimal solution to that of $L$-Greedy is at most $\frac{L+2}{L+1}$.

When an edge/vertex arrives, the algorithm checks if there is an augmenting path. By the well-known Hopcroft-Karp algorithm [17], it can be done in $O(n \cdot n^{2.5}) = O(n^{3.5})$-time, where $n$ is the number of vertices in the final graph.    □

Since it was shown that to achieve 1.5-competitiveness, every vertex incurs at most 2 recourse, we consider a target competitive ratio $1 < t < 2$ and have the following theorem. Note that $1 < t^* < 2$, thus $0 < \frac{t^*-1}{3-t^*} < 1$.

**Theorem 5.** *For the* MAXIMUM CARDINALITY MATCHING *problem in the vertex/edge-arrival model, the L-Greedy algorithm is t-competitive for any $1 < t < 2$ and incurs at most $\frac{(2-t^*)}{(t^*-1)(3-t^*)} + \frac{t^*-1}{3-t^*}$ amortized recourse, where $t^*$ is the largest number such that $t^* \leq t$ and $t^* = 1 + \frac{1}{j}$ for some integer j.*

*Proof* (**Ideas**). Consider the connected components generated by the union of edges chosen by $L$-Greedy or by OPT. By Observation 1, we prove this theorem by showing that for any component in the graph, the total recourse incurred at this component divided by its size is at most $\frac{(2-t^*)}{(t^*-1)(3-t^*)} + \frac{t^*-1}{3-t^*}$.

By selecting $L = \lceil \frac{2-t}{t-1} \rceil$, the path eliminations only happen at odd-size components with length from 3 to $2 \cdot (\lceil \frac{1}{t-1} \rceil - 1) + 1$ (note that $\lceil \frac{1}{t-1} \rceil \geq 2$ since $1 < t < 2$). Moreover, for such a $(2k+1)$-edge augmenting path, the total recourse incurred by the $2k + 1$ elements in the path is at most $1 + \sum_{k=1}^{\lceil \frac{1}{t-1} \rceil - 1} 2k$. Hence, the amount of amortized recourse incurred by this component is at most $\frac{(\lceil \frac{1}{t-1} \rceil - 1) \cdot \lceil \frac{1}{t-1} \rceil + 1}{2\lceil \frac{1}{t-1} \rceil - 1}$. Thus, the theorem is proven if $t = 1 + \frac{1}{j}$ for some integer $j$.

For the case in which there is no integer $j$ such that $t = 1 + \frac{1}{j}$, the proof can be adapted by rounding down $t$ to the largest $t^* \leq t$ such that $t^* = 1 + \frac{1}{j}$ for some integer $j$. By eliminating all augmenting paths that have length at most $\frac{2}{t^*-1} - 1$, the amount of incurred amortized recourse is at most $\frac{2-t^*}{(t^*-1)(3-t^*)} + \frac{t^*-1}{3-t^*}$, and the algorithm attains a competitive ratio of $t^* \leq t$.    □

**Theorem 6.** *No deterministic t-competitive online algorithm can incur amortized recourse less than $\frac{(2-t^*)}{(t^*-1)(3-t^*)}$ in the worst case.*

*Proof* (**Ideas**). Given that $\frac{n+2}{n+1} \leq t < \frac{n+1}{n}$ for some integer $n \geq 1$, consider the adversarial instance that releases a sequence of $2n + 1$ edges that form a path. More specifically, given the current instance which is an $\ell$-path, the adversary releases a new edge that is incident to one of the endpoints of the $\ell$-path and forms a $(\ell + 1)$-path. For any $1 \leq k \leq n$, the following invariants hold for any $t$-competitive algorithm:

(**I1**) For a path with length $2k+1$, a $t$-competitive algorithm has to accept $k+1$ edges.

(**I2**) For a path with length $2k$, a $t$-competitive algorithm has to accept $k$ edges.

(**I3**) When an instance is increased from a $2(k-1)+1$ path to a $2k+1$ path, a $t$-competitive algorithm incurs at least $2k$ amount of recourse.

Given the invariants **I1**, **I2**, and **I3**, the $(2n+1)$-length path instance incurs recourse with total amount at least $\sum_{k=1}^{n}(2k) = n \cdot (n+1)$. Therefore, any $t$-competitive algorithm incurs at least $\frac{n\cdot(n+1)}{2n+1}$ amortized recourse for this instance. Let $t^* = \frac{n+2}{n+1} \leq t$. It follows that $n = \frac{2-t^*}{t^*-1}$. Therefore, the amortized recourse is at least $\frac{n\cdot(n+1)}{2n+1} = \frac{(2-t^*)}{(t^*-1)(3-t^*)}$. $\qquad\square$

## 5   Minimum Vertex Cover

In this section, we propose a special version of the TaS$_t$ algorithm, Duo-Halve, that attains a competitive ratio of $2 - \frac{2}{\mathsf{OPT}}$ for the MINIMUM VERTEX COVER problem with optimal vertex cover size OPT in polynomial time.

The Duo-Halve algorithm uses an optimal solution as the yardstick with $t = 2 - \frac{2}{\mathsf{OPT}}$. However, the computation of the optimal solution of VERTEX COVER is very expensive. Thus, we maintain a maximal matching greedily (as the well-known 2-approximation algorithm for VERTEX COVER) on the current input graph and only select vertices that are saturated by the matching. Intuitively, if the Duo-Halve algorithm rejects two of these saturated vertices, the competitive ratio is at most $2 - \frac{2}{\mathsf{OPT}}$. We show that either we can remove up to 2 carefully chosen vertices from the matching-based solution without violating its feasibility, or the optimal solution contains at least one vertex more than the number of edges in the maximal matching (Lemma 3 and Theorem 7). In either case, the constructed solution is $(2 - \frac{2}{\mathsf{OPT}})$-competitive. We can refine the choice of vertices to be removed so that they are incident to one of the two last edges added to the matching. This restriction allows us to show that maintaining this constructed solution needs only a constant amount of amortized recourse, and polynomial time (Lemma 2).

In the following discussion, we use some terminology. Let ME1 and ME2 be the last and second-to-last edges added to the matching respectively. Note that ME1 and ME2 change over the course of the input sequence as more vertices are revealed and edges are added to the matching. Also, let $V_{M(\mathcal{X})}$ be the vertices saturated by the maximal matching $M(\mathcal{X})$. The DH algorithm partitions the vertices into three groups: **Group-1**: the endpoints of ME1 or ME2, **Group-2**: the vertices in $V_M$ but not in Group-1, and **Group-3**: the vertices in $V\backslash V_M$.

**Duo-Halve Algorithm (DH).** When a new vertex $v$ arrives, if an edge $(p,v)$ is added to $M(\mathcal{X})$, then it introduces a new ME1 (namely $(p,v)$). The algorithm first accepts all **Group-2** vertices that are adjacent to $v$. Then, the algorithm decides the assignment of ME1 and ME2 and minimizes the number of accepted endpoints of ME1 and ME2. If there is a tie, we apply the one that accepts fewer endpoints in ME1 and/or incurs less recourse. (See Algorithm 2).

---

**Algorithm 2.** Duo-Halve algorithm (DH) for Minimum Vertex Cover Problem

---

ME1 ← ∅, ME2 ← ∅, $V_M$ ← ∅
**while** new vertex $v$ arrives **do**
    **if** there is a vertex $p \in N(v) \cup (V \backslash V_M)$ **then**
        ME2 ← ME1
        ME1 ← $(p, v)$    ▷ $(p, v)$ is a new matched edge. If there is more than one $p$,
choose one arbitrarily.
        Add $p$ and $v$ into $V_M$
        LateAccept all rejected vertices in $(V_M \backslash \{\text{vertices in ME1 or ME2}\}) \cap N(v)$
        HalveBoth(ME1, ME2)
    **else**
        LateAccept all rejected vertices in $V_M \cap N(v)$
        HalveBoth(ME1, ME2)
    **end if**
**end while**

**Function** HalveBoth(matched edge ME1, matched edge ME2)
Among accept/reject configurations of ME1 and ME2 that yield a valid vertex cover, return one that maximizes the number of half edges among ME1 and ME2 with the minimum number of late operations. If there is a tie, prioritize ME1 (see Fig. 1 for details).
**end Function**

---

The DH algorithm returns a feasible solution in $O(n^3)$ time, where $n$ is the number of vertices in the graph. Intuitively, the algorithm maintains a valid solution as it greedily covers edges using vertices in the maximal matching, with the exception of ME1 and ME2, where it carefully ensures that a feasible configuration is chosen. Furthermore, the most computationally-expensive component of the DH algorithm, which checks the validity of a constant number of configurations by looking at the neighborhoods of ME1 and ME2, runs in $O(n^2)$ time for each new element.

**Lemma 2.** *The DH algorithm always returns a valid vertex cover in $O(n^3)$ time, where $n$ is the number of vertices in the graph.*

We first show that if DH fails to produce a solution where it accepts only one vertex of ME1, then OPT $\geq |M| + 1$. The intuition is that if DH has to accept both endpoints of ME1, there must be at least one **Group-3** vertex in each of the endpoints' neighborhoods. Therefore, the optimal solution has to cover the corresponding edges with at least two vertices.

**Lemma 3.** *In the assignment of DH, if both endpoints of ME1 are selected, then the optimal solution must contain at least two vertices in $ME1 \cup (V \backslash V_M)$, and $\frac{DH}{OPT} \leq 2 - \frac{2}{OPT}$.*

**Theorem 7.** *The DH algorithm is $(2 - \frac{2}{OPT})$-competitive.*

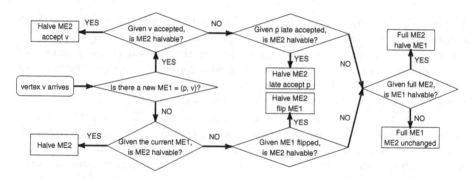

**Fig. 1.** An illustration of the flow of `HalveBoth` (ME1, ME2).

*Proof* **(Ideas).** We define an edge as being *half* if exactly one of its endpoints is accepted by DH, and *full* if both of its endpoints are accepted by DH. In any possible solution provided by DH, there are three states based on the configuration of ME1 and ME2: 1) both ME1 and ME2 are half, 2) ME1 is half and ME2 is full, and 3) ME1 is full. In state 1, we can directly show that the bound holds since DH $\leq 2|M| - 2$. The bound holds for state 3 by Lemma 3.

State 2 requires more involved analysis. If an endpoint of ME2 has a rejected **Group-2** neighbor, then DH rejects at least two vertices in $V_M$ (this **Group-2** neighbor and 1 ME1 vertex) and DH $\leq 2|M| - 2$. Otherwise, if at least one endpoint of ME2 has no **Group-3** neighbor, then we can show that there is no solution based on the maximal matching containing only 2 **Group-1** vertices. This means that OPT must contain either a **Group-3** vertex or 3 **Group-1** vertices, and thus OPT $\geq |M| + 1$. Finally, if each endpoint of ME2 has a **Group-3** neighbor, then OPT must either select a **Group-3** vertex or both endpoints of ME2, and OPT $\geq |M| + 1$. □

For a single newly-revealed vertex, the amount of recourse incurred can be up to $O(n)$. Even if we restrict our consideration to ME1 and ME2, a single new vertex can incur recourse at most 4. However, this cannot happen at every input. We use a potential function to show that the amortized recourse incurred by DH is at most 3.33.

**Theorem 8.** *The amortized recourse incurred by DH is at most $\frac{10}{3}$.*

*Proof* **(Ideas).** We prove the theorem by using a potential function. To this end, we define an edge $(u, v)$ as being *free* if there exist feasible assignments both by either accepting $u$ or by accepting $v$. Also, we define a matched edge with only one endpoint selected as being *expired* if it is neither ME1 nor ME2. Finally, we define $A$ as the set of vertices accepted by DH. Using these terms, we define the potential function $\Phi$ as

$$\Phi := |\{(u,v)|(u,v) \text{ expired}\}| + \frac{1}{3}|A \cap (\text{ME1} \cup \text{ME2})| + \frac{2}{3} \cdot \mathbb{1}[\text{ME2 is free}]$$

Furthermore, at any given moment in the input sequence where the matching constructed by DH contains at least 2 edges, the status of ME1 and ME2 is characterized by one of 6 states according to their possible combinations of selection statuses of their endpoints. We also differentiate between the two half possibilities for ME1, since the newly-revealed vertex in ME1 can be accepted without incurring a late operation when there is a new ME1.

We show that, for any possible state transition triggered by a newly-revealed vertex, the number of incurred late operations LO added to the change in potential $\Delta\Phi$ is bounded above by $\frac{10}{3}$. Note that, for any newly-revealed vertex $v$, $v$ may be adjacent to $k \geq 0$ rejected vertices that are matched by some expired edge. This incurs $k$ late operations, but also decreases $\Phi$ by $k$, so this may be ignored when computing $LO + \Delta\Phi$. Since $\Phi_0 = 0$ and $\Phi_i \geq 0$, this allows us to conclude the statement of our theorem.                                                      $\square$

Moreover, we can show a lower bound by constructing a family of instances that alternates between incurring a late accept on a **Group-2** vertex, and 4 late operations on ME1 and ME2. This is illustrated in Fig. 2.

**Lemma 4.** *For any $\varepsilon > 0$, there exists an instance such that DH incurs amortized recourse strictly greater than $\frac{5}{2} - \varepsilon$.*

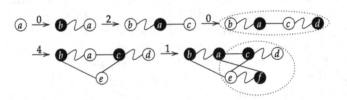

**Fig. 2.** Adversarial instance for VERTEX COVER such that DH incurs asymptotic amortized recourse $\frac{5}{2}$. Each arrow's number denotes the number of late operations incurred by the next vertex's reveal. The dotted ovals highlight the repeating structure.

Finally, we show that the analysis in Theorem 7 is tight for a class of online algorithms where its solution only contains vertices saturated by the matching maintained throughout the process in an incremental manner. In other words, no online algorithm in this class achieves a lower competitive ratio, no matter how much amortized recourse it uses.

**Definition 4.** *An algorithm for vertex cover is* incremental matching-based *if it maintains a maximal matching throughout the process in an incremental manner, and its solution only contains vertices saturated by the matching.*

**Theorem 9.** *No deterministic incremental matching-based algorithm achieves a competitive ratio smaller than $2 - \frac{2}{OPT}$.*

*Proof.* Consider the instance which first reveals $k$ disconnected edges via their endpoints. For any incremental matching-based algorithm, each of these $k$ edges will be added to the matching, and at least one vertex from each pair will be accepted. Then, the instance reveals a final vertex that is adjacent to all previously-revealed vertices (See Fig. 3). The incremental matching-based algorithm will not accept this vertex, as it is not matched, but must accept all other vertices for a vertex cover of size $2k$. However, the optimal solution consists of the last revealed vertex, and one endpoint from each of the $k$ edges. Therefore, no incremental matching-based algorithm can achieve a competitive ratio smaller than $\frac{2k}{k+1} = 2 - \frac{2}{\mathsf{OPT}}$. □

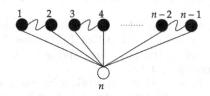

**Fig. 3.** Adversarial instance for VERTEX COVER such that any incremental matching-based algorithm is exactly $(2 - \frac{2}{\mathsf{OPT}})$-competitive. Vertices are labeled by their release order. Any such algorithm must accept $n - 1$ vertices, whereas the optimal solution contains $\frac{n-1}{2} + 1$ vertices.

## 6   Concluding Remarks

In this paper, we propose a general Target-and-Switch algorithm for online problems that allow recourse. We prove that for any monotone-sum graph problem, the algorithm attains a competitive ratio of $t > 1$ while incurring $w_{\max} \cdot (1 + \frac{1}{t-1})$ amortized recourse. Many interesting problems remain open. A major future direction is extending the analysis to non-monotone problems such as DOMINATING SETS or monotone-max problems such as COLORING.

**Acknowledgement.** We wish to thank the anonymous referees for their comments and suggestions on a previous version of this paper. In particular, we thank them for helping us complete the monotone problem definition.

## References

1. Albers, S., Schraink, S.: Tight bounds for online coloring of basic graph classes. In: Pruhs, K., Sohler, C. (eds.) 25th Annual European Symposium on Algorithms, ESA 2017, 4–6 September 2017, Vienna, Austria. LIPIcs, vol. 87, pp. 7:1–7:14. Schloss Dagstuhl - Leibniz-Zentrum für Informatik (2017). https://doi.org/10. 4230/LIPIcs.ESA.2017.7
2. Angelopoulos, S., Dürr, C., Jin, S.: Online maximum matching with recourse. J. Comb. Optim. **40**(4), 974–1007 (2020). https://doi.org/10.1007/s10878-020-00641-w

3. Avitabile, T., Mathieu, C., Parkinson, L.H.: Online constrained optimization with recourse. Inf. Process. Lett. **113**(3), 81–86 (2013). https://doi.org/10.1016/j.ipl. 2012.09.011

4. Azar, Y., Panigrahi, D., Touitou, N.: Online graph algorithms with predictions. In: Naor, J.S., Buchbinder, N. (eds.) Proceedings of the 2022 ACM-SIAM Symposium on Discrete Algorithms, SODA 2022, Virtual Conference, Alexandria, VA, USA, 9–12 January 2022, pp. 35–66. SIAM (2022). https://doi.org/10.1137/1. 9781611977073.3

5. Azar, Y., Touitou, N.: Beyond tree embeddings - a deterministic framework for network design with deadlines or delay. In: Irani [18], pp. 1368–1379 (2020). https:// doi.org/10.1109/FOCS46700.2020.00129

6. Bernstein, A., Holm, J., Rotenberg, E.: Online bipartite matching with amortized $O(\log^2 n)$ replacements. J. ACM **66**(5), 37:1–37:23 (2019). https://doi.org/10. 1145/3344999

7. Bienkowski, M., Kraska, A., Liu, H.-H., Schmidt, P.: A primal-dual online deterministic algorithm for matching with delays. In: Epstein, L., Erlebach, T. (eds.) WAOA 2018. LNCS, vol. 11312, pp. 51–68. Springer, Cham (2018). https://doi. org/10.1007/978-3-030-04693-4_4

8. Boppana, R.B., Halldórsson, M.M.: Approximating maximum independent sets by excluding subgraphs. BIT **32**(2), 180–196 (1992). https://doi.org/10.1007/ BF01994876

9. Bosek, B., Disser, Y., Feldmann, A.E., Pawlewicz, J., Zych-Pawlewicz, A.: Recoloring interval graphs with limited recourse budget. In: Albers, S. (ed.) 17th Scandinavian Symposium and Workshops on Algorithm Theory, SWAT 2020, 22–24 June 2020, Tórshavn, Faroe Islands. LIPIcs, vol. 162, pp. 17:1–17:23. Schloss Dagstuhl - Leibniz-Zentrum für Informatik (2020). https://doi.org/10.4230/LIPIcs.SWAT. 2020.17

10. Boyar, J., Favrholdt, L.M., Kotrbčík, M., Larsen, K.S.: Relaxing the irrevocability requirement for online graph algorithms. Algorithmica **84**, 1916–1951 (2022). https://doi.org/10.1007/s00453-022-00944-w

11. Cygan, M., Czumaj, A., Mucha, M., Sankowski, P.: Online facility location with deletions. In: Azar, Y., Bast, H., Herman, G. (eds.) 26th Annual European Symposium on Algorithms, ESA 2018, 20–22 August 2018, Helsinki, Finland. LIPIcs, vol. 112, pp. 21:1–21:15. Schloss Dagstuhl - Leibniz-Zentrum für Informatik (2018). https://doi.org/10.4230/LIPIcs.ESA.2018.21

12. Gu, A., Gupta, A., Kumar, A.: The power of deferral: Maintaining a constant-competitive steiner tree online. SIAM J. Comput. **45**(1), 1–28 (2016). https://doi. org/10.1137/140955276

13. Gupta, A., Guruganesh, G., Kumar, A., Wajc, D.: Fully-dynamic bin packing with limited repacking. CoRR abs/1711.02078 (2017). http://arxiv.org/abs/1711.02078

14. Gupta, A., Levin, R.: Fully-dynamic submodular cover with bounded recourse. In: Irani [18], pp. 1147–1157 (2020). https://doi.org/10.1109/FOCS46700.2020.00110

15. Halperin, E.: Improved approximation algorithms for the vertex cover problem in graphs and hypergraphs. SIAM J. Comput. **31**(5), 1608–1623 (2002). https://doi. org/10.1137/S0097539700381097

16. Harutyunyan, H.A., Pankratov, D., Racicot, J.: Online domination: the value of getting to know all your neighbors. In: Bonchi, F., Puglisi, S.J. (eds.) 46th International Symposium on Mathematical Foundations of Computer Science, MFCS 2021, 23–27 August 2021, Tallinn, Estonia. LIPIcs, vol. 202, pp. 57:1– 57:21. Schloss Dagstuhl - Leibniz-Zentrum für Informatik (2021). https://doi.org/ 10.4230/LIPIcs.MFCS.2021.57

17. Hopcroft, J.E., Karp, R.M.: An $n^{5/2}$ algorithm for maximum matchings in bipartite graphs. SIAM J. Comput. **2**(4), 225–231 (1973). https://doi.org/10.1137/0202019
18. Irani, S. (ed.): 61st IEEE Annual Symposium on Foundations of Computer Science, FOCS 2020, Durham, NC, USA, 16–19 November 2020. IEEE (2020). https://doi.org/10.1109/FOCS46700.2020
19. Karakostas, G.: A better approximation ratio for the vertex cover problem. ACM Trans. Algorithms **5**(4), 41:1–41:8 (2009). https://doi.org/10.1145/1597036.1597045
20. Khot, S., Regev, O.: Vertex cover might be hard to approximate to within 2-epsilon. J. Comput. Syst. Sci. **74**(3), 335–349 (2008). https://doi.org/10.1016/j.jcss.2007.06.019
21. Megow, N., Nölke, L.: Online minimum cost matching with recourse on the line. In: Byrka, J., Meka, R. (eds.) Approximation, Randomization, and Combinatorial Optimization. Algorithms and Techniques, APPROX/RANDOM 2020, 17–19 August 2020, Virtual Conference. LIPIcs, vol. 176, pp. 37:1–37:16. Schloss Dagstuhl - Leibniz-Zentrum für Informatik (2020). https://doi.org/10.4230/LIPIcs.APPROX/RANDOM.2020.37
22. Megow, N., Skutella, M., Verschae, J., Wiese, A.: The power of recourse for online MST and TSP. SIAM J. Comput. **45**(3), 859–880 (2016). https://doi.org/10.1137/130917703
23. Monien, B., Speckenmeyer, E.: Ramsey numbers and an approximation algorithm for the vertex cover problem. Acta Inform. **22**(1), 115–123 (1985). https://doi.org/10.1007/BF00290149
24. Wang, Y., Wong, S.C.: Two-sided online bipartite matching and vertex cover: beating the greedy algorithm. In: Halldórsson, M.M., Iwama, K., Kobayashi, N., Speckmann, B. (eds.) ICALP 2015, Part I. LNCS, vol. 9134, pp. 1070–1081. Springer, Heidelberg (2015). https://doi.org/10.1007/978-3-662-47672-7_87

# An Improved Algorithm for Open Online Dial-a-Ride

Júlia Baligács[ID], Yann Disser[ID], Nils Mosis[ID], and David Weckbecker[(✉)][ID]

TU Darmstadt, Darmstadt, Germany
{baligacs,disser,mosis,weckbecker}@mathematik.tu-darmstadt.de

**Abstract.** We consider the open online dial-a-ride problem, where transportation requests appear online in a metric space and need to be served by a single server. The objective is to minimize the completion time until all requests have been served. We present a new, parameterized algorithm for this problem and prove that it attains a competitive ratio of $1 + \varphi \approx 2.618$ for some choice of its parameter, where $\varphi$ is the golden ratio. This improves the best known bounds for open online dial-a-ride both for general metric spaces as well as for the real line. We also give a lower bound of 2.457 for the competitive ratio of our algorithm for any parameter choice.

**Keywords:** Online optimization · Dial-a-ride · Competitive analysis

## 1 Introduction

In the online dial-a-ride problem, transportation requests appear over time in a metric space $(M, d)$ and need to be transported by a single server. Each request is of the form $r = (a, b; t)$, appears at its starting position $a \in M$ at its release time $t \geq 0$, and needs to be transported to its destination $b \in M$. The server starts at a distinguished point $O \in M$, called the origin, can move at unit speed, and has a capacity $c \in \mathbb{N} \cup \{\infty\}$ that bounds the number of requests it is able to carry simultaneously. Importantly, the server only learns about request $r$ when it appears at time $t$ during the execution of the server's algorithm. Moreover, the total number of requests is initially unknown and the server cannot tell upon arrival of a request whether it is the last one.[1] Requests do not have to be served in the same order in which they appear.

The objective of the open dial-a-ride problem is to minimize the time until all requests have been served, by loading each request $r = (a, b; t)$ at point $a$ no earlier than time $t$, transporting it to point $b$ and unloading it there. We consider the non-preemptive variant of the problem, meaning that requests may only be

---

[1] If the server can distinguish the last request, it can start an optimal schedule once all requests are released, achieving a completion time of at most twice the optimum.

Supported by DFG grant DI 2041/2.

P. Chalermsook and B. Laekhanukit (Eds.): WAOA 2022, LNCS 13538, pp. 154–171, 2022.
https://doi.org/10.1007/978-3-031-18367-6_8

unloaded at their respective destinations. Note that, in contrast to the closed variant of the problem, we do not require the server to return to the origin after serving the last request.

As usual, we measure the quality of a (deterministic) online algorithm in terms of competitive analysis. That is, we compare the completion time $\text{ALG}(\sigma)$ of the algorithm to an offline optimum completion time $\text{OPT}(\sigma)$ over all request sequences $\sigma$. Here, the offline optimum is given by the best possible completion time that can be achieved if all requests are known (but not released) from the start. The (strict) competitive ratio of the algorithm is given by $\rho := \sup_\sigma \text{ALG}(\sigma)/\text{OPT}(\sigma).$[2] Note that, in particular, the running time of an algorithm does not play a role in its competitive analysis.

***Our Results.*** We present a parameterized online algorithm $\text{LAZY}(\alpha)$ and show that it improves on the best known upper bound for open online dial-a-ride for $\alpha = \varphi$, where $\varphi = \frac{1+\sqrt{5}}{2}$ denotes the golden ratio. We also show a lower bound on potential improvements for other parameter choices. More precisely, we show the following.

**Theorem 1.** $\text{LAZY}(\varphi)$ *has competitive ratio* $1 + \varphi \approx 2.618$ *for the open online dial-a-ride problem on general metric spaces for any server capacity* $c \in \mathbb{N} \cup \{\infty\}$. *For every* $\alpha \geq 1$ *and any* $c \in \mathbb{N} \cup \{\infty\}$, $\text{LAZY}(\alpha)$ *has competitive ratio at least* $\max\{1 + \alpha, 2 + 2/(3\alpha)\}$, *even if the metric space is the real line.*

In particular, we obtain a lower bound on the competitive ratio of our algorithm, independent of $\alpha$.

**Corollary 1.** *For every* $\alpha \geq 0$, $\text{LAZY}(\alpha)$ *has competitive ratio at least* $3/2 + \sqrt{11/12} \approx 2.457$ *for open online dial-a-ride on the line.*

Our lower bound also narrows the range of parameter choices that could allow improved competitive ratios.

**Corollary 2.** *For* $\alpha \notin (2\varphi/3, \varphi) \approx (1.078, 1.618)$, $\text{LAZY}(\alpha)$ *has competitive ratio at least* $1 + \varphi$ *for open online dial-a-ride on the line.*

Our upper bound improves the best known upper bound of 2.70 for general metric spaces [8], and even the best known upper bound of 2.67 for the real line [10]. Figure 1 gives an overview over the previous upper bounds for open online dial-a-ride. Note that, in contrast to previous results, $\text{LAZY}(\alpha)$ is not a so-called schedule-based algorithm as defined in [8], because it interrupts schedules.

We note that an upper bound of $\varphi + 1 \approx 2.618$ was already claimed in [26] for the Wait-or-Ignore algorithm, but the proof in [26] is inconclusive. While the general idea of our algorithm is similar to Wait-or-Ignore, our implementation is more involved and avoids issues in the analysis that are not being addressed

---

[2] We adopt a strict definition of the competitive ratio that requires a bounded ratio for all request sequences, i.e., we do not allow an additive constant.

in [26]. In particular, Wait-or-Ignore only waits at the origin, while our algorithm crucially also waits at other locations.

***Related Work.*** As listed in Fig. 1, the best previously known upper bound for open online dial-a-ride of 2.70 was shown by Birx [8] and a slightly better bound of 2.67 for the line was shown by Birx et al. [10]. In this paper, we improve both bounds to $1 + \varphi \approx 2.618$. A better upper bound of $1 + \sqrt{2} \approx 2.41$ is known for the preemptive variant of the problem, due to Bjelde et al. [11]. The TSP problem is an important special case of dial-a-ride, where $a = b$ for every request $(a, b; t)$, i.e., requests just need to be visited. Bonifaci and Stougie gave an upper bound for open online TSP on general metric spaces of 2.41. Bjelde et al. [11] were able to show a tight bound of 2.04 for open online TSP on the line. Birx et al. [10] showed that open online dial-a-ride is strictly more difficult than open online TSP by providing a slightly larger lower bound of 2.05. Weaker lower bounds for the half-line were given by Lippmann [26].

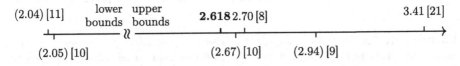

**Fig. 1.** Overview of the state-of-the-art for the open online dial-a-ride problem. Bounds in parentheses were shown for the real line. Note that lower bounds on the real line carry over to general metric spaces and the converse is true for upper bounds. In particular, our upper bound also holds on the line.

The competitive analysis of the closed online dial-a-ride problem on general metric spaces has proven to be structurally much simpler and conclusive results are known: The best possible competitive ratio of 2 is achieved by the conceptually clean Smartstart algorithm, as shown by Ascheuer et al. [2] and Feuerstein and Stougie [14]. Ausiello et al. [4] gave a matching lower bound already for TSP. The situation is more involved on the line. Bjelde et al. [11] gave a sophisticated algorithm for closed online TSP on the line that tightly matches the lower bound of 1.64 shown by Ausiello et al. [4]. Birx [8] separated closed online dial-a-ride on the line by giving a lower bound of 1.76. No better upper bound than 2 is known in this setting, not even for preemptive algorithms. Blom et at. [12] gave a tight bound of 1.5 for the half-line, and the best known lower bound of 1.71 for closed dial-a-ride on the half-line is due to Ascheuer et al. [2].

Clearly, most variants of the online dial-a-ride problem have resisted tight competitive analysis for many years. As a remedy, several authors have resorted to considering restricted classes of algorithms, restricted adversary models, or resource augmentation. In that vein, Blom et al. [12] considered "zealous" (or "diligent") algorithms that do not stay idle if there are unserved requests, and Birx [8] derived stronger lower bounds for "schedule-based" algorithms that subdivide the execution into schedules that may not be interrupted. Examples of restricting the adversary include "non-abusive" or "fair" models introduced by Krumke et al. [22] and Blom et al. [12], that force the optimum solution to stay in the convex hull of all released requests. In the same spirit, Hauptmeier et al. [16] adopted a "reasonable load" model, which requires that the length of an optimum schedule for serving all requests revealed up to time $t$ is bounded by a function of $t$. In terms of resource augmentation, Allulli et al. [1] and Ausiello et al. [5] considered a model with "lookahead", where the algorithm learns about requests before they are released. In contrast, Lippmann et al. [25] considered a restricted information model where the server learns the destination of a request only upon loading it. Bonifaci and Stougie [13] and Jaillet and Wagner [19] considered resource augmentation regarding the number of servers, their speeds, and their capacities.

While we concentrate on minimizing completion time, other objectives have been studied: Krumke [21] presented first results for randomized algorithms minimizing expected completion time, Krumke et al. [24] and Bienkowski et al. [6,7] minimized the sum of completion times, Krumke et al. [22,23] and Hauptmeier et al. [16] minimized the flow time, and Yi and Tian [27] maximized the number of served requests (with deadlines). Regarding other metric spaces, Jawgal et al. [20] considered online TSP on a circle. Various generalizations of online dial-a-ride have been investigated: Ausiello et al. [3] introduced the online quota TSP, where only a minimum weighted fraction of requests need to be served, and, similarly, Jaillet and Lu [17,18] adopted a model where requests can be rejected for a penalty in the objective. Jaillet and Wagner [19] and Hauptmeier et al. [15] allowed precedence constraints between requests.

## 2  Notation and Definition of the Algorithm

Let $\sigma = (r_1, \ldots, r_n)$ be a sequence of requests $r_i = (a_i, b_i; t_i)$ with release times $0 < t_1 < \cdots < t_n$. Note that we do not allow multiple requests to appear at the same time or a request to appear at time 0 but this is not a restriction as the release times can differ by arbitrarily small values. We let $\mathrm{OPT}(t)$ denote the completion time of the offline optimum over all requests released not later than $t$. A *schedule* is a sequence of actions of the server, specifying when requests are collected and unloaded, how the server moves, and, in particular, when the server stays stationary. Let $\mathrm{OPT}[t]$ denote an optimal schedule with completion time $\mathrm{OPT}(t)$. We say that a server *visits* point $p \in M$ at time $t \geq 0$ if the server is in position $p$ at time $t$.

The rough idea of our algorithm is to wait until we gather several requests and then start a schedule serving them. If a new request arrives during the execution of a schedule, it would be desirable to include it in the server's plan. Therefore, we check whether we can "reset" the server's state in a reasonable time, i.e., deliver all currently loaded requests and return to the origin, so that we can compute a new schedule. If this is not possible, we keep following the current schedule and consider the new requests later.

We introduce some notation to capture this more formally. Let $R$ be a set of requests and $x \in M$. Then, the schedule $S(R, x)$ is the shortest schedule starting from point $x$ and serving all requests in $R$. Note that this schedule can ignore the release times of the requests as we will only compute it after all requests in $R$ are released. As it is not beneficial to wait at some point during the execution of a schedule, the walked distance in $S(R, x)$ is the same as the time needed to complete it. We denote its length by $|S(R, x)|$.

Now, we can describe our algorithm. The factor $\alpha \geq 1$ will be a measure of how long we wait before starting a schedule. A precise description of the algorithm is given below (cf. Algorithm 1). In short, whenever a new request $r = (a, b; t)$ arrives, we determine whether it is possible to serve all loaded requests and return to the origin in time $\alpha \cdot \text{OPT}(t)$. If this is possible, we do so. In this case, we say that the schedule was *interrupted*. Otherwise, we ignore the request and consider it in the next schedule. Before starting a new schedule, we wait at least until time $\alpha \cdot \text{OPT}(t)$.

In the following, the algorithm $\text{LAZY}(\alpha)$ with waiting parameter $\alpha \geq 1$ is described. The first part of the algorithm is invoked whenever a new request $r = (a, b; t)$ is released, and the second part of the algorithm is invoked whenever the algorithm becomes idle, i.e., when the server has finished waiting or finished a schedule. We denote by $t$ the current time, by $R_t$ the set of unserved requests at time $t$ and by $p_t$ the position of the server at time $t$. There are three commands that can be executed, namely DELIVER_AND_RETURN, WAIT_UNTIL($t'$), and FOLLOW_SCHEDULE($S$). Whenever one of these commands is invoked, the server aborts what it is currently doing and executes the new command. The command DELIVER_AND_RETURN instructs the server to deliver all loaded requests and return to the origin in an optimal way. The command WAIT_UNTIL($t'$) orders the server to remain at its position until time $t'$ and the command FOLLOW_SCHEDULE($S$) tells the server to execute schedule $S$. Once the server completes the execution of a command, it becomes *idle*.

---

**Algorithm 1:** $\text{LAZY}(\alpha)$

---

initialize: $i \leftarrow 0$

---

*upon receiving request $r = (a, b; t)$:*
**if** *server can serve loaded requests and return to $O$ until time $\alpha \cdot \text{OPT}(t)$* **then**
    ⌞ **execute** DELIVER_AND_RETURN         `// interrupt` $S^{(i)}$

---

*upon becoming idle:*
**if** $t < \alpha \cdot \text{OPT}(t)$ **then**
    | **execute** WAIT_UNTIL($\alpha \cdot \text{OPT}(t)$)
**else if** $R_t \neq \emptyset$ **then**
    | $i \leftarrow i + 1$, $R^{(i)} \leftarrow R_t$, $t^{(i)} \leftarrow t$, $p^{(i)} \leftarrow p_t$
    | $S^{(i)} \leftarrow S(R^{(i)}, p^{(i)})$
    ⌞ **execute** FOLLOW_SCHEDULE($S^{(i)}$)

---

# 3 Analysis of LAZY

In this section, we analyze LAZY($\alpha$) and show that LAZY($\alpha$) is $1 + \alpha$ competitive for $\alpha \geq \varphi = \frac{1+\sqrt{5}}{2}$. This implies in particular that LAZY($\varphi$) is $(1+\varphi)$-competitive, i.e., that the first part of Theorem 1 holds.

**Theorem 2.** *For $\alpha \geq \varphi \approx 1.618$, LAZY($\alpha$) is $(1 + \alpha)$-competitive for the open dial-a-ride problem on general metric spaces for any server capacity $c \in \mathbb{N} \cup \{\infty\}$.*

*Proof.* For a given request sequence $(r_1, \ldots, r_n)$, we denote the number of schedules started by LAZY($\alpha$) by $k \leq n$. Let $S^{(i)}$, $t^{(i)}$, $p^{(i)}$, and $R^{(i)}$ be as defined in the algorithm, i.e., $S^{(i)}$ is the $i$-th schedule started by LAZY($\alpha$), $t^{(i)}$ is its starting time, $p^{(i)}$ its starting position, and $R^{(i)}$ is the set of requests served by $S^{(i)}$. Observe that some schedules might be interrupted so that $R^{(1)}, \ldots, R^{(k)}$ are not necessarily disjoint. Also observe that we have $p_1 = O$ and, for $i > 1$, $p^{(i)}$ is either the ending position of $S^{(i-1)}$ or $O$ if $S^{(i-1)}$ was interrupted.

We show by induction on $i$ that, for all $i \in \{1, \ldots, k\}$,

a) $|S^{(i)}| \leq \mathrm{OPT}(t^{(i)})$, and
b) $t^{(i)} + |S^{(i)}| \leq (1 + \alpha) \cdot \mathrm{OPT}(t^{(i)})$.

Note that this completes the proof since the last schedule is completed at time $t^{(k)} + |S^{(k)}|$ and since $\mathrm{OPT}(t^{(k)})$ is the completion time of the offline optimum over all requests.

Before starting the induction, let us make some observations. Since the server does not start a schedule at time $t$ if $t < \alpha \cdot \mathrm{OPT}(t)$, we have

$$t^{(i)} \geq \alpha \cdot \mathrm{OPT}(t^{(i)}) \tag{1}$$

for all $i \in \{1, \ldots, k\}$. Further, for every request $r = (a, b; t) \in R^{(i+1)} \backslash R^{(i)}$, we have $t > t^{(i)}$ because $R^{(i)}$ contains all unserved requests released until time $t^{(i)}$. Moreover, $R^{(i+1)} \backslash R^{(i)} \neq \emptyset$ because otherwise $S^{(i)}$ is not interrupted and we have $R^{(i+1)} = \emptyset$, contradicting that the algorithm starts $S^{(i+1)}$. Therefore, for all $i \in \{1, \ldots, k-1\}$,

$$\mathrm{OPT}(t^{(i+1)}) > t^{(i)} \overset{(1)}{\geq} \alpha \cdot \mathrm{OPT}(t^{(i)}). \tag{2}$$

Now, let us start the induction.

*Base Case:* a) Since $\mathrm{OPT}[t^{(1)}]$ is a schedule serving all requests in $R^{(1)}$ starting from $O$, possibly with additional waiting times, we have

$$|S^{(1)}| = |S(R^{(1)}, p^{(1)})| = |S(R^{(1)}, O)| \leq \mathrm{OPT}(t^{(1)}).$$

b) Consider the time $t^{(1)}$ at which schedule $S^{(1)} = S(R^{(1)}, O)$ is started, and let $t' \leq t^{(1)}$ denote the largest release time of a request in $R^{(1)}$. In particular, no requests are released in the time period $(t', t^{(1)}]$ and thus $\mathrm{OPT}(t') = \mathrm{OPT}(t^{(1)})$.

When the request at time $t'$ is released, the server is in $O$ so that the command DELIVER_AND_RETURN is completed immediately. Therefore, the server becomes idle at time $t'$ and the waiting time is set to $\alpha \cdot \text{OPT}(t') = \alpha \cdot \text{OPT}(t^{(1)})$. Observe that $\alpha \cdot \text{OPT}(t') > \text{OPT}(t') \geq t'$. The server becomes idle again and starts schedule $S^{(1)}$ precisely at time $t^{(1)} = \alpha \cdot \text{OPT}(t^{(1)})$. We obtain

$$t^{(1)} + |S^{(1)}| \overset{\text{a)}}{\leq} (1+\alpha) \cdot \text{OPT}(t^{(1)}).$$

*Induction Step:* Assume that a) and b) hold for some $i \in \{1, \ldots, k-1\}$. We show that this implies that a) and b) also hold for $i + 1$.

First, consider the case that the schedule $S^{(i)}$ is interrupted. Then, we have $p^{(i+1)} = O$ and $t^{(i+1)} = \alpha \cdot \text{OPT}(t^{(i+1)})$. It immediately follows that $S^{(i+1)} = |S(R^{(i+1)}, p^{(i+1)})| \leq \text{OPT}(t^{(i+1)})$ because $\text{OPT}[t^{(i+1)}]$ serves all requests in $R^{(i+1)}$ (among others) and starts in $O$. With this, we obtain

$$t^{(i+1)} + |S^{(i+1)}| \leq (1+\alpha) \cdot \text{OPT}(t^{(i+1)}).$$

Therefore, a) and b) hold for $i+1$ if $S^{(i)}$ is interrupted. For the rest of the proof, assume that $S^{(i)}$ is not interrupted.

Assume that $\text{OPT}[t^{(i+1)}]$ visits $p^{(i+1)}$ before collecting any request in $R^{(i+1)}$. Then, by definition of $S^{(i+1)} = S(R^{(i+1)}, p^{(i+1)})$, we immediately see that a) holds for $i + 1$ because $\text{OPT}[t^{(i+1)}]$ needs to serve all requests in $R^{(i+1)}$ after visiting point $p^{(i+1)}$. Thus, for the proof of a), it suffices to consider the case that $\text{OPT}[t^{(i+1)}]$ collects some request in $R^{(i+1)}$ before visiting $p^{(i+1)}$. We denote the first request in $R^{(i+1)}$ collected by $\text{OPT}[t^{(i+1)}]$ by $r = (a, b; t)$.

Since $S^{(i)}$ is not interrupted, we have $R^{(i+1)} \cap R^{(i)} = \emptyset$ and thus $t > t^{(i)}$. Together with (1), this implies that $\text{OPT}[t^{(i+1)}]$ collects $r$ at $a$ not earlier than time $t > t^{(i)} \geq \alpha \cdot \text{OPT}(t^{(i)})$. By definition of $r$ and $S(R^{(i+1)}, a)$, this implies

$$\text{OPT}(t^{(i+1)}) \geq t + |S(R^{(i+1)}, a)| > \alpha \cdot \text{OPT}(t^{(i)}) + |S(R^{(i+1)}, a)|. \tag{3}$$

Further, since we assumed that $\text{OPT}[t^{(i+1)}]$ visits $p^{(i+1)}$ after visiting $a$ later than $\alpha \cdot \text{OPT}(t^{(i)})$ and since the server needs at least time $d(a, p^{(i+1)})$ to get from $a$ to $p^{(i+1)}$, we have

$$\text{OPT}(t^{(i+1)}) \geq \alpha \cdot \text{OPT}(t^{(i)}) + d(a, p^{(i+1)}). \tag{4}$$

Let $t_\ell \leq t^{(i+1)}$ denote the largest release time of a request in $R^{(i+1)}$. No requests appears in the, possibly empty, time interval $(t_\ell, t^{(i+1)}]$. Thus, we have $\text{OPT}(t_\ell) = \text{OPT}(t^{(i+1)})$. Recall that the schedule $S^{(i)}$ is not interrupted. In particular, it is not interrupted at time $t_\ell$, i.e., at time $t_\ell$, the server cannot serve all loaded requests and return to the origin until time $\alpha \cdot \text{OPT}(t_\ell)$. At time $t_\ell$, the server can trivially serve all loaded requests in time $t^{(i)} + |S^{(i)}|$ by following the current schedule, which ends in $p^{(i+1)}$. This yields

$$t^{(i)} + |S^{(i)}| + d(p^{(i+1)}, O) > \alpha \cdot \text{OPT}(t_\ell) = \alpha \cdot \text{OPT}(t^{(i+1)}). \tag{5}$$

Recall that $p^{(i+1)}$ is the ending position of $S^{(i)}$ and, therefore, it is the destination of a request in $R^{(i)}$. Since $\text{OPT}[t^{(i)}]$ needs to visit $p^{(i+1)}$ and starts in $O$, we have $d(p^{(i+1)}, O) \leq \text{OPT}(t^{(i)})$. Further, by the induction hypothesis, we have $t^{(i)} + |S^{(i)}| \leq (1 + \alpha) \cdot \text{OPT}(t^{(i)})$. This yields

$$(2 + \alpha) \cdot \text{OPT}(t^{(i)}) \geq t^{(i)} + |S^{(i)}| + d(p^{(i+1)}, O)$$
$$\overset{(5)}{>} \alpha \cdot \text{OPT}(t^{(i+1)})$$
$$\overset{(4)}{\geq} \alpha^2 \cdot \text{OPT}(t^{(i)}) + \alpha \cdot d(a, p^{(i+1)}),$$

so that

$$d(a, p^{(i+1)}) < \frac{1}{\alpha} \cdot (2 + \alpha - \alpha^2) \cdot \text{OPT}(t^{(i)}). \tag{6}$$

The schedule $S^{(i+1)}$ starts in $p^{(i+1)}$ and needs to serve all requests in $R^{(i+1)}$. By applying the triangle inequality, we can conclude that

$$|S^{(i+1)}| \leq d(p^{(i+1)}, a) + |S(R^{(i+1)}, a)|$$
$$\overset{(3)}{<} d(p^{(i+1)}, a) + \text{OPT}(t^{(i+1)}) - \alpha \cdot \text{OPT}(t^{(i)})$$
$$\overset{(6)}{<} \left( \frac{2}{\alpha} + 1 - 2\alpha \right) \cdot \text{OPT}(t^{(i)}) + \text{OPT}(t^{(i+1)})$$
$$\leq \text{OPT}(t^{(i+1)}),$$

where the last inequality holds because we have $\left( \frac{2}{\alpha} + 1 - 2\alpha \right) \leq 0$ if $\alpha \geq \frac{1 + \sqrt{17}}{4} \approx 1.2808$.

It remains to show that b) also holds for $i + 1$. If the schedule $S^{(i)}$ is completed before time $\alpha \cdot \text{OPT}(t^{(i+1)})$, the schedule $S^{(i+1)}$ is started precisely at time $t^{(i+1)} = \alpha \cdot \text{OPT}(t^{(i+1)})$. Together with part a), this yields the assertion. Therefore, assume that $S^{(i)}$ is not completed before time $\alpha \cdot \text{OPT}(t^{(i+1)})$. Then, the schedule $S^{(i+1)}$ is started as soon as $S^{(i)}$ is completed. Together with the induction hypothesis, this implies $t^{(i+1)} = t^{(i)} + |S^{(i)}| \leq (1 + \alpha) \cdot \text{OPT}(t^{(i)})$. Hence, the schedule $S^{(i+1)}$ can be completed in time

$$t^{(i+1)} + |S^{(i+1)}| \leq (1 + \alpha) \cdot \text{OPT}(t^{(i)}) + |S^{(i+1)}|$$
$$\overset{(2),a)}{\leq} (1 + \alpha) \cdot \frac{1}{\alpha} \cdot \text{OPT}(t^{(i+1)}) + \text{OPT}(t^{(i+1)})$$
$$= \left( \frac{1}{\alpha} + 2 \right) \cdot \text{OPT}(t^{(i+1)})$$
$$\leq (1 + \alpha) \cdot \text{OPT}(t^{(i+1)}),$$

where the last inequality holds because we have $\frac{1}{\alpha} + 2 \leq 1 + \alpha$ if $\alpha \geq \frac{1 + \sqrt{5}}{2} = \varphi$.

# 4    Lower Bound for LAZY

In this section, we provide lower bounds on the competitive ratio of $\text{LAZY}(\alpha)$. We give a lower bound construction for $\alpha \geq 1$ and a separate construction for $\alpha < 1$. Together they show that $\text{LAZY}(\alpha)$ cannot be better than $(3/2 + \sqrt{11/12})$-competitive for all $\alpha \geq 0$, i.e., that Corollary 1 holds. Furthermore, they narrow the range of parameter choices that would lead to an improvement over the competitive ratio of $\varphi + 1$.

In the following constructions, we let the metric space $(M, d)$ be the real line, i.e., $M = \mathbb{R}$, $O = 0$, and $d(a, b) = |a - b|$. Note that lower bounds on the line trivially carry over to general metric spaces. Moreover, our constructions work for any given server capacity $c \in \mathbb{N} \cup \{\infty\}$ because a larger server capacity does neither change the behavior of the optimum solution nor the behavior of $\text{LAZY}$.

First, observe that, for any $\alpha \geq 0$, $\text{LAZY}(\alpha)$ has a competitive ratio of at least $1 + \alpha$. This can be easily seen by observing the request sequence consisting of the single request $r_1 = (1, 1; \frac{1}{2})$. In this case, the offline optimum has completed the sequence by time 1, whereas $\text{LAZY}(\alpha)$ waits in $O$ until time $\max(\alpha, \frac{1}{2})$ and then moves to 1 and serves $r_1$ not earlier than $1 + \alpha$.

**Lemma 1.** *For any $\alpha \geq 0$, $\text{LAZY}(\alpha)$ has a competitive ratio of at least $1 + \alpha$ for the open online dial-a-ride problem on the line for any capacity $c \in \mathbb{N} \cup \{\infty\}$.*

Now, we give a construction for the case $\alpha \geq 1$.

**Proposition 1.** *For $\alpha \geq 1$, $\text{LAZY}(\alpha)$ has a competitive ratio of at least $2 + \frac{2}{3\alpha}$ for the open online dial-a-ride problem on the line.*

*Proof.* First, observe that, for $\alpha \geq 1/2 + \sqrt{11/12}$, we have $2 + \frac{2}{3\alpha} \leq 1 + \alpha$ so that the assertion follows from Lemma 1. Therefore, let $\alpha \in [1, 1/2 + \sqrt{11/12})$ and let $\varepsilon > 0$ be small enough such that $3\alpha + 2 > 3\alpha^2 + \alpha\varepsilon$. Note that this is possible because $3\alpha + 2 > 3\alpha^2$ for $\alpha \in [1, (1/2) + \sqrt{11/12})$.

We construct an instance of the open online dial-a-ride problem, where the competitive ratio of $\text{LAZY}(\alpha)$ converges to $2 + \frac{2}{3\alpha}$ for $\varepsilon \to 0$ (cf. Fig. 2). We define the instance by giving the requests

$$r_1 = (0, 1; \varepsilon), \; r_2 = (0, -1; 2\varepsilon), \text{ and } r_3 = (2 - 3\alpha - \varepsilon, 2 - 3\alpha - \varepsilon; 3\alpha + \varepsilon).$$

One solution is to first serve $r_1$ and then $r_2$. This is possible in 3 time units and, after this, the server is in position $-1$. Then, the server can reach point $2 - 3\alpha - \varepsilon$ by time $3 + (3\alpha + \varepsilon - 2 - 1) = 3\alpha + \varepsilon$. At this point in time, $r_3$ is released and can immediately be served. Thus, we have

$$\text{OPT} := \text{OPT}(3\alpha + \varepsilon) = 3\alpha + \varepsilon.$$

We now analyze what $\text{LAZY}(\alpha)$ does on this request sequence. We have $\text{OPT}(2\varepsilon) = 3$. Thus, the server waits in $O$ until time $3\alpha$. Since no new request arrives until this time, the server starts an optimal schedule serving $r_1$ and $r_2$. Without loss of generality, we can assume that $\text{LAZY}(\alpha)$ starts by serving $r_2$, because the starting positions and destinations of $r_1$ and $r_2$ are symmetrical.

At time $3\alpha + \varepsilon$, request $r_3$ is released, and the server has currently loaded $r_2$. Delivering $r_2$ and returning to the origin takes the server until time $3\alpha + 2$. By definition of $\alpha$ and $\varepsilon$, we have

$$3\alpha + 2 > 3\alpha^2 + \alpha\varepsilon = \alpha\text{OPT}. \qquad (7)$$

This implies that the server is not interrupted in its current schedule. It continues serving $r_2$ and then serves $r_1$ at time $3\alpha + 3$. Together with (7), it follows that, after serving $r_1$, the server immediately starts serving the remaining request $r_3$. Moving from 1 to $2 - 3\alpha - \varepsilon$ takes $3\alpha - 1 + \varepsilon$ time units, i.e., the server serves $r_3$ at time $(3\alpha + 3) + (2 - 3\alpha - \varepsilon) = 6\alpha + 2 + \varepsilon$. Thus, the competitive ratio is at least

$$\frac{6\alpha + 2 + \varepsilon}{\text{OPT}} = \frac{6\alpha + 2 + \varepsilon}{3\alpha + \varepsilon} = 2 + \frac{2 - \varepsilon}{3\alpha + \varepsilon}.$$

The statement follows by taking the limit $\varepsilon \to 0$.

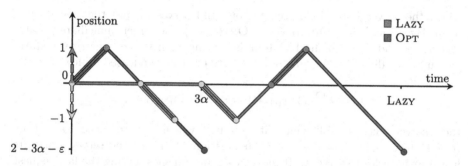

**Fig. 2.** Instance of the open online dial-a-ride problem on the line where $\text{LAZY}(\alpha)$ has a competitive ratio of at least $2 + \frac{2}{3\alpha}$ for all $\alpha \geq 1$.

Next, we give a lower bound construction for $\alpha < 1$.

**Proposition 2.** *For $\alpha \in [0,1)$, the algorithm $\text{LAZY}(\alpha)$ has a competitive ratio of at least $1 + \frac{3}{\alpha+1}$ for the open dial-a-ride problem on the line.*

*Proof.* Let $\alpha \in [0,1)$ and $\varepsilon \in (0, \min\{\frac{\alpha}{2}, \frac{1}{\alpha} - \alpha, 1 - \alpha\})$. We construct an instance of the open dial-a-ride problem, where the competitive ratio of $\text{LAZY}(\alpha)$ converges to $1 + \frac{3}{\alpha+1}$ for $\varepsilon \to 0$ (cf. Fig. 3). We define the instance by giving the requests

$$r_1 = \left(\frac{\varepsilon}{2}, \frac{1}{2}; \frac{\varepsilon}{2}\right), \ r_2 = (1, 1; \varepsilon), \ r_3 = (0, 0; \alpha + \varepsilon),$$

$$r_4 = \left(\frac{1}{2} + \varepsilon, 1; \alpha + 2\varepsilon\right), \text{ and } r_5 = (1, 1; \alpha + 1 + \varepsilon).$$

One solution is to first wait in $O$ until time $\alpha + \varepsilon$ and serve $r_3$. Then, the server can move to $\varepsilon/2$, pick up $r_1$ and deliver it. Then, we can move to $\frac{1}{2} + \varepsilon$, pick up $r_4$ and deliver it. This can be done by time $\alpha + 1 + \varepsilon$. Now, the server is in position 1 and can thus immediately serve $r_5$. It finishes serving all request in time $\alpha + 1 + \varepsilon$. Since the last request is released at time $\alpha + 1 + \varepsilon$, we have

$$\text{OPT} := \text{OPT}(\alpha + 1 + \varepsilon) = \alpha + 1 + \varepsilon. \tag{8}$$

We now analyze what $\text{LAZY}(\alpha)$ does on this request sequence. We have $\text{OPT}(\varepsilon/2) = 1/2$ so that $\alpha \cdot \text{OPT}(\varepsilon/2) = \alpha/2 > \varepsilon$ and the server does not start moving before $r_2$ is released. Then, we have $\text{OPT}(\varepsilon) = 1$. Hence, the server waits in $O$ until time $\alpha$. Since no new requests arrive until this time, the server starts an optimal schedule serving $r_1$ and $r_2$, i.e., it moves to $\varepsilon/2$ and picks up $r_1$. At times $\alpha + \varepsilon$ and $\alpha + 2\varepsilon$, $r_3$ and $r_4$ are released. We have $\text{OPT}(\alpha + \varepsilon) = \text{OPT}(\alpha + 2\varepsilon) = \alpha + 1 + \varepsilon$. Serving the loaded request $r_1$ and returning to 0 would take the server until time

$$\alpha + 1 \overset{\varepsilon < \frac{1}{\alpha} - \alpha}{>} \alpha + (\alpha^2 + \alpha\varepsilon) = \alpha(\alpha + 1 + \varepsilon) = \alpha \cdot \text{OPT}(\alpha + \varepsilon). \tag{9}$$

Thus, the server keeps following its tour and serves $r_1$ and then $r_2$ at time $\alpha + 1$. By (9) and since $\text{OPT}(\alpha + 1) = \text{OPT}(\alpha + \varepsilon)$, the server immediately starts serving $r_3$ and $r_4$. The shortest tour is serving $r_4$ first, i.e., the server starts moving towards $\frac{1}{2} + \varepsilon$. At time $\alpha + 1 + \varepsilon$, request $r_5$ is released. Since

$$\alpha + 1 + \varepsilon \overset{(8)}{=} \text{OPT}(\alpha + 1 + \varepsilon) > \alpha \cdot \text{OPT}(\alpha + 1 + \varepsilon),$$

the server keeps following its tour, which is finished at time $(\alpha + 1) + (1 - 2\varepsilon) + 1 = 3 + \alpha - 2\varepsilon$ in position $O$. Then, the server starts its last tour in order to serve $r_5$. It moves to 1 and finishes serving the last request at time $4 + \alpha - 2\varepsilon$. Thus, the competitive ratio is

$$\frac{4 + \alpha - 2\varepsilon}{\text{OPT}} = \frac{4 + \alpha - 2\varepsilon}{\alpha + 1 + \varepsilon} = 1 + \frac{3 - 3\varepsilon}{\alpha + 1 + \varepsilon}.$$

The statement follows by taking the limit $\varepsilon \to 0$.

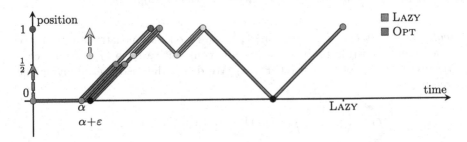

**Fig. 3.** Instance of the open online dial-a-ride problem on the line where $\text{LAZY}(\alpha)$ has a competitive ratio of at least $1 + \frac{3}{\alpha + 1}$ for all $\alpha \in [0, 1)$.

We now give another lower bound construction which is stonger than the previous one for large $\alpha < 1$.

**Proposition 3.** *For $\alpha \in [0, 1)$, the algorithm* LAZY$(\alpha)$ *has a competitive ratio of at least $2 + \alpha + \frac{1-\alpha}{2+3\alpha}$ for the open online dial-a-ride problem on the line.*

*Proof.* Let $\varepsilon > 0$ be small enough. We construct an instance of the open online dial-a-ride problem on the line, where the competitive ratio of LAZY$(\alpha)$ converges to $2 + \alpha + \frac{1-\alpha}{2+3\alpha}$ for $\varepsilon \to 0$ (cf. Fig. 4). We distinguish between three cases.

*Case 1 ($\alpha \in [0, 2/3)$):* We define the instance by giving the requests

$$r_1 = (0, 1; \varepsilon),$$
$$r_2 = (-\alpha, -\alpha; \alpha + \varepsilon),$$
$$r_3 = (2 + \alpha - \varepsilon, 2 + \alpha - \varepsilon; \alpha + 2\varepsilon),$$
$$r_4 = (2 + \alpha - \varepsilon, 2 + \alpha - \varepsilon; 2 + 3\alpha).$$

One solution is to move to $-\alpha$ and wait there until time $\alpha + \varepsilon$. Then, $r_2$ can be served, and the server can move to $0$ where it arrives at time $2\alpha + \varepsilon$. It picks up $r_1$ and delivers it at time $2\alpha + 1 + \varepsilon$ at $1$. It keeps moving to position $2 + \alpha - \varepsilon$ and serves $r_3$ and $r_4$ there at time $2 + 3\alpha$. Since the last request is released at time $2 + 3\alpha$, we have

$$\text{OPT} := \text{OPT}(2 + 3\alpha) = 2 + 3\alpha.$$

We now analyze what LAZY$(\alpha)$ does. We have $\text{OPT}(\varepsilon) = 1 + \varepsilon$. Thus, the server starts waiting in $0$ until time $\alpha(1 + \varepsilon)$. At time $\alpha(1 + \varepsilon)$, the server starts an optimal schedule over all unserved requests, i.e., over $\{r_1\}$. It picks up $r_1$ and starts moving towards $1$. At time $\alpha + \varepsilon > \alpha(1 + \varepsilon)$, request $r_2$ arrives. We have $\text{OPT}(\alpha + \varepsilon) = 2\alpha + 1 + \varepsilon$. Serving the loaded request $r_1$ and returning to the origin would take the server until time

$$\alpha(1 + \varepsilon) + 2 > \alpha(2\alpha + 1 + \varepsilon) = \alpha\text{OPT}(\alpha + \varepsilon),$$

i.e., the server continues its current schedule. At time $\alpha + 2\varepsilon$, $r_3$ is released. We have $\text{OPT}(\alpha + 2\varepsilon) = 2 + 3\alpha$. Serving the loaded request and returning to the origin would still take until time

$$\alpha(1 + \varepsilon) + 2 > 2\alpha + 3\alpha^2 = \alpha\text{OPT}(\alpha + 2\varepsilon),$$

where the inequality follows from the fact that $\alpha < \frac{2}{3}$. Thus, the server continues the current schedule, which is finished at time $\alpha(1 + \varepsilon) + 1$ in position $1$. The server waits until time $\max\{1 + \alpha, \alpha\text{OPT}(\alpha + 2\varepsilon)\}$ and then starts the next schedule. Since $\alpha\text{OPT}(\alpha + 2\varepsilon) = 2\alpha + 3\alpha^2 < 2 + 3\alpha$, the next schedule is thus started before $r_4$ is released. It is faster to serve $r_3$ before $r_2$ in this schedule because the server starts from point $1$. At time $2 + 3\alpha$, request $r_4$ is released. Since this does not change the completion time of the optimum and because the server started the current schedule not earlier than time $\alpha\text{OPT}(\alpha + 2\varepsilon)$, it continues the current schedule. The second schedule takes $(1+\alpha-\varepsilon)+(2+2\alpha-\varepsilon) = 3+3\alpha-2\varepsilon$ time units and ends in $-\alpha$. The last schedule, in which $r_4$ is served, is started immediately and takes $2 + 2\alpha - \varepsilon$ time units. Hence, LAZY$(\alpha)$ takes at least

$$(\alpha(2 + 3\alpha)) + (3 + 3\alpha - 2\varepsilon) + (2 + 2\alpha - \varepsilon) = 5 + 7\alpha + 3\alpha^2 - 3\varepsilon$$

time units to serve all requests.

*Case 2* $(\alpha \in [\frac{2}{3}, \frac{\sqrt{37-12\varepsilon}-1}{6}))$: We define the instance by giving the requests

$$r_1 = (0, 1; \varepsilon),$$
$$r_2 = (-\alpha, -\alpha; \alpha + \varepsilon),$$
$$r_3 = \left(\frac{2}{\alpha} + 1 - 2\alpha - \varepsilon, \frac{2}{\alpha} + 1 - 2\alpha - \varepsilon; \alpha + 2\varepsilon\right),$$
$$r_4 = (2 + \alpha - \varepsilon, 2 + \alpha - \varepsilon; 2 + \alpha - \varepsilon),$$
$$r_5 = (2 + \alpha - \varepsilon, 2 + \alpha - \varepsilon; 2 + 3\alpha).$$

One solution is to move to $-\alpha$ and wait there until time $\alpha + \varepsilon$. Then, $r_2$ can be served, and the server can move to 0 where it arrives at time $2\alpha + \varepsilon$. It picks up $r_1$ and delivers it at time $2\alpha + 1 + \varepsilon$ at 1. It keeps moving to position $\frac{2}{\alpha} + 1 - 2\alpha - \varepsilon$, where it arrives at time $\frac{2}{\alpha} + 1$ and immediately serves $r_3$. It continues to move to $2 + \alpha - \varepsilon$ and serves $r_4$ and $r_5$ there at time $2 + 3\alpha$. Since the last request is released at time $2 + 3\alpha$, we have

$$\text{OPT} = \text{OPT}(2 + 3\alpha) = 2 + 3\alpha.$$

We now analyze what LAZY$(\alpha)$ does. We have OPT$(\varepsilon) = 1 + \varepsilon$. Thus, the server starts waiting in 0 until time $\alpha(1 + \varepsilon)$. At time $\alpha(1 + \varepsilon)$, the server starts an optimal schedule over all unserved requests, i.e., over $\{r_1\}$. It picks up $r_1$ and starts moving towards 1. At time $\alpha + \varepsilon > \alpha(1 + \varepsilon)$, request $r_2$ arrives. We have OPT$(\alpha + \varepsilon) = 2\alpha + 1 + \varepsilon$. Serving the loaded request $r_1$ and returning to the origin would take the server until time

$$\alpha(1 + \varepsilon) + 2 > \alpha(2\alpha + 1 + \varepsilon) = \alpha\text{OPT}(\alpha + \varepsilon),$$

i.e., the server continues its current schedule. At time $\alpha + 2\varepsilon$, $r_3$ is released. We have OPT$(\alpha + 2\varepsilon) = \frac{2}{\alpha} + 1$. Serving the loaded request and returning to the origin would still take until time

$$\alpha(1 + \varepsilon) + 2 > 2 + \alpha = \alpha\text{OPT}(\alpha + 2\varepsilon).$$

Thus, the server continues the current schedule which is finished at time $1 + \alpha$ in position 1. Since $1 + \alpha < 2 + \alpha = \alpha\text{OPT}(\alpha + 2\varepsilon)$, the server starts waiting in 1 until time $2 + \alpha$. At time $2 + \alpha - \varepsilon$, request $r_4$ is released. We have OPT$(2 + \alpha - \varepsilon) = 2 + 3\alpha$. It would take the server until time

$$3 + \alpha - \varepsilon > 2\alpha + 3\alpha^2 = \alpha\text{OPT}(2 + \alpha - \varepsilon)$$

to return to the origin, where the inequality follows from the fact that $\alpha < \frac{\sqrt{37-12\varepsilon}-1}{6}$. Thus, the server starts waiting until time $\alpha\text{OPT}(2+\alpha-\varepsilon) = 2\alpha + 3\alpha^2$. After it finished waiting, it starts the second schedule and tries to serve $r_3$, $r_4$ and then $r_2$. At time $2 + 3\alpha$, request $r_5$ is released. Since this does not change the completion time of the optimum and because the server started the current schedule at time $\alpha\text{OPT}(\alpha+2\varepsilon)$, it continues its current schedule. The second schedule takes $(1 + \alpha - \varepsilon) + (2 + 2\alpha - \varepsilon) = 3 + 3\alpha - 2\varepsilon$ time units and ends in $-\alpha$. The last schedule, in which $r_5$ is served, is started immediately and takes $2 + 2\alpha - \varepsilon$ time units. Hence, $\text{LAZY}(\alpha)$ takes until time

$$(\alpha(2 + 3\alpha)) + (3 + 3\alpha - 2\varepsilon) + (2 + 2\alpha - \varepsilon) = 5 + 7\alpha + 3\alpha^2 - 3\varepsilon$$

to serve all requests.

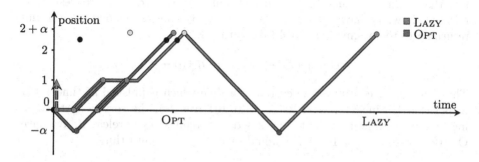

**Fig. 4.** Instance of the open online dial-a-ride problem on the line where $\text{LAZY}(\alpha)$ has a competitive ratio of at least $2 + \alpha\frac{1-\alpha}{2+3\alpha}$ for all $\alpha \in [0, 1)$. This is the construction of Case 2 in the proof of Theorem 3.

*Case 3 ($\alpha \in [\frac{\sqrt{37-12\varepsilon}-1}{6}, 1)$):* We define the instance by giving the requests

$$r_1 = (0, 1; \varepsilon),$$
$$r_2 = (-\alpha, -\alpha; \alpha + \varepsilon),$$
$$r_3 = \left(\frac{2}{\alpha} + 1 - 2\alpha - \varepsilon, \frac{2}{\alpha} + 1 - 2\alpha - \varepsilon; \alpha + 2\varepsilon\right),$$
$$r_4 = \left(\frac{3}{\alpha} + 1 - 2\alpha - \frac{(2+\alpha)\varepsilon}{\alpha}, \frac{3}{\alpha} + 1 - 2\alpha - \frac{(2+\alpha)\varepsilon}{\alpha}; 2 + \alpha - \varepsilon\right),$$
$$r_5 = (2 + \alpha - \varepsilon, 2 + \alpha - \varepsilon; 3 + \alpha - 3\varepsilon),$$
$$r_6 = (2 + \alpha - \varepsilon, 2 + \alpha - \varepsilon; 2 + 3\alpha).$$

One solution is to move to $-\alpha$ and wait there until time $\alpha + \varepsilon$. Then, $r_2$ can be served, and the server can move to $0$ where it arrives at time $2\alpha + \varepsilon$. It picks up $r_1$ and delivers it at time $2\alpha + 1 + \varepsilon$ at $1$. It keeps moving to position $\frac{2}{\alpha} + 1 - 2\alpha - \varepsilon$, where it arrives at time $\frac{2}{\alpha} + 1$ and immediately serves $r_3$. It continues to move to $\frac{3}{\alpha} + 1 - 2\alpha - \frac{(2+\alpha)\varepsilon}{\alpha}$, where it arrives at time $\frac{3}{\alpha} + 1 - \frac{2\varepsilon}{\alpha}$ and immediately serves $r_4$. Lastly, it moves to $2 + \alpha - \varepsilon$, where it arrives at time $2 + 3\alpha$ and serves $r_4$ and $r_5$ there. Since the last request is released at time $2 + 3\alpha$, we have

$$\text{OPT} = \text{OPT}(2 + 3\alpha) = 2 + 3\alpha.$$

We now analyze what $\text{LAZY}(\alpha)$ does. We have $\text{OPT}(\varepsilon) = 1 + \varepsilon$. Thus, the server starts waiting in 0 until time $\alpha(1 + \varepsilon)$. At time $\alpha(1 + \varepsilon)$, the server starts an optimal schedule over all unserved requests, i.e., over $\{r_1\}$. It picks up $r_1$ and starts moving towards 1. At time $\alpha + \varepsilon > \alpha(1 + \varepsilon)$, request $r_2$ arrives. We have $\text{OPT}(\alpha + \varepsilon) = 2\alpha + 1 + \varepsilon$. Serving the loaded request $r_1$ and returning to the origin would take the server until time

$$\alpha(1 + \varepsilon) + 2 > \alpha(2\alpha + 1 + \varepsilon) = \alpha\text{OPT}(\alpha + \varepsilon),$$

i.e., the server continues its current schedule. At time $\alpha + 2\varepsilon$, request $r_3$ is released. Now, we have $\text{OPT}(\alpha + 2\varepsilon) = \frac{2}{\alpha} + 1$. Serving the loaded request and returning to the origin would still take until time

$$\alpha(1 + \varepsilon) + 2 > 2 + \alpha = \alpha\text{OPT}(\alpha + 2\varepsilon).$$

Thus, the server continues the current schedule which is finished at time $1 + \alpha$ in position 1. Since $1 + \alpha < 2 + \alpha = \alpha\text{OPT}(\alpha + 2\varepsilon)$, the server starts waiting in 1 until time $2 + \alpha$. At time $2 + \alpha - \varepsilon$, request $r_4$ is released. We have $\text{OPT}(2 + \alpha - \varepsilon) = \frac{3}{\alpha} + 1 - \frac{2\varepsilon}{\alpha}$. It would take the server until time

$$3 + \alpha - \varepsilon > 3 + \alpha - 2\varepsilon = \alpha\text{OPT}(2 + \alpha - \varepsilon),$$

to return to the origin, i.e., the server starts waiting until time $\alpha\text{OPT}(2 + \alpha - \varepsilon) = 3 + \alpha - 2\varepsilon$. At time $3 + \alpha - 3\varepsilon$, request $r_5$ is released. We have $\text{OPT}(3 + \alpha - 3\varepsilon) = 2 + 3\alpha$. It would take the server until time

$$4 + \alpha - 3\varepsilon > 2\alpha + 3\alpha^2 = \alpha\text{OPT}(3 + \alpha - 3\varepsilon)$$

to return to the origin, where the inequality follows from the fact that $\alpha < 1$ and that $\varepsilon$ is small. Thus, the server waits in 1 until time $\alpha\text{OPT}(3 + \alpha - 3\varepsilon) = 2\alpha + 3\alpha^2$ and then starts its second schedule to serve requests $r_3$, $r_4$, $r_5$ and then $r_2$. At time $2 + 3\alpha$, request $r_6$ is released. Since this does not change the completion time of the optimum and because the server started the current schedule at time $\alpha\text{OPT}(3 + \alpha - 3\varepsilon)$, it continues its current schedule. The second schedule takes $(1 + \alpha - \varepsilon) + (2 + 2\alpha - \varepsilon) = 3 + 3\alpha - 2\varepsilon$ time units and ends in $-\alpha$. The last schedule, in which $r_6$ is served, is started immediately and takes $2 + 2\alpha - \varepsilon$ time units. Hence, $\text{LAZY}(\alpha)$ takes until time

$$(\alpha(2 + 3\alpha)) + (3 + 3\alpha - 2\varepsilon) + (2 + 2\alpha - \varepsilon) = 5 + 7\alpha + 3\alpha^2 - 3\varepsilon$$

to serve all requests.

In all three cases, the optimal solution is $2 + 3\alpha$ and the algorithm takes at least $5 + 7\alpha + 3\alpha^2 - 3\varepsilon$ time units. Thus, the competitive ratio of LAZY($\alpha$) is at least

$$\frac{5 + 7\alpha + 3\alpha^2 - 3\varepsilon}{2 + 3\alpha} = 2 + \alpha + \frac{1 - \alpha - 3\varepsilon}{2 + 3\alpha}.$$

The statement follows by taking the limit $\varepsilon \to 0$.

Now, we combine our results for the lower bounds (cf. Fig. 5). Combining Lemma 1 and Proposition 1, we obtain that, for $\alpha \geq 1$, LAZY($\alpha$) has a competitive ratio of at least $\max\{1 + \alpha, 2 + 2/3\alpha\}$, which proves the lower bound of Theorem 1. Minimizing over $\alpha \geq 1$ yields a competitive ratio of at least $3/2 + \sqrt{11/12} > 2.457$ in that domain. For the case $\alpha < 1$, we have seen in Proposition 2 that the algorithm LAZY($\alpha$) has a competitive ratio of at least $1 + 3/(\alpha + 1) > 5/2$. Together, this proves Corollary 1.

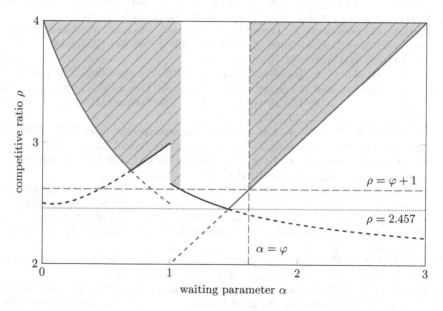

**Fig. 5.** Lower bounds on the competitive ratio of LAZY($\alpha$) depending on $\alpha$. The lower bound of Lemma 1 is depicted in green, the lower bound of Proposition 1 in red, the lower bound of Proposition 2 in orange, and the lower bound of Proposition 3 in blue. The highlighted area over the plot indicates the domain of $\alpha$ for which no improvement over $1 + \varphi$ is possible. (Color figure online)

Moreover, we conclude that the results above narrow the range for $\alpha$ in which LAZY($\alpha$) might have a competitive ratio better than $\varphi + 1$. By Lemma 1, it follows that LAZY($\alpha$) cannot have a better competitive ratio than $\varphi + 1$ for any $\alpha > \varphi \approx 1.618$. By Proposition 1, we obtain that LAZY($\alpha$) has competitive ratio at least $\varphi + 1$ for any $\alpha$ with $1 \leq \alpha \leq \frac{2\varphi}{3} \approx 1.079$. Proposition 2 yields that, for $0 \le \alpha \le 0.695$, the competitive ratio of LAZY($\alpha$) is at least $2.768 > \varphi + 1$. Lastly, for $0.695 < \alpha < 1$, Proposition 3 gives a lower bound of 2.768 on the competitive ratio of LAZY($\alpha$). To summarize, an improvement of the competitive ratio of LAZY($\alpha$) might only be possible for some $\alpha \in (2\varphi/3, \varphi) \approx [1.08, 1.618)$, which proves Corollary 2.

# References

1. Allulli, L., Ausiello, G., Laura, L.: On the power of lookahead in on-line vehicle routing problems. In: Wang, L. (ed.) COCOON 2005. LNCS, vol. 3595, pp. 728–736. Springer, Heidelberg (2005). https://doi.org/10.1007/11533719_74
2. Ascheuer, N., Krumke, S.O., Rambau, J.: Online dial-a-ride problems: minimizing the completion time. In: Reichel, H., Tison, S. (eds.) STACS 2000. LNCS, vol. 1770, pp. 639–650. Springer, Heidelberg (2000). https://doi.org/10.1007/3-540-46541-3_53
3. Ausiello, G., Demange, M., Laura, L., Paschos, V.: Algorithms for the on-line quota traveling salesman problem. Inf. Process. Lett. **92**(2), 89–94 (2004)
4. Ausiello, G., Feuerstein, E., Leonardi, S., Stougie, L., Talamo, M.: Algorithms for the on-line travelling salesman. Algorithmica **29**(4), 560–581 (2001). https://doi.org/10.1007/s004530010071
5. Ausiello, G., Allulli, L., Bonifaci, V., Laura, L.: On-line algorithms, real time, the virtue of laziness, and the power of clairvoyance. In: Cai, J.-Y., Cooper, S.B., Li, A. (eds.) TAMC 2006. LNCS, vol. 3959, pp. 1–20. Springer, Heidelberg (2006). https://doi.org/10.1007/11750321_1
6. Bienkowski, M., Kraska, A., Liu, H.: Traveling repairperson, unrelated machines, and other stories about average completion times. In: Bansal, N., Merelli, E., Worrell, J. (eds.) Proceedings of the 48th International Colloquium on Automata, Languages, and Programming (ICALP), pp. 1–20 (2021)
7. Bienkowski, M., Liu, H.: An improved online algorithm for the traveling repairperson problem on a line. In: Proceedings of the 44th International Symposium on Mathematical Foundations of Computer Science (MFCS), pp. 6:1–6:12 (2019)
8. Birx, A.: Competitive analysis of the online dial-a-ride problem. Ph.D. Thesis, TU Darmstadt (2020)
9. Birx, A., Disser, Y.: Tight analysis of the smartstart algorithm for online dial-a-ride on the line. SIAM J. Discret. Math. **34**(2), 1409–1443 (2020)
10. Birx, A., Disser, Y., Schewior, K.: Improved bounds for open online dial-a-ride on the line. In: Proceedings of the 22nd International Workshop on Approximation Algorithms for Combinatorial Optimization Problems (APPROX), vol. 145, p. 21(22) (2019)
11. Bjelde, A., et al.: Tight bounds for online TSP on the line. ACM Transact. Algorithms **17**(1), 1–58 (2020)
12. Blom, M., Krumke, S.O., de Paepe, W.E., Stougie, L.: The online TSP against fair adversaries. INFORMS J. Comput. **13**(2), 138–148 (2001)
13. Bonifaci, V., Stougie, L.: Online k-server routing problems. Theor. Comput. Syst. **45**(3), 470–485 (2008)
14. Feuerstein, E., Stougie, L.: On-line single-server dial-a-ride problems. Theoret. Comput. Sci. **268**(1), 91–105 (2001)
15. Hauptmeier, D., Krumke, S., Rambau, J., Wirth, H.C.: Euler is standing in line dial-a-ride problems with precedence-constraints. Discret. Appl. Math. **113**(1), 87–107 (2001)

16. Hauptmeier, D., Krumke, S.O., Rambau, J.: The online dial-a-ride problem under reasonable load. In: Bongiovanni, G., Petreschi, R., Gambosi, G. (eds.) CIAC 2000. LNCS, vol. 1767, pp. 125–136. Springer, Heidelberg (2000). https://doi.org/10.1007/3-540-46521-9_11

17. Jaillet, P., Lu, X.: Online traveling salesman problems with service flexibility. Networks **58**(2), 137–146 (2011)

18. Jaillet, P., Lu, X.: Online traveling salesman problems with rejection options. Networks **64**(2), 84–95 (2014)

19. Jaillet, P., Wagner, M.R.: Generalized online routing: new competitive ratios, resource augmentation, and asymptotic analyses. Oper. Res. **56**(3), 745–757 (2008)

20. Jawgal, V.A., Muralidhara, V.N., Srinivasan, P.S.: Online travelling salesman problem on a circle. In: Gopal, T.V., Watada, J. (eds.) TAMC 2019. LNCS, vol. 11436, pp. 325–336. Springer, Cham (2019). https://doi.org/10.1007/978-3-030-14812-6_20

21. Krumke, S.O.: Online optimization competitive analysis and beyond. Habilitation thesis, Zuse Institute Berlin (2001)

22. Krumke, S.O., et al.: Non-abusiveness helps: an O(1)-competitive algorithm for minimizing the maximum flow time in the online traveling salesman problem. In: Jansen, K., Leonardi, S., Vazirani, V. (eds.) APPROX 2002. LNCS, vol. 2462, pp. 200–214. Springer, Heidelberg (2002). https://doi.org/10.1007/3-540-45753-4_18

23. Krumke, S.O., de Paepe, W.E., Poensgen, D., Lipmann, M., Marchetti-Spaccamela, A., Stougie, L.: On minimizing the maximum flow time in the online dial-a-ride problem. In: Erlebach, T., Persinao, G. (eds.) WAOA 2005. LNCS, vol. 3879, pp. 258–269. Springer, Heidelberg (2006). https://doi.org/10.1007/11671411_20

24. Krumke, S.O., de Paepe, W.E., Poensgen, D., Stougie, L.: News from the online traveling repairman. Theoret. Comput. Sci. **295**(1–3), 279–294 (2003)

25. Lipmann, M., Lu, X., de Paepe, W.E., Sitters, R.A., Stougie, L.: On-line dial-a-ride problems under a restricted information model. Algorithmica **40**(4), 319–329 (2004)

26. Lippmann, M.: On-line routing. Ph.D. thesis, Technische Universiteit Eindhoven (2003)

27. Yi, F., Tian, L.: On the online dial-a-ride problem with time-windows. In: Megiddo, N., Xu, Y., Zhu, B. (eds.) AAIM 2005. LNCS, vol. 3521, pp. 85–94. Springer, Heidelberg (2005). https://doi.org/10.1007/11496199_11

# Stochastic Graph Exploration with Limited Resources

Ilan Reuven Cohen$^{(\boxtimes)}$ (iD)

Faculty of Engineering, Bar-Ilan University, Ramat Gan, Israel
`ilan-reuven.cohen@biu.ac.il`

**Abstract.** In recent years, the explosion of research on large-scale networks has been fueled to a large extent by the increasing availability of large, detailed network data sets. Specifically, exploration of social networks constitutes a growing field of research, as they generate a huge amount of data on a daily basis and are the main tool for networking, communications, and content sharing. Exploring these networks is resource-consuming (time, money, energy, etc.). Moreover, uncertainty is a crucial aspect of graph exploration since links costs are unknown in advance, e.g., creating a positive influence between two people in social networks. One approach to model this problem is the stochastic graph exploration problem [4], where, given a graph and a source vertex, rewards on vertices, and distributions for the costs of the edges. The goal is to probe a subset of the edges, so the total cost of the edges is at most some prespecified budget, and the subgraph is connected, containing the source vertex, and maximizes the total reward of the spanned vertices. In this stochastic setting, an optimal probing strategy is likely to be adaptive, i.e., it may determine the next edge to probe based on the realized costs of the already probed edges. As computing such adaptive strategies is intractable [15], we focus on developing non-adaptive strategies, which fix a list of edges to probe in advance. A non-adaptive strategy would not be competitive versus the optimal adaptive one unless it uses a budget augmentation. The current results demand an augmentation factor, which depends logarithmically on the number of nodes. Such a factor is unrealistic in large-scale network scenarios. In this paper, we provide constant competitive non-adaptive strategies using only a constant budget augmentation for various scenarios.

**Keywords:** Stochastic optimization · Graph exploration · Non-adaptive strategies

## 1 Introduction

Network exploration is a fundamental paradigm for discovering information available at the nodes of a network. The rise of social networks increased the number of nodes and links dramatically, and exploring them demands many resources. A network exploration algorithm defines a probing strategy that decides at each stage of the process which edges to probe. For the exploration of social networks, the network structure is known in

This research was supported by the Israel Science Foundation (grant No. 1737/21).

© The Author(s), under exclusive license to Springer Nature Switzerland AG 2022
P. Chalermsook and B. Laekhanukit (Eds.): WAOA 2022, LNCS 13538, pp. 172–189, 2022.
https://doi.org/10.1007/978-3-031-18367-6_9

advance, e.g., followers on Twitter, friends on Facebook, etc. However, the aggregation of information in this network is uncertain, e.g., whether tweets/posts of a user would be retweeted/shared by their followers. Most of the recent work [23,26,27] deals with real-world networks' exploration when a limited budget is available, but they do not provide a comprehensive theoretical study of these problems. In this work, we continue the theoretical study of exploring networks, initiated by [4].

The main difficulty in designing effective probing strategies, other than their enormous size, is that they must perform well in an uncertain environment, where the amount of resource (cost) associated with a specific link (edge) is unknown in advance. A common assumption, instead, is that the distributions of the edges' costs may be well-estimated by using various properties of the connecting nodes. Accordingly, a good probing strategy might have an adaptive nature; it may determine the future network portion to be explored according to the realized cost of the edges already explored. Unfortunately, finding such optimal adaptive strategies is often intractable [15]. Moreover, implementing an efficient adaptive strategy might be impossible. In various exploration process applications, many machines work in parallel, and the updated adaptive strategy must be communicated to the machines participating in the process. As a result, the communication cost required by an adaptive strategy may also be high. Therefore, we are interested in devising non-adaptive probing strategies that are simple and that define the sequence of probes in advance before the process is started. We continue the recent line of research [8,21,22] developing polynomial computable non-adaptive strategies that are competitive against the optimal adaptive strategy.

In this work, we extended the work in [4] where they considered exploring a network from a root node. The graph has deterministic rewards on nodes and costs on edges, where each edge cost is drawn independently from a known distribution. They mainly focused on designing non-adaptive strategies for exploring the graph, where they demonstrated that in order to achieve a reasonable approximation guaranteeing a budget augmentation is mandatory, so a crucial aspect is the augmentation factor used by the non-adaptive strategy. Unfortunately, they only provide algorithms that use budget augmentation, which depends on the logarithm of the number of nodes in the network (and the maximal revenue from a node). As the main motivation is exploring large-scale networks where the number of nodes is immense, those algorithms are, in fact, impractical. In this paper, we develop non-adaptive algorithms for strategies that use a considerably less amount of resources compared to the currently known algorithms.

**Problem Definition.** Given an instance $\mathcal{I} = (V, E, r, \pi, \mathcal{R}, B)$, the underlying graph is $G(V, E)$ and $r \in V$ is the source vertex, and $n = |V|$ is the number of vertices. The edge costs $C : E \to \mathbb{R}_{\geq 0}$ are drawn independently according to $\pi(e)$, for $e \in E$, and deterministic rewards of vertices $\mathcal{R} : V \to \mathbb{R}_{\geq 0}$. (The model can be easily extended to rewards distributed according to independent random variables.) A graph-exploration process constructs a set of edges $F \subseteq E$ that it probes. All vertices of the subgraph of $G$ spanned by $F$ must be connected to $r$ via the edges of $F$. The process probes one by one the edges and adds them to $F$. The actual cost of an edge $e$, drawn independently from the distribution $\pi(e)$, is revealed only when the edge is probed. The objective is to maximize the expected total reward from the vertices spanned by the edge set $F$, while the total cost of the edges in $F$ remains bounded by a prespecified budget $B$. As soon as the total cost of $F$ exceeds $B$, the process terminates. Our goal is to design a polynomial computable non-adaptive strategy, where, given an instance $\mathcal{I}$, it computes an ordered list of the edges in advance, where every prefix of the list induces a subtree that contains $r$. The expected gain of a list strategy is the sum of the vertex's reward times the probability that the vertex is successfully added, i.e., that the total cost of the vertex's list prefix does not exceed the budget. We compare this gain to an *adaptive strategy* expected gain, which may decide on the next edge to be probed after the cost of all previously probed edges is revealed. Note that the adaptive algorithm does not know the realization of the edges' costs in advance, and once it probes an edge, it must be added to its set. As mentioned, there exist simple instances where the ratio between any non-adaptive expected reward to an adaptive reward is $\Omega(n)$. Therefore, we allow the non-adaptive one to use a limited amount of budget augmentation. Accordingly, we call an algorithm $(\alpha, \beta)$-approximate if it computes a strategy which uses budget $\beta \cdot B$, and obtains an expected reward of at least $1/\alpha$ times the optimal reward (obtained by an adaptive algorithm). In this work, we will focus on algorithms that are $(O(1), O(1))$-approximate, i.e., algorithms for strategies that use a constant factor of budget augmentation and ensure a constant fraction of the optimal adaptive reward.

## 1.1 Summary of Results and Techniques

Our main contribution is developing non-adaptive competitive algorithms for the stochastic graph exploration problem, which uses only a constant amount of budget augmentations. We provide positive results for two important scenarios: spider graphs and bounded-weighted-depth trees.

**Spider Graphs.** We study a spider-tree graph where all the vertices except the root have an out-degree of at most 1. These graphs are natural extensions for job scheduling applications for the stochastic knapsack, introduced in [15]. In this application, the goal is to schedule a maximum-value subset of $n$ jobs with uncertain duration within a fixed amount of time, and captures the reality of job scheduling, where we cannot go back in time, in case a scheduled job has taken too long to complete. The spider graph instances apply to scenarios where before processing a certain job, it must process some chain of jobs in advance, and the duration of each job in the chain is stochastic. Note that for scenarios where the reward is just when completing a complete job chain (i.e., the reward is zero on non-leafs vertices), the spider-tree formulation gives extra power to the adaptive strategies versus the stochastic knapsack formulation. This arises from the fact that in such instances, an adaptive strategy may abort a chain in the middle, while in the knapsack setting, the entire chain must be represented as a single edge (so if begun, it must be completed). On the other hand, in both formulations, the non-adaptive strategies must complete the entire chain of jobs once started.

**Bounded Weighted Depth Trees.** For a general tree structure, we achieved a constant approximation and augmentation for bounded weighted depth trees instances. In those instances, the expected cost of the path from the root to each node is bounded (with respect to the entire budget). The structure of the social network (as in *Six degrees of separation assumption* [30]) implies that this scenario is extremely practical.

**Our Techniques.** Given an instance $\mathcal{I}$, we bound the value of the optimal adaptive strategy using a linear program (LP) formulation $\Phi_{\mathcal{I}}$. This extends the LP formulation of [15] and captures the dependence between the probing probabilities of different edges in the graph. In addition, we define a relaxation of this LP, namely $\hat{\Phi}_{\mathcal{I}}$, and characterize the structure of an optimal solution for it. In order to solve *spider graphs* instances, we divide the set of nodes into two, the risky nodes, where the expected cost from the root to those nodes is more than half of the entire budget, and the rest of the nodes (which are a connected component that contains $r$). For the risky nodes, using the solution of the LP $\Phi_{\mathcal{I}}$, we prove that a non-adaptive strategy that probes a single leg and uses budget augmentation is a constant competitive. For the non-risky nodes, we use the solution's structure of $\hat{\Phi}_{\mathcal{I}}$ and present a competitive set strategy. Combining the two results implies a constant approximation algorithm that uses at most constant augmentation. Our algorithm for *bounded-weighted-depth trees* instances also uses the solution of $\hat{\Phi}_{\mathcal{I}}$. However, the resulting structure of the optimal solution is more complex, and the total expected cost of the edges in support of the solution might be much higher than the budget. To address this issue, we introduce the *Tree Decomposition* algorithm, which can decompose such a solution to a bounded number of subtrees, where each subtree contains $r$ and has a bounded total expected cost. We prove that sampling one of those subtrees (entirely) is a competitive non-adaptive strategy. Finally, we prove the limitation of the proposed LP $\Phi_{\mathcal{I}}$, by presenting an instance where the gap between the LP and an adaptive solution (even with budget augmentation) is unbounded, which raises questions about future directions on how to bound a general instance of this problem.

## 1.2  Related Work

The first work on the adaptivity gap of stochastic problems has been studied for the knapsack problem [10,15]. Note that the stochastic knapsack problem is a special case of the problem we study, where the underlying graph is a star-graph. As mentioned, the model studied in this paper has been introduced in [4]. In [4], they show an $\Omega(n)$ adaptivity gap for this problem, and in order to circumvent the impossibility result, they allow a limited amount of resource augmentation. They presented polynomial time computable non-adaptive strategies in a graph of $n$ vertices and where $R$ is maximum reward of a vertex, which are $(O(1), O(\log nR))$-approximate for trees and $(O(\log(nR)), O(\log(nR)))$-approximate for general graphs.

Another set of problems related to the stochastic graph exploration problem are various *stochastic orienteering* problems [8,19,28]. In stochastic orienteering, the set of traversed edges must form a path in a metric graph with deterministic costs on the edges, while the time spent on a node is a random variable, which follows an a priori known distribution. In stochastic graph exploration, the random variables are the costs of the edges of the graph, but we cannot ensure that the costs on the edges form a metric since the random variables are independent.

The adaptivity gap has also been studied for budgeted multi-armed bandits [17, 18,24] by resorting to suitable linear programming relaxation. Unlike previous work on budgeted multi-armed bandit problems, we consider the setting in which new arms appear after some arms are pulled. Stochastic probing problems have also been studied for matching [1,7,13], motivated by kidney exchange and for more general classes of matroid optimization problems [20, 21]. Various other stochastic problems are recently studied, such as variations on the Pandora's box [5, 9, 11, 12, 25], stochastic $k$-TSP [22], scheduling [2, 3, 16], probing [14, 21, 29], stochastic vertex cover in (hyper-)graphs [6], etc.

## 2 Preliminaries

**Notations.** We assume without loss of generality that $B = 1$, and we focus of on tree instances. For instance, $\mathcal{I} = (V, E, r, \pi, \mathcal{R})$, where $r \in V$ is the root of the directed tree $G = (V, E)$, we assume that the edges are pointed away from the root. Accordingly, we denote $Parent(v)$ as the parent of a vertex $v$, the vertex connected to $v$ on its path to the root, $\langle Parent(v), v \rangle \in E$, and let $e_v = \langle Parent(v), v \rangle \in E$. For ease of notation, we denote $C(v) = C(e_v)$ ($C(r) = 0$), and $\pi(v) = \pi(e_v)$. Let $Path(u, v)$ be the set of vertices on the path from $u$ to $v$ (including $u, v$), and let $D(v) = Path(r, v)$. For a subtree $F \subseteq V$, let $C(F)$ be the total realized cost of its edges, i.e., $C(F) = \sum_{e \in F} C(e)$ and let $\mathcal{R}(F)$ be the total reward of its vertices, i.e. $\mathcal{R}(F) = \sum_{v \in F} \mathcal{R}(v)$.

**Types of Strategies.** An *adaptive strategy* determines for any subtree $F \subseteq V, r \in F$, and remaining budget $1 - C(F)$ which adjacent edge to $F$ to probe, (i.e., edge $\langle u, v \rangle$ such that $u \in F$ and $v \notin F$). Note that if $C(F) > 1$, the strategy must halt. A vertex $v$ is *successfully added* if $e_v$ is probed and $C(F) \leq 1$ immediately afterward ($\mathcal{R}(v)$ is added to the total gain of the strategy). For a fixed adaptive strategy, we denote $x_v$ the probability that this strategy probes the edge $e_v$, and $y_v$ the probability that this strategy successfully add the vertex $v$ to its connected component. The expected gain of this strategy is:

$$\sum_{v \in V} y_v \cdot \mathcal{R}(v).$$

In Sect. 3, we bound the optimal solution value via a set of constraints on $\{x_v, y_v : v \in V\}$.

The two main non-adaptive strategies are: the *list strategy* and the *set strategy*. A *list strategy* specifies an ordered list of the vertices $L = (v_1 = r, \ldots, v_n)$, where for any $k$, $(v_1, \ldots, v_k)$ is a connected subtree. The expected gain of a list strategy $L$ with a budget $\mathcal{B}$ is:

$$\sum_{k=1}^{n} \mathbf{Pr}\left[\sum_{i=1}^{k} C(v_i) \leq \mathcal{B}\right] \cdot \mathcal{R}(v_k).$$

A *set strategy* (all-or-nothing) specifies a subtree $F$ ($r \in F$) in advance, and either gains the entire content of $F$, if the total cost of $F$ did not overflow the budget, or gains nothing otherwise. The expected gain of a set strategy $F$ with a budget $\mathcal{B}$ is:

$$\mathbf{Pr}[C(F) \leq \mathcal{B}] \cdot \mathcal{R}(F).$$

## 3  Bounding the Optimal Adaptive Policy

In this section, we bound the value of an adaptive strategy given an instance $\mathcal{I}$. Recall that for a fixed strategy, we denote $x_v$, the probability that the strategy probes the edge $e_v$, and $y_v$, the probability that the strategy successfully add vertex $v$ to its connected component. We first note that we can bound the probability that a strategy probes an edge by the probability it probes its successor edge, i.e., for any $v \in V$ we have $x_v \leq x_{Parent(v)}$.

A simple Example 16, demonstrated that this condition is not sufficient. Instead, we bound $y_v$ (the probability of successfully adding $v$) by the probability it probed one of its predecessors edge $e_u \in D(v)$, times the probability the cost of the path between $u, v$ is less than the adaptive budget (1).

**Lemma 1.** *Given a tree-instance $\mathcal{I}$, and $v, u \in V$ s.t. $u \in D(v)$, then*

$$y_v \leq \mathbf{Pr}[C(Path(u, v)) \leq 1] \cdot x_u.$$

*Proof.* Given a pair of vertices $v, u$, where $u \in D(v)$, by the definition of a feasible adaptive strategy, it may probe the path $Path(u, v)$ only if it probes the edge $e_u$ first, and the total cost of $Path(u, v)$ is independent of the probability of sampling $e_u$. Finally, in order that $v$ would be successfully added to $F$, the total cost of this path must be at most 1, i.e., $C(Path(u, v)) \leq 1$.

Our second lemma bounds the set of all vertices that an adaptive policy tries to insert. We extend the lemma in [15] to deal with budget augmentation. The main idea is to exploit the property of irrevocable decisions, which forces the adaptive policy to keep a vertex even if its size turns out to be large. Let $\mu_{\mathcal{B}}(v) = \mathbb{E}[\min\{\mathcal{B}, C(v)\}]$ the truncated expected size with respect to the augmented budget $\mathcal{B}$, and $\mu_{\mathcal{B}}(F) = \sum_{v \in F} \mu_{\mathcal{B}}(v)$. For ease of notation, we denote $\mu(v) = \mu_1(v)$.

**Lemma 2.** *For $\mathcal{B} \geq 1$ we have $\sum_{v \in V} x_v \mu_{\mathcal{B}}(v) \leq 1 + \mathcal{B}$.*

*Proof.* Our proof extends lemma 2 in [15]. For an adaptive policy, let $S_t$ denote the set of the first $t$ vertices chosen by it. Once the size of $S_t$ overflows, no further vertices are added to $S_t$, and $S_t$ remains constant after this. Let $C_{\mathcal{B}}(v) = \min\{C(v), \mathcal{B}\}$, We define a series $X_t = \sum_{i \in S_t} C_{\mathcal{B}}(v_i) - \mu_{\mathcal{B}}(v_i)$. It is easy to verify $X_t$ is a martingale, and since $X_0 = 0$, we have $\mathbb{E}[X_t] = 0$, which yields $\mathbb{E}[\sum_{i \in S_t} C_{\mathcal{B}}(v_i)] = \mathbb{E}[\sum_{i \in S_t} \mu_{\mathcal{B}}(v_i)]$. The process stops when $C(S_t) > 1$ or we have no more vertices left. Since $\sum_{i \in S_t} C_{\mathcal{B}}(v_i) \leq C(S_t)$ and each $C_{\mathcal{B}}(v_i) \leq \mathcal{B}$, we get $\sum_{i \in S_t} C_{\mathcal{B}}(v_i) \leq 1 + \mathcal{B}$ for any $t > 0$. The mean size of all the vertices inserted by the policy (including the first one which exceeds knapsack capacity) is $\mu(S) = \lim_{t \to \infty} \mu(S_t)$ and therefore $\mathbb{E}[\mu(S)] = \lim_{t \to \infty} \mathbb{E}[\mu(S_t)] \leq 1 + \mathcal{B}$.

Given an instance $\mathcal{I}$, we define $\Phi_{\mathcal{I},\mathcal{B}}(t)$ as the optimal solution value for the following linear program (Fig. 1):

Using Lemmas 1, 2, we have:

**Theorem 3.** *Given a tree-instance $\mathcal{I}$, the expected gain of any adaptive policy with budget $\mathcal{B}$ is at most $\Phi_{\mathcal{I},\mathcal{B}}(1 + \mathcal{B})$.*

$$\max \sum_v y_v \cdot \mathcal{R}(v)$$

$$s.t. : \sum_v x_v \cdot \mu_{\mathcal{B}}(v) \leq t$$

$$y_v \leq \mathbf{Pr}[C(Path(u,v)) \leq 1] \cdot x_u \qquad \text{for } u \in D(v) \qquad (1)$$

$$0 \leq x_v, y_v \leq 1 \qquad \text{for } v \in V$$

**Fig. 1.** $\Phi_{\mathcal{I},\mathcal{B}}(t)$ - the linear program for bounding the optimal adaptive policy.

Next, we provide several characterizations for the solution of $\Phi(t)$ (we omit the subscripts $\mathcal{I},\mathcal{B}$ for a fixed set of parameters). First, we prove that $\Phi_{\mathcal{I},\mathcal{B}}(t)$ is a concave function.

*Claim.* For $\gamma \in [0,1]$ we have, $\Phi_{\mathcal{I},\mathcal{B}}(\gamma \cdot t) \geq \gamma \cdot \Phi_{\mathcal{I},\mathcal{B}}(t)$.

*Proof.* Given an optimal solution $y_v, x_v$ for $\Phi(t)$, then $\gamma \cdot x_v, \gamma \cdot y_v$ is a feasible solution for $\Phi(\gamma \cdot t)$ and its values is $\gamma \cdot \Phi_{\mathcal{I},\mathcal{B}}(t)$.

The next claim states that the constraint $x_u \leq x_{Parent(u)}$ for $u \in V \setminus \{r\}$ implied by the other constraints.

*Claim.* There exists an optimal solution of $\Phi_{\mathcal{I},\mathcal{B}}(t)$, such for $u \in D(v)$ that

$$x_v \leq x_u \cdot \mathbf{Pr}[C(Path(u, Parent(v))) \leq 1].$$

*Proof.* Given a solution, and $v, u \in D(v)$ such that $x_v > x_u \cdot \mathbf{Pr}[C(Path(u, Parent(v))) \leq 1]$, we show that $x_v$ is not binding in any constraint and can be reduced. For a constraint $v, w$ ($v \in D(w)$):

$$x_v \cdot \mathbf{Pr}[C(Path(v,w)) \leq 1] > x_u \cdot \mathbf{Pr}[C(Path(u, Parent(v))) \leq 1] \cdot \mathbf{Pr}[C(Path(v,w)) \leq 1]$$
$$\geq x_u \cdot \mathbf{Pr}[C(Path(u,w)) \leq 1] \geq y_w,$$

where the first inequality is derived from our assumption, the second inequality arises from: $Path(u,w) = Path(u, Parent(v)) \cup Path(v,w)$, and the last inequality is a by the definition of $\Phi_{\mathcal{I}}$.

**Corollary 4.** *There exists an optimal solution of $\Phi_{\mathcal{I},\mathcal{B}}(t)$, such that, for $v \in V$ we have* $x_v \leq x_{Parent(v)}$.

Next, we define a linear program $\hat{\Phi}_{\mathcal{I},\mathcal{B}}(t)$, a relaxation for the original linear program $\Phi_{\mathcal{I},\mathcal{B}}(t)$, and characterize an optimal solution for it.

Using Corollary 4, and the fact that in $\Phi_{\mathcal{I},\mathcal{B}}$, we have $y_v \leq x_v$, we conclude:

**Observation 5.** *For any instance $\mathcal{I}$, and for any $t > 0$, we have* $\hat{\Phi}_{\mathcal{I},\mathcal{B}}(t) \geq \Phi_{\mathcal{I},\mathcal{B}}(t)$.

While $\hat{\Phi}_{\mathcal{I},\mathcal{B}}$ might not have a constant gap for general instances, see Example 16, we will use it for sub-instances where the probability of probing any vertex in this sub-instance is constant. As mentioned, the next lemma would characterize a possible optimal solution.

$$\max \sum_{v \in V} x_v \cdot \mathcal{R}(v)$$

$$s.t. : \sum_{v \in V} x_v \cdot \mu_B(v) \leq t \tag{2}$$

$$x_v \leq x_{Parent(v)} \qquad\qquad \text{for } v \in V \tag{3}$$

$$0 \leq x_v \leq 1 \qquad\qquad \text{for } v \in V$$

**Fig. 2.** $\hat{\Phi}_{I,B}$ - relaxation for the linear program for bounding the optimal adaptive policy.

**Lemma 6.** *There exists a solution for $\hat{\Phi}(t)$, where for each pair of vertices $v_1, v_2 \in V$, such that $0 < x_{v_1}, x_{v_2} < 1$, then for all vertices $w \in Path(v_1, v_2)$ we have $x_w = x_{v_2}$.*

*Proof.* We show a constructive proof that starts from an optimal solution $x$ and computes a modified optimal solution $x'$ until the condition holds.

We omit all vertices $v$ and their corresponding edges such that $x_v = 0$. For $i \in \{1, 2\}$, let $v_i \in V$ such that $0 < x_{v_i} < 1$ and $T_i = \{w : x_u = x_{v_i}$ for all $u \in Path(w, v_i)\}$, such that $T_1 \neq T_2$ (if such a pair does not exist the condition on $x$ holds). Note that $T_i$ is a maximal connected subtree, which contains $v_i$ and includes vertices $u$ such that $x_{v_i} = x_u$. Assume w.l.o.g that $v_1, v_2$ are the roots of their corresponding subtrees (the closet vertex to $r$ in their corresponding subtrees).

If $\mathcal{R}(T_1) = 0$ (or $\mathcal{R}(T_2) = 0$), we may set $x'_u = x_u - \epsilon$ for $u \in T_1$ and $x'_u = x_u$. Otherwise, without decreasing the value of the solution, for large enough $\epsilon$, some vertices will join $T_1$ connected component, or the value will reach 0. Similarly, if $\mu(T_1) = 0$ (or $\mu(T_2) = 0$), we may set $x'_u = x_u + \epsilon$ for $u \in T_1$ and $x'_u = x_u$, otherwise $x'$ is feasible without decreasing the value of the solution. For large enough $\epsilon$, some vertices will join $T_1$ connected component, or this value will reach 1.

Let $g_1 = \frac{\mathcal{R}(T_1)}{\mu(T_1)}$ and $g_2 = \frac{\mathcal{R}(T_2)}{\mu(T_2)}$, we prove that in an optimal solution $x$, $g_1 = g_2$. We show that there exists $\epsilon' > 0$ such that for $\epsilon \in \{-\epsilon', \epsilon'\}$, the following modified solution $x'$ is feasible.

$$x'_v = \begin{cases} x_v + \epsilon, & \text{for } v \in T_1 \\ x_v - \epsilon \cdot \frac{\mu(T_1)}{\mu(T_2)}, & \text{for } v \in T_2 \\ x_v, & \text{otherwise} \end{cases}$$

First, Constraint 2 holds since:

$$\sum_{v \in v} x'_v \mu(v) = \sum_{v \in V} x'_v \mu(v) + \epsilon \cdot \mu(T_1) - \epsilon \cdot \mu(T_2) \cdot \frac{\mu(T_1)}{\mu(T_2)} = \sum_{v \in V} x_v \mu(v).$$

Second, note that, by definition $x_{Parent(v_i)} > x_{v_i}$ and for all descendants of $u$ of $T_i$, $x_u < x_{v_i}$, therefore there exists a small enough $\epsilon$ for which Constraint 3 will still hold. Finally, the objective function for $x'$ equals to:

$$\sum_{\ell \in L} x'_{v_\ell} \cdot \mathcal{R}(v_\ell) - \sum_{\ell \in L} x_{v_\ell} \cdot \mathcal{R}(v_\ell) = \epsilon \cdot \mathcal{R}(T_1) - \epsilon \cdot \mathcal{R}(T_2) \cdot \frac{\mu(T_1)}{\mu(T_2)} = \epsilon \cdot \mu(T_1) \cdot (g_1 - g_2),$$

and if $g_1 \neq g_2$, the value of the objective for $x'$ for $\epsilon = \epsilon'$ or $\epsilon = -\epsilon'$ is higher than the value of the objective for $x$, which contradicts its optimality. Therefore, by setting $\epsilon = \min\{x_{Parent(v_1)} - x_{v_1}, x_{v_2} \cdot \frac{\mu(T_2)}{\mu(T_1)}\}$, we have that $x'$ is an optimal feasible solution and a progress has been made (either $x_{v_1}$ is 1, $x_{v_2}$ is 0, or another vertex added to $T_1$).

Using Lemma 6, we conclude:

**Corollary 7.** *There exists an optimal solution $x$ for $\hat{\Phi}(t)$ where there exists sub-trees $T_1, T_2 \subseteq V$ and a value $\zeta \geq 0$ such that for $v \in T_1$, $x_v = 1$ and for $v \in T_2$, $x_v = \zeta$, and $x_v = 0$ otherwise. The value of this solution is $\mathcal{R}(T_1) + \zeta \cdot \mathcal{R}(T_2)$, and $\mu(T_1) + \zeta \cdot \mu(T_2) \leq t$.*

Finally, for a subtree $F$, we prove a lower bound on the probability of the realized cost of $F$ to be less than (a fraction of) the budget as a function of the subtree truncated expected size.

**Lemma 8.** *For a subtree $F$, and $\gamma \in [0, 1]$, we have $\mathbf{Pr}[C(F) < \gamma \cdot \mathcal{B}] \geq 1 - \frac{\mu_{\mathcal{B}}(F)}{\gamma \cdot \mathcal{B}}$.*

*Proof.*

$$\mathbf{Pr}[C(F) \geq \gamma \cdot \mathcal{B}] = \mathbf{Pr}[\min\{C(F), \mathcal{B}\} \geq \gamma \cdot \mathcal{B}]$$

$$\leq \frac{\mathbb{E}[\min\{C(F), \mathcal{B}\}]}{\gamma \cdot \mathcal{B}} \leq \frac{\mathbb{E}\left[\sum_{v \in F} \min\{C(v), \mathcal{B}\}\right]}{\gamma \cdot \mathcal{B}} = \frac{\mu_{\mathcal{B}}(F)}{\gamma \cdot \mathcal{B}},$$

where the first equality is due to $\gamma \leq 1$, the first inequality is given by Markov's inequality, and the second equality arises from the definition of $\mu$.

## 4    Spider Graphs

Our first objective is to to develop a constant approximation algorithm for *spider graphs* instances using a constant augmentation. In spider graphs, all the vertices except the root have an out-degree of at most 1. Let $L$ be the number of legs in the graph, we denote in leg $i \in [L]$, $v_{i,j}$ as the level $j$ vertex, i.e. a vertex in leg $i$ where its distance from the root is $j$, note that this vertex is uniquely defined. Let $C_i(j, k) = C(Path(v_{i,j}, v_{i,k}))$. Example 14 (Lemma 1 in [4]) demonstrates that even for this simple graph structure a budget augmentation is necessary to achieve a constant competitiveness. In addition, Example 15 demonstrates that even with budget augmentation there might not be a competitive set-strategy. Nevertheless, we show that using a constant budget augmentation, the adaptivity gap is bounded, by proving there exists a suitable list strategy.

Algorithm SpiderNoAdaptive divides the vertices into two sets, one set contains the risky vertices, where the expected cost of the path to them is at least half the budget, and the other set contains the rest of the vertices. The algorithm computes a constant approximation non-adaptive strategy for each set, and the maximum of those two strategies yield a constant non-adaptive strategy for the entire instance. For the risky vertices, the algorithm outputs a single arm, we prove that a non-adaptive list strategy that probes this arm and uses a constant budget augmentation has a constant competitive ratio. For the non-risky vertices, the algorithm computes a solution to $\hat{\Phi}_{\mathcal{I},1}(0.5)$ according to the structure of Corollary 7. The algorithm uses this solution to compute a fixed set of vertices. We prove that a non-adaptive set strategy that probes this set has a constant competitive ratio.

## 4.1   Non-adaptive Algorithm

For a constant $0 < \epsilon < 1$, we show that using $(1 + \epsilon)$ budget augmentation, there exists non-adaptive strategy and its expected gain is $O(\epsilon)$ factor of the optimal adaptive gain. The algorithm is composed of two parts, the first part would address the "risky" vertices, vertices with expected cost of the path from the root to them is at least 0.5, and the second part will deal with the rest of the vertices. Formally, let $Risky = \{v_{i,j} : \mu(D(v_{i,j})) > 0.5\}$. Accordingly, to the spider graph's notation, let $x_{i,j}$ the probability that the optimal adaptive strategy probed the edge $(v_{i,j-1}, v_{i,j})$ and let $y_{i,j}$ the probability that $v_{i,j}$ is successfully added to the probed sub-tree. The LP bound for adaptive strategy $\Phi_{\mathcal{I},1}(t)$ reduced to:

$$\max \sum_{i,j} y_{i,j} \cdot \mathcal{R}(v_{i,j})$$

$$s.t. : \sum_v x_{i,j} \mu_1(v_{i,j}) \le t$$

$$y_{i,j} \le \mathbf{Pr}[C_i(k,j) \le 1] \cdot x_{i,k} \qquad \text{for } k \le j$$

$$0 \le x_{i,j}, y_{i,j} \le 1 \qquad \text{for } v \in V$$

---

**Data:** Spider leg tree instance $\mathcal{I}(V, E, r, \pi, \mathcal{R})$
**Result:** Non-adaptive list strategy
**Procedure**   Risky($\mathcal{I}$)
 $x, y \leftarrow$ Solve $\Phi_{\mathcal{I},1}(2)$
 $L_i \leftarrow \sum_j x_{i,j} \cdot \mu(v_{i,j})$, for all $i \in L$
 $L^* \leftarrow L_i$ with probability $L_i/2$
 **return** $L^*$
**Procedure**   NonRisky($\mathcal{I}$)
 $x \leftarrow$ Solve $\hat{\Phi}_{\mathcal{I},1}(0.5)$ // A solution according to Corollary 7
 $T_1 \leftarrow \{(i,t) : x_{i,t} = 1\}, T_2 \leftarrow \{(i,t) : x_{i,t} = \zeta\}$
 **return** $\arg\max\{\mathcal{R}(T_1), \mathcal{R}(T_2)\}$
$\mathcal{I}_{\text{Risky}} \leftarrow \mathcal{I}, \mathcal{R}(v_{i,t}) = 0 :$ for $v_{i,t} \notin Risky$
$\mathcal{I}_{\text{Non}} \leftarrow \mathcal{I}, \mathcal{R}(v_{i,t}) = 0 :$ for $v_{i,t} \in Risky$
$T_r \leftarrow$   Risky($\mathcal{I}_{\text{Risky}}$)
$T_n \leftarrow$   NonRisky($\mathcal{I}_{\text{Non}}$)
**return** $\arg\max\{NonAdpative(T_r), NonAdpative(T_n)\}$

**Algorithm SpiderNoAdaptive:** Non-adaptive algorithm for spider trees

## 4.2  Risky Vertices

**Lemma 9.** *For instance $\mathcal{I}$ where $\mathcal{R}(v_{i,j}) = 0$, for $v_{i,j} \notin$ Risky, Procedure* `Risky(`$\mathcal{I}$`)` *outputs a list strategy which gains at least $\epsilon/8$ factor of the optimal adaptive policy using $(1 + \epsilon)$ budget augmentation.*

*Proof.* First note that the procedure is well defined since $\sum_i L_i \leq 2$, by $\Phi(2)$'s definition. The probability of vertex $v_{i,t}$ to successfully being probed is $L_i/2 \cdot \mathbf{Pr}[C_i(1,t) \leq 1 + \epsilon]$. For a leg $i$, let $k$ the first index s.t. $\sum_{t=1}^{k} \mu(v_{i,t}) \geq \epsilon/2$.

Note that, $\mathcal{R}(v_{i,k'}) = 0$ for for $k' < k$. Since by the definition of $k$, $\sum_{t=1}^{k'} \mu(v_{i,t}) < \epsilon/2 \leq 0.5$ and therefore, $v_{i,k'} \notin$ *Risky*, and $\mathcal{R}(v_{i,k'}) = 0$ by our assumption. Therefore, it is sufficient to bound the probability of vertices being successfully probed just for $v_{i,h}$ where $h \geq k$ and compare it to $y_{i,h}$ the corresponding adaptive probability. We first show that for a leg $i$, $L_i \geq \epsilon/2 \cdot x_{i,k}$.

$$L_i = \sum_{t=1}^{k} x_{i,t} \cdot \mu(v_{i,t}) \geq \sum_{t=1}^{k} x_{i,t} \cdot \mu(v_{i,t}) \geq x_{i,k} \sum_{t=1}^{k} \mu(v_{i,t}) \geq \epsilon/2 \cdot x_{i,k}, \quad (4)$$

where the first inequality is by removing positive terms, the second inequality is by Corollary 4, and the third inequality is by $k$'s definition.

Finally, the probability that $v_{i,t}$ (for $t \geq k$) is successfully probed:

$$\frac{L_i}{2} \cdot \mathbf{Pr}[C_i(1,t) \leq 1 + \epsilon] \geq \frac{L_i}{2} \cdot \mathbf{Pr}[C_i(1, k - 1) \leq \epsilon] \cdot \mathbf{Pr}[C_i(k, t) \leq 1]$$

$$\geq \frac{L_i}{4} \cdot \mathbf{Pr}[C_i(k, t) \leq 1] \geq \frac{x_{i,k} \cdot \epsilon}{8} \cdot \mathbf{Pr}[C_i(k, t) \leq 1] \geq \frac{y_{i,k} \cdot \epsilon}{8},$$

where the first inequality is by the decomposition of the path, the second inequality is by Lemma 8 ($\gamma = \epsilon$) $\mathbf{Pr}[C_i(1, k - 1) \leq \epsilon] \geq 1/2$ since $\sum_{j=1}^{k-1} \mu_{i,j} < \epsilon/2$ by $k$'s definition, the third inequality is by 4 and the last inequality is by (3) in the definition of $\Phi_{\mathcal{I}}$. $\qquad\square$

## 4.3  Non-risky Vertices

**Lemma 10.** *For instance $\mathcal{I}$ where $\mathcal{R}(v_{i,j}) = 0$, for $v_{i,j} \in$ Risky, Procedure* `NonRisky(`$\mathcal{I}$`)` *outputs a set strategy which gains at least a $1/16$ fraction of the optimal strategy gain without budget augmentation.*

*Proof.* Let $x$ be a solution which fulfill the conditions of Corollary 7, and $T_1, T_2$ according to the algorithm's definition. Note that, since $x_r = 1$, therefore $T_2$ is a single arm. By Observation 5 and by Theorem 3, the optimal gain is bounded by $\Phi_{\mathcal{I},1}(2) \leq \Phi_{\mathcal{I},1}(0.5) \cdot 4 \leq (\mathcal{R}(T_1) + \mathcal{R}(T_2)) \cdot 4 \leq \max\{\mathcal{R}(T_1), \mathcal{R}(T_2)\} \cdot 8$. Note that, we have $\mu(T_1) \leq 0.5$, since for $v_{i,t} \in T_1$, $x_{i,t} = 1$ and $\sum x_{i,t}\mu(v_{i,t}) \leq 0.5$ by $\hat{\Phi}_{\mathcal{I},1}(0.5)$ definition. Second, note that $\mu(T_2) \leq 0.5$, since it's a single arm, and we omitted the risky vertices. Therefore $\mathbf{Pr}[C(T_i) \leq 1] \geq 0.5$ and the gain of the strategy is at least $0.5 \cdot \max\{\mathcal{R}(T_1), \mathcal{R}(T_2)\}$.

## 4.4  Putting Things Together

**Theorem 11.** *For spider graphs instances, and a constant $\epsilon < 1$, there exist a $(24/\epsilon, 1 + \epsilon)$-approximate non-adaptive strategy.*

*Proof.* Clearly, half of the gain is from the risky vertices or from the non-risky vertices. If most of the gain is from the risky vertices, by Lemma 9, there exists a $\epsilon/8$ competitive list strategy using $1 + \epsilon$ augmentation. In most of the gain is from non-risky vertices, then by Lemma 10, there exists $1/16$ competitive set strategy without augmentation. Therefore, if $\gamma \in [0, 1]$ fraction of the optimal gain is out of non-risky vertices, the maximum reward out of these two strategies, is at least $\max\{\gamma/16, (1 - \gamma) \cdot \epsilon/8\}$ fraction, which is at least $24/\epsilon$ fraction for any $\gamma$.

# 5  Bounded Weighted Depth Trees Instances

We now focus on the practical scenario where the depth of the tree's instance is bounded. We define the weighted depth of a graph as the maximum over all vertices of the total expected cost of the path from the root to them. Formally, an instance is $(\beta, \mathcal{B})$ weighted depth bounded, if for all $v \in V$, $\mu_{\mathcal{B}}(D(v)) \leq \beta$. We prove that for $(\mathcal{B} \cdot (1 - \epsilon), \mathcal{B})$ bounded weighted depth instances, for some constants $\epsilon > 0, \mathcal{B} \geq 1$, there exists a constant competitive set strategy with a budget $\mathcal{B}$. Note that for $(1 - \epsilon, 1)$ bounded weighted-depth instances, our result implies there exists a constant competitive set strategy without budget augmentation, i.e., $(O(1), 1)$-approximate strategy. We observe that, given a solution $x$ for $\hat{\Phi}_{\mathcal{T}, \mathcal{B}}(t)$, and let $T$ be the support tree of $x$, i.e. $v \in T$ if $x_v > 0$, then unlike for spider graphs, $\mu(T)$ can be much higher than $t$); see Example 18. Nevertheless, we show that, given a tree $T$ ($r \in T$), such that $\mu(D(v)) \leq \beta$ (we omit the subscript $\mathcal{B}$ it is clear from the context) for $v \in T$. It is possible to decompose $T$ to $\mathcal{S} = \{S_1, \ldots, S_k\}$, where for all $v \in T$ there exists $i \in [k]$ such that $v \in S_i$. For all $i \in [k]$, $S_i$ is a subtree that contains $r$, and $\mu(S_i) \leq \alpha$, and the number of subtrees is bounded by $k \leq \lceil \frac{2\mu(T)}{\alpha - \beta} \rceil$. Given a subtree $T'$, denote $S_{T'}(v) = \{u : v \in D(u)\} \cap T'$, the subtree of $v$ with respect to $T'$.

Algorithm TreeDecompose works in iterations. At each iteration, it locates a proper set of subtrees, where the parent of the root of each of those subtrees is the same. It adds the subtree containing this set and its entire path to the root to the set of sub-trees, and removes this subset from the current tree. Specifically, it denotes $T'$ as the current sub-tree. If $\mu(T') \leq \alpha$, then $T'$ is a proper sub-tree, and the algorithm adds it to the set of sub-trees and terminates. Otherwise, it locates a subset of vertices $S'$, where $S'$ contains several sub-trees with the same parent $w$, and $\mu(S') \geq \frac{\alpha - \beta}{2}, \mu(D(w) \cup S') \leq \alpha$ holds. The algorithm adds $D(w) \cup S'$ to the set of sub-trees and omits $S'$ from the graph. To locate such $S'$, in each iteration, it starts from the root and iteratively proceeds to one of his children $u$ with the maximum value of $\mu(S_{T'}(u))$, until it reaches a vertex $v$ such that $\mu(D(v) \cup S_{T'}(v)) > \alpha$, the algorithm locates $S'$ for the vertex $w$, the parent of the vertex $v$.

**Lemma 12.** *Algorithm TreeDecompose outputs $\mathcal{S} = \{S_1, \ldots, S_k\}$ a feasible tree decomposition and $k \leq \lceil \frac{2\mu(T)}{\alpha - \beta} \rceil$.*

> **Input** : A $(\beta, 1)$ weighted depth bounded instance, edge weights $\mu$, and $\alpha > \beta$
> **Output:** A tree decomposition $\mathcal{S} = \{S_1, \ldots, S_k\}$, where $\mu(S_i) \leq \alpha$ for $i \in [k]$
>
> $\mathcal{S} \leftarrow \emptyset, T' \leftarrow T$
> **while** $\mu(T') > \alpha$ **do**
> $\quad v \leftarrow r$
> $\quad$ **while** $\mu(D(v) \cup S_{T'}(v)) > \alpha$ **do**
> $\quad\quad v \leftarrow \arg\max_{(v,u) \in E} \mu(S_{T'}(u))$
> $\quad$ **end**
> $\quad w \leftarrow Parent(v)$
> $\quad$ Let $(u_1, \ldots, u_h)$ such that $(w, u_i) \in T'$ and $\mu(S_{T'}(u_i)) \leq \mu(S_{T'}(u_{i+1}))$ for
> $\quad\quad i \in [h-1]$
> $\quad h^* \leftarrow \arg\min\{j \in [h] : \mu(D(w)) + \sum_{i=j}^{h} \mu(S_{T'}(u_i)) \leq \alpha\}$
> $\quad S' \leftarrow \bigcup_{j=h^*}^{h} S_{T'}(u_j)$
> $\quad \mathcal{S} \leftarrow \mathcal{S} \cup \{(D(w) \cup S')\}, T' \leftarrow T' \setminus S'$
> **end**
> return $\mathcal{S} \cup \{T'\}$

**Algorithm TreeDecompose:** Decomposition of a tree to bounded weight rooted subtrees.

*Proof.* First, we show that the algorithm is well-defined, and $T'$ is a connected subtree of $T$ and contains the root at any step, which follows from the fact that the algorithm only omits a vertex with its entire subtree from $T'$. Next, we observe that the inner loop is well-defined, and $v$ would not be a leaf since for any leaf $\ell$, we have $\mu(D(\ell)) + \mu(S_{T'}(\ell)) = \mu(D(\ell)) \leq \beta \leq \alpha$. Therefore, the inner loop always halts at vertex $v \neq r$ such that, for $w = Parent(v)$ then $v = u_h$ since $v = \arg\max_{(w,u) \in E} \mu(S_{T'}(u))$. Therefore, $\mu(D(w) \cup S_{T'}(w)) > \alpha$ and $\mu(D(u_h) \cup S_{T'}(u_h)) \leq \alpha$. Note that, $h^*$ is well-defined since for $j = h$, we have: $\mu(D(w)) + \mu(S_{T'}(u_h)) = \mu(D(u_h) \cup S_{T'}(u_h)) \leq \alpha$. Next, we have that $h^* > 1$ since for $j = 1$ we have $\mu(D(w)) + \sum_{i=1}^{h} \mu(S_{T'}(e_i)) = \mu(D(w) \cup S_{T'}(w)) > \alpha$.

We observe that $\mathcal{S} = \{S_1, \ldots, S_k\}$ is the tree decomposition of $T$, since the algorithm terminates only after covering all the vertices. Additionally, each subtree added to $\mathcal{S}$ is $(D(w) \cup S')$, where $S' = \bigcup_{j=h^*}^{h} S_{T'}(u_j)$ is a collection of subtrees and their path to the root, and by the condition on $h^*$, we have $\mu(D(w) \cup S') \leq \alpha$.

In order to complete the proof, we need to show that $k \leq \frac{2\mu(T)}{\alpha - \beta}$. For any iteration where $\mu(T') > \alpha$, let $w, v, h^*$ be the corresponding values of a main loop iteration, we have:

$$\alpha < \mu(D(w)) + \sum_{i=h^*-1}^{h} \mu(S_{T'}(u_i)) \leq \beta + \sum_{i=h^*-1}^{h} \mu(S_{T'}(u_i)) \leq \beta + 2\sum_{i=h^*}^{h} \mu(S_{T'}(u_i)),$$

where the first inequality is since $h^* > 1$, the second inequality is due to $\beta \geq D(w)$ for all $w \in T$, and the last inequality is a result of $\mu(S_{T'}(u_{h^*-1})) \leq \mu(S_{T'}(u_{h^*})) \leq \sum_{i=h^*}^{h} \mu(S_{T'}(u_i))$ since the vertices $u_i$ are sorted accordingly. Therefore, the decrease in $\mu(T')$ in each such iteration is at least $\mu(S') \geq \frac{\alpha - \beta}{2}$, the number of iterations (until

$\mu(T') \leq \alpha)$ is at most $\lceil \frac{2(\mu(T)-\alpha)}{\alpha-\beta} \rceil$, and the number of subtrees is at most $\lceil \frac{2(\mu(T)-\alpha)}{\alpha-\beta} + 1 \rceil \leq \lceil \frac{2\mu(T)}{\alpha-\beta} \rceil$ as required.

**Theorem 13.** *For $(\mathcal{B} \cdot (1-\epsilon), \mathcal{B})$ bounded weighted depth instances, where $\epsilon > 0, \mathcal{B} \geq 1$, there exists $(\frac{\epsilon^2}{16 \cdot (\mathcal{B}+1)}, \mathcal{B})$-approximate non-adaptive strategy.*

*Proof.* Let $x$ be a solution to $\Phi_{\mathcal{I},\mathcal{B}}(\mathcal{B}+1)$, which fulfills the conditions of Corollary 7, and let $T_1, T_2, \zeta$ be the corresponding integral and fractional trees and the fractional tree assignment value, respectively. Note that, since $x_r = 1$, we have $r \in T_1$. Let $r_2$ be the root of $T_2$, and let $T = T_2 \cup D(r_2)$. Assume w.l.o.g. that $\zeta \cdot \mathcal{R}(T) \geq \mathcal{R}(T_1)$. Note that by $\hat{\Phi}_{\mathcal{I}}$ definition, we have $\mu(T) \leq (\mathcal{B}+1)/\zeta$. By Lemma 12, and $\alpha = \mathcal{B} \cdot (1 - \epsilon/2)$ we have a $\mathcal{S} = \{S_1, \ldots, S_k\}$, a feasible tree decomposition and $k \leq \lceil \frac{2\mu(T)}{\alpha-\beta} \rceil \leq \frac{4 \cdot (\mathcal{B}+1)}{\epsilon \cdot \zeta}$. Therefore, by choosing a uniform subtree $S^*$ out of the $k$ subtrees, the probability that a vertex $v \in T$ would be successfully added to the strategy is:

$$\mathbf{Pr}[v \in S^*] \cdot \mathbf{Pr}[C(S^*) \leq \mathcal{B}] \geq \frac{1}{k} \cdot (1 - \frac{\alpha}{\mathcal{B}}) \geq \frac{\epsilon \cdot \zeta}{4 \cdot (\mathcal{B}+1)} \cdot \frac{\epsilon}{2} = \zeta \cdot \frac{\epsilon^2}{8 \cdot (\mathcal{B}+1)}.$$

By summing over all the vertices, we find that the non-adaptive gain is at least $\zeta \cdot \mathcal{R}(T) \cdot \epsilon^2/(8 \cdot (\mathcal{B}+1))$, while the gain of the optimal adaptive policy is at most $2 \cdot \zeta \cdot \mathcal{R}(T)$.

## 6   Examples

In this section, we provide several of examples that demonstrate various tree graph instances' properties. Let $Be(s, p)$ denote a Bernoulli distribution, i.e., for a randomized variable $x \sim Be(s, p)$, then $x = s$ has a probability $p$, and $x = 0$ otherwise. Let $1^+$ denote a large constant, i.e., $1^+ \gg \mathcal{B}$. Note that, for $e \sim Be(1^+, p)$, $\mu_{\mathcal{B}}(e) = p \cdot \mathcal{B}$.

First, for completeness, we state again the example in [4], which demonstrates that any competitive non-adaptive strategy must use budget augmentation even in spider graphs (Fig. 3).

*Example 14.* A spider graph with $L$ legs, each leg $i$ contains two vertices, $v_{i,1}$, and $v_{i,2}$, $\mathcal{R}(v_{i,1}) = 0$ and $\mathcal{R}(v_{i,2}) = 1$ for all $i \in [L]$, $\pi(v_{i,1}) \sim Be(2^{-i}, 1 - \frac{1}{L})$, and $\pi(v_{i,2}) = (1 - 2^{-i} + 2^{-L}, 1)$.

*Claim (Lemma 1 in [4]).* The adaptivity gap of stochastic graph exploration is $\Omega(n)$ (without budget augmentation) even in spider graphs.

*Proof.* Consider the spider graph in Example 14. Note that, for any $i \neq j \in [L]$, we have, $C(\{v_{i,2}, v_{j,2}\}) > 1$ with probability 1; therefore, any strategy would gain at most 1. Given a list strategy, let $v_{i,2}$ be the first second-level vertex in the list, the probability that this list strategy will gain is at most $\mathbf{Pr}[C(\{v_{i,1}, v_{i,2}\} \leq 1] = \frac{1}{L}$; therefore, the expected gain of any non-adaptive strategy is at most $1/L$.

On the other hand, an adaptive strategy probes $v_{i,1}$ sequentially from $L$ to 1 until $C(v_{i,1}) = 0$ for some $i \in [L]$, and if such $i$ exists, it probes $v_{i,2}$ and halts. Note that, in the case that such $i$ exists, this strategy successfully probes $v_{i,2}$ (with probability 1). Therefore, the expected gain of this strategy is $(1 - 1/L)^L \approx 0.36$ and the gap is unbounded.

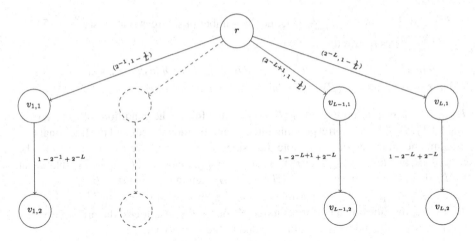

**Fig. 3.** Instance of spider graph demonstrating that the budget augmentation is necessary, see Example 14.

The next example demonstrates that even with budget augmentation, a *set* strategy is not competitive versus an *adaptive* strategy.

**Fig. 4.** An example demonstrating that a set strategy cannot approximate an adaptive strategy even with budget augmentation.

*Example 15.* A spider graph with 1 legs, the leg contains $k$ vertices, $v_{1,j}$ for $j \in [k]$. $\mathcal{R}(v_{1,k}) = 2^k$, and $\pi(v_{i,1}) \sim Be(1^+, 1/2)$. See also Fig. 4.

*Claim.* The gap between a *set* strategy and an *adaptive* strategy is unbounded, even when the set strategy uses a constant budget augmentation.

*Proof.* Consider the graph in Example 15. The probability that a list strategy which probes the single leg will successfully probe vertex $v_{i,h}$ is $2^{-h}$; therefore its expected gain is $\sum_{h=1}^{k} 2^{-h} \cdot 2^h = k$. While a set strategy which contains vertices with a prefix of $v_{1,j}$ for $j \in [k]$ would gain $\sum_{h=1}^{j} 2^h \le 2^{j+1}$ and the probability it gains is $2^{-j}$, there it's expected gain is at most $2^{j+1} \cdot 2^{-j} = 2$ and the gap is unbounded.

*Example 16.* A spider graph with 1 legs, the leg contains 2 vertices, $v_{1,1}$ for and $v_{1,2}$. $\mathcal{R}(v_{1,1}) = 0, \mathcal{R}(v_{1,2}) = 1, \pi(v_{1,1}) \sim Be(1^+, 1 - 1/k), \pi(v_{1,2}) \sim Be(0, 1)$. The value of $\hat{\Phi}_{\mathcal{I},1}(1) = 1$, while the value of any adaptive policy (even with budget augmentation) is at most $1/k$.

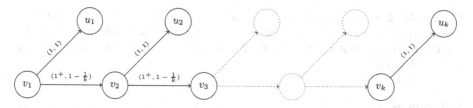

**Fig. 5.** An example demonstrates that for general instances, there is a non-constant gap between the value of the solution of $\Phi_I$ and the gain of an adaptive strategy, even with a constant budget augmentation.

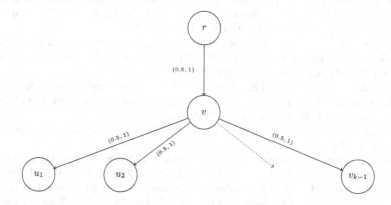

**Fig. 6.** An instance where $\hat{\Phi}_\mathcal{I}$ has an unbounded weight support tree.

Next, we demonstrate that for general instances, we cannot use the value of the solution of $\Phi_\mathcal{I}$ for the lower bound of an adaptive strategy, even with budget augmentation.

*Example 17.* Consider a graph with $2 \cdot k$ vertices, $v_1 = r, v_i, u_i$ for $i \in [k]$, $Parent(v_i) = v_{i-1}), Parent(u_i) = v_i$, and $\mathcal{R}(u_i) = k^{i-1}, \mathcal{R}(v_i) = 0, \pi(v_i) \sim Be(1^+, 1 - 1/k), \pi(u_i) \sim Be(1, 1)$. See Fig. 5

*Claim.* There exists an instance $\mathcal{I}$, such that the gap between $\Phi_{\mathcal{I},\mathcal{B}}(2)$ and the value of any adaptive policy on $\mathcal{I}$ is unbounded, even with budget augmentation.

*Proof.* Consider the instance of Example 17; first note that for $i \in [k]$, $x_{v_i} = k^{-i+1}/2$ and $y_{u_i} = x_{u_i} = k^{-i}/2$ is a feasible solution and its value is $\sum_{i=1}^k y_{u_i} \cdot k^{i-1} = \sum_{i=1}^k k^{-i}/2 \cdot k^{i-1} = k/2$. Considering an adaptive strategy with budget $\mathcal{B}$, we can assume w.l.o.g that it is deterministic, since the only random variables are $C(v_i)$, and if $C(v_i) \neq 0$, the algorithm must halt. Let $h_1, \ldots, h_\mathcal{B}$, where the algorithm decides to probe $u_{h_i}$ for $i \in \mathcal{B}$ (clearly there must be at most $\mathcal{B}$ since the cost of any of them is deterministically 1 ). The probability it gains, $u_{h_j}$, is at most $k^j$, and therefore it's expected gain is at most $\sum_{j=1}^\mathcal{B} k^{-j} \cdot k^j = \mathcal{B}$, while as we have shown $\Phi_{\mathcal{I},\mathcal{B}} \geq \Phi_{\mathcal{I},1} = k/2$.

188    I. R. Cohen

*Example 18.* Consider a graph with $k+1$ vertices, $r, v, u_i$ for $i \in [k-1]$, $Parent(v) = r$, $Parent(u_i) = v$, and $\mathcal{R}(u_i) = 1, \mathcal{R}(v) = 0, \pi(v_i) \sim Be(0.5, 1), \pi(u_i) \sim Be(1, 1)$. See Fig. 6. The optimal solution for $\hat{\Phi}_{\mathcal{I}}(2)$ will set $x_v = x_{u_i} = 4/k$, and therefore $\mu(T) = (k-1)/2$.

# References

1. Adamczyk, M.: Improved analysis of the greedy algorithm for stochastic matching. Inf. Process. Lett. **111**(15), 731–737 (2011)
2. Albers, S., Eckl, A.: Explorable uncertainty in scheduling with non-uniform testing times. In: Kaklamanis, C., Levin, A. (eds.) WAOA 2020. LNCS, vol. 12806, pp. 127–142. Springer, Cham (2021). https://doi.org/10.1007/978-3-030-80879-2_9
3. Albers, S., Eckl, A.: Scheduling with testing on multiple identical parallel machines. In: Lubiw, A., Salavatipour, M. (eds.) WADS 2021. LNCS, vol. 12808, pp. 29–42. Springer, Cham (2021). https://doi.org/10.1007/978-3-030-83508-8_3
4. Anagnostopoulos, A., Cohen, I.R., Leonardi, S., Lacki, J.: Stochastic graph exploration. In: 46th International Colloquium on Automata, Languages, and Programming (ICALP 2019). Schloss Dagstuhl-Leibniz-Zentrum fuer Informatik (2019)
5. Aouad, A., Ji, J., Shaposhnik, Y.: Pandora's box problem with sequential inspections. SSRN 3726167 (2020)
6. Bampis, E., Dürr, C., Erlebach, T., Santos de Lima, M., Megow, N., Schlöter, J.: Orienting (hyper)graphs under explorable stochastic uncertainty. In: Mutzel, P., Pagh, R., Herman, G. (eds.) 29th Annual European Symposium on Algorithms, ESA 2021, Lisbon, Portugal, 6–8 September 2021 (Virtual Conference). LIPIcs, vol. 204 pp. 10:1–10:18. Schloss Dagstuhl - Leibniz-Zentrum für Informatik (2021)
7. Bansal, N., Gupta, A., Li, J., Mestre, J., Nagarajan, V., Rudra, A.: When LP is the cure for your matching woes: improved bounds for stochastic matchings. Algorithmica **63**(4), 733–762 (2012)
8. Bansal, N., Nagarajan, V.: On the adaptivity gap of stochastic orienteering. In: Lee, J., Vygen, J. (eds.) IPCO 2014. LNCS, vol. 8494, pp. 114–125. Springer, Cham (2014). https://doi.org/10.1007/978-3-319-07557-0_10
9. Beyhaghi, H., Kleinberg, R.: Pandora's problem with nonobligatory inspection. In: Proceedings of the 2019 ACM Conference on Economics and Computation, pp. 131–132 (2019)
10. Bhalgat, A., Goel, A., Khanna, S.: Improved approximation results for stochastic knapsack problems. In: Proceedings of the Twenty-Second Annual ACM-SIAM Symposium on Discrete Algorithms, pp. 1647–1665. SIAM (2011)
11. Boodaghians, S., Fusco, F., Lazos, P., Leonardi, S.: Pandora's box problem with order constraints. In: Proceedings of the 21st ACM Conference on Economics and Computation, pp. 439–458 (2020)
12. Chawla, S., Gergatsouli, E., Teng, Y., Tzamos, C., Zhang, R.: Pandora's box with correlations: learning and approximation. In: 2020 IEEE 61st Annual Symposium on Foundations of Computer Science (FOCS), pp. 1214–1225. IEEE (2020)
13. Chen, N., Immorlica, N., Karlin, A.R., Mahdian, M., Rudra, A.: Approximating matches made in heaven. In: Albers, S., Marchetti-Spaccamela, A., Matias, Y., Nikoletseas, S., Thomas, W. (eds.) ICALP 2009. LNCS, vol. 5555, pp. 266–278. Springer, Heidelberg (2009). https://doi.org/10.1007/978-3-642-02927-1_23
14. Chugg, B., Maehara, T.: Submodular stochastic probing with prices. In: 2019 6th International Conference on Control, Decision and Information Technologies (CoDIT), pp. 60–66. IEEE (2019)

15. Dean, B.C., Goemans, M.X., Vondrák, J.: Approximating the stochastic knapsack problem: the benefit of adaptivity. Math. Oper. Res. **33**(4), 945–964 (2008)
16. Dürr, C., Erlebach, T., Megow, N., Meißner, J.: An adversarial model for scheduling with testing. Algorithmica **82**(12), 3630–3675 (2020)
17. Guha, S., Munagala, K.: Approximation algorithms for budgeted learning problems. In: Proceedings of the Thirty-Ninth Annual ACM Symposium on Theory of Computing, STOC 2007, San Diego, California, USA, pp. 104–113. ACM, New York (2007)
18. Gupta, A., Krishnaswamy, R., Molinaro, M., Ravi, R.: Approximation algorithms for correlated knapsacks and non-martingale bandits. In: IEEE 52nd Annual Symposium on Foundations of Computer Science, FOCS 2011, Palm Springs, CA, USA, 22–25 October 2011, pp. 827–836 (2011)
19. Gupta, A., Krishnaswamy, R., Nagarajan, V., Ravi, R.: Running errands in time: approximation algorithms for stochastic orienteering. Math. Oper. Res. **40**(1), 56–79 (2015)
20. Gupta, A., Nagarajan, V.: A stochastic probing problem with applications. In: Goemans, M., Correa, J. (eds.) IPCO 2013. LNCS, vol. 7801, pp. 205–216. Springer, Heidelberg (2013). https://doi.org/10.1007/978-3-642-36694-9_18
21. Gupta, A., Nagarajan, V., Singla, S.: Algorithms and adaptivity gaps for stochastic probing. In: Proceedings of the Twenty-Seventh Annual ACM-SIAM Symposium on Discrete Algorithms, SODA 2016, Arlington, Virginia, pp. 1731–1747. Society for Industrial and Applied Mathematics, Philadelphia (2016)
22. Jiang, H., Li, J., Liu, D., Singla, S.: Algorithms and adaptivity gaps for stochastic $k$-TSP. arXiv preprint arXiv:1911.02506 (2019)
23. Laishram, R., Areekijseree, K., Soundarajan, S.: Predicted max degree sampling: sampling in directed networks to maximize node coverage through crawling. In: 2017 IEEE Conference on Computer Communications Workshops (INFOCOM WKSHPS), pp. 940–945 (2017)
24. Ma, W.: Improvements and generalizations of stochastic knapsack and multi-armed bandit approximation algorithms. In: Proceedings of the Twenty-Fifth Annual ACM-SIAM Symposium on Discrete Algorithms, pp. 1154–1163. SIAM (2014)
25. Singla, S.: The price of information in combinatorial optimization. In: Proceedings of the Twenty-Ninth Annual ACM-SIAM Symposium on Discrete Algorithms, pp. 2523–2532. SIAM (2018)
26. Soundarajan, S., Eliassi-Rad, T., Gallagher, B., Pinar, A.: MaxReach: reducing network incompleteness through node probes. In: 2016 IEEE/ACM International Conference on Advances in Social Networks Analysis and Mining (ASONAM), pp. 152–157 (2016)
27. Soundarajan, S., Eliassi-Rad, T., Gallagher, B., Pinar, A.: $\epsilon$WGXX: adaptive edge probing for enhancing incomplete networks. In: Proceedings of the 2017 ACM on Web Science Conference, WebSci 2017, New York, NY, USA, pp. 161–170. ACM, New York (2017)
28. Thayer, T.C., Carpin, S.: An adaptive method for the stochastic orienteering problem. IEEE Robot. Autom. Lett. **6**(2), 4185–4192 (2021)
29. Wang, W., Gupta, A., Williams, J.: Probing to minimize. arXiv preprint arXiv:2111.01955 (2021)
30. Watts, D.J., Strogatz, S.H.: Collective dynamics of 'small-world' networks. Nature **393**(6684), 440–442 (1998)

# Adaptivity Gaps for the Stochastic Boolean Function Evaluation Problem

Lisa Hellerstein[1], Devorah Kletenik[2], Naifeng Liu[3], and R. Teal Witter[1]([⊠])

[1] NYU Tandon, Brooklyn, NY 11201, USA
rtealwitter@nyu.edu
[2] Brooklyn College, Brooklyn, NY 11210, USA
[3] CUNY Graduate Center, New York, NY 10016, USA

**Abstract.** We consider the Stochastic Boolean Function Evaluation (SBFE) problem where the task is to efficiently evaluate a known Boolean function $f$ on an unknown bit string $x$ of length $n$. We determine $f(x)$ by sequentially testing the variables of $x$, each of which is associated with a cost of testing and an independent probability of being true. If a strategy for solving the problem is adaptive in the sense that its next test can depend on the outcomes of previous tests, it has lower expected cost but may take up to exponential space to store. In contrast, a non-adaptive strategy may have higher expected cost but can be stored in linear space and benefit from parallel resources. The adaptivity gap, the ratio between the expected cost of the optimal non-adaptive and adaptive strategies, is a measure of the benefit of adaptivity. We present lower bounds on the adaptivity gap for the SBFE problem for popular classes of Boolean functions, including read-once DNF formulas, read-once formulas, and general DNFs. Our bounds range from $\Omega(\log n)$ to $\Omega(n/\log n)$, contrasting with recent $O(1)$ gaps shown for symmetric functions and linear threshold functions.

## 1 Introduction

We consider the question of determining adaptivity gaps for the Stochastic Boolean Function Evaluation (SBFE) problem, for different classes of Boolean formulas. In an SBFE problem, we are given a (representation of a) Boolean function $f : \{0,1\}^n \to \{0,1\}$, a positive cost vector $c = [c_1, \ldots, c_n]$, and a probability vector $p = [p_1, \ldots, p_n]$. The problem is to determine the value $f(x)$ on an initially unknown random input $x \in \{0,1\}^n$. The value of each $x_i$ can only be determined by performing a test, which incurs a cost of $c_i$. Each $x_i$ is equal to 1 (is true) with independent probability $p_i$. Tests are performed sequentially and continue until $f(x)$ can be determined. We say $f(x)$ is determined by a set of tests if $f(x) = f(x')$ for all $x' \in \{0,1\}^n$ such that $x'_i = x_i$ for every $i$ in the set of tests.

For example, if $f(x) = x_1 \vee \ldots \vee x_n$ then testing continues until a test is performed on some $x_i$ such that $x_i = 1$ at which point we know $f(x) = 1$, or until all $n$ tests have been performed with outcome $x_i = 0$ for each $x_i$ so we

P. Chalermsook and B. Laekhanukit (Eds.): WAOA 2022, LNCS 13538, pp. 190–210, 2022.
https://doi.org/10.1007/978-3-031-18367-6_10

know $f(x) = 0$. The problem is to determine the order to perform tests that minimizes the total expected cost of the tests.

We will call a testing order a strategy which we can think of as a decision tree for evaluating $f$. A strategy can be *adaptive*, meaning that the choice of the next test $x_i$ can depend on the outcome of previous tests. In some practical settings, however, it is desirable to consider only *non-adaptive* strategies. Non-adaptive strategies often take up less space than adaptive strategies, and they may be able to be evaluated more quickly if tests can be performed in parallel [20], such as in the problem of detecting network faults [23] or in group testing for viruses, such as the coronavirus [28]. A non-adaptive testing strategy is a permutation of the tests where testing continues in the order specified by the permutation until the value of $f(x)$ can be determined from the outcomes of the tests performed so far. A non-adaptive strategy also corresponds to a decision tree where all non-leaf nodes on the same level contain the same test $x_i$.

The *adaptivity gap* measures how much benefit can be obtained by using an adaptive strategy. Consider a class $F$ of $n$-variable functions $f : \{0,1\}^n \to \{0,1\}$. Let $\mathsf{OPT}_\mathcal{N}(f,c,p)$ be the expected evaluation cost of the optimal non-adaptive strategy on function $f$ under costs $c$ and probabilities $p$. Similarly, $\mathsf{OPT}_\mathcal{A}(f,c,p)$ is the expected evaluation cost of the optimal adaptive strategy on $f$ under $c$ and $p$. The adaptivity gap of the function class $F$ is

$$\max_{f \in F} \sup_{c,p} \frac{\mathsf{OPT}_\mathcal{N}(f,c,p)}{\mathsf{OPT}_\mathcal{A}(f,c,p)}.$$

The SBFE problem for a class of Boolean formulas $F$ restricts the evaluated $f$ to be a member of $F$. In this paper we prove bounds on the adaptivity gaps for the SBFE problem on read-once DNF formulas, DNF formulas, and read-once formulas. (See Sect. 1.3 for definitions.) A summary of our results can be found in Table 1.

All the bounds in the table have a dependence on $n$, meaning that none of the listed SBFE problems has a constant adaptivity gap. This contrasts with recent work of Ghuge et al. [14], which shows that the adaptivity gaps for the SBFE problem for symmetric Boolean functions and linear threshold functions are $O(1)$.

For any SBFE problem, the non-adaptive strategy of testing the $x_i$ in increasing order of $c_i$ has an expected cost that is within a factor of $n$ of the optimal adaptive strategy [25]. Thus $n$ is an upper bound on the adaptivity gap for all SBFE problems.

**Outline:** We present our results on formula classes in increasing order of generality. In Sect. 2, we warm up with a variety of results on read-once DNF formulas in different settings. In Sect. 3, we prove our main technical result for read-once formulas, drawing on branching process identities and concentration inequalities. In Sect. 4, we prove our most general results on DNF formulas (the bounds also apply to the restricted class of DNF formulas with a linear number of terms). Note that we state our lower bound results in the most restricted context because they of course apply to more general settings. Due to space constraints, we defer some proofs to Appendix A.

**Table 1.** A summary of our results. We also prove an $O(\sqrt{n})$ upper bound for tribes formulas i.e., read-once DNFs with unit costs where every term has the same number of variables. We say all probabilities are equal if $p_1 = p_2 = \ldots = p_n$.

| Formula class | Adaptivity gap |
|---|---|
| Read-once DNF | $\Theta(\log n)$ for unit costs, uniform distribution |
| | $\Omega(\sqrt{n})$ for unit costs |
| | $\Omega(n^{1-\epsilon}/\log n)$ for uniform distribution |
| Read-once | $\Omega(\epsilon^3 n^{1-2\epsilon/\ln(2)})$ for unit costs, equal probabilities |
| DNF | $\Omega(n/\log n)$ for unit costs, uniform distribution |
| | $\Theta(n)$ for uniform distribution |

## 1.1    Connection to $st$-Connectivity in Uncertain Networks

Our result for read-once formulas has implications for a problem of determining $st$-connectivity in an uncertain network, studied by Fu et al. [13]. The input is a multi-graph with a source node $s$ and a destination node $t$. Each edge corresponds to a variable $x_i$ indicating whether it is usable, which is true with probability $p_i$. Testing the usability of edge $i$ costs $c_i$. The $st$-connectivity function for the multi-graph is true if and only if there is a path of usable edges from $s$ to $t$. The problem is to find a strategy to evaluate the $st$-connectivity function that has minimum expected cost.

The $st$-connectivity function associated with a multi-graph can be represented by a read-once formula if and only if the multi-graph is a two-terminal series-parallel graph. This type of graph has two distinguished nodes, $s$ and $t$, and is formed by recursively combining disjoint series-parallel graphs either in series, or in parallel (see [12] for the precise definitions).[1] Fu et al. performed experiments with both adaptive and non-adaptive strategies for this problem, comparing their performance, but did not prove theoretical adaptivity gap bounds. Since the $st$-connectivity function on a series-parallel graph is a read-once formula, our lower bound on the adaptivity gap for read-once formulas applies to the problem of $st$-connectivity.

## 1.2    Related Work

It is well-known that the SBFE problem for the Boolean OR function given by $f(x) = x_1 \vee \ldots \vee x_n$ has a simple solution: test the variables $x_i$ in increasing order of the ratio $c_i/p_i$ until a test reveals a variable set to true, or until all variables

---

[1] The term *series-parallel circuits* (systems) refers to a set of parallel circuits that are connected in series (see, e.g., [11,31]). Viewed as graphs, they correspond to the subset of two-terminal series-parallel graphs whose $st$-connectivity functions correspond to read-once CNF formulas. We note that Kowshik used the term "series-parallel graph" in a non-standard way to refer only to this subset; Fu et al. in citing Kowshik, used the term the same way [13,27].

are tested and found to be false (cf. [30]). This strategy is non-adaptive, meaning that the SBFE problem for the Boolean OR function has an adaptivity gap of 1. That is, there is no benefit to adaptivity.

Gkenosis et al. [16] introduced the Stochastic Score Classification problem, which generalizes the SBFE problem for both symmetric Boolean functions and for linear threshold functions. Ghuge et al. [14] showed that the Stochastic Score Classification problem has an adaptivity gap of $O(1)$. In the unit-cost case, Gkenosis et al. showed the gap is at most 4 for symmetric Boolean functions, and at most $\phi$ (the golden ratio) for the not-all-equal function [16].

Adaptivity gaps were introduced by Dean et al. [7] in the study of the stochastic knapsack problem which, in contrast to the SBFE problem for Boolean functions, is a maximization problem. It has an adaptivity gap of 4 [7,9]. Adaptivity gaps have also been shown for other stochastic maximization problems (e.g., [2,5,8,19,24]). Notably, the problem of maximizing a monotone submodular function with stochastic inputs, subject to any class of prefix-closed constraints, was shown to have an $O(1)$ adaptivity gap [5,20].

Adaptivity gaps have also been shown for stochastic covering problems, which, like SBFE, are minimization problems. Goemans et al. [17] showed that the adaptivity gap for the Stochastic Set Cover problem, in which each item can only be chosen once, is $\Omega(d)$ and $O(d^2)$, where $d$ is the size of the target set to be covered. If the items can be used repeatedly, the adaptivity gap is $\Theta(\log d)$.

Agarwal et al. [1] and Ghuge et al. [15] proved bounds of $\Omega(Q)$ and $O(Q \log Q)$ respectively, on the adaptivity gap for the more abstract Stochastic Submodular Cover Problem in which each item can only be used once. Applied to the special case of Stochastic Set Cover, the upper bound is $O(d \log d)$, which improves the above $O(d^2)$ bound. They also gave bounds parameterized by the number of rounds of adaptivity allowed. We note that, as shown by Deshpande et al. [10], one approach to solving SBFE problems is to reduce them to special cases of Stochastic Submodular Cover. However, this approach does not seem to have interesting implications for SBFE problem adaptivity gaps.

## 1.3   Preliminaries

Consider a Boolean function $f : \{0,1\}^n \to \{0,1\}$, a positive cost vector $c = [c_1,\ldots,c_n]$, and probability vector $p = [p_1,\ldots,p_n]$. We assume $c_i > 0$ and $0 < p_i < 1$ for $i \in [n]$ where $[n]$ denotes the set $\{1,\ldots,n\}$. Let strategy $S$ be a decision tree for evaluating $f(x)$ on an unknown input $x \in \{0,1\}^n$. We define $\mathrm{cost}_c(f,x,S)$ as the total cost of the variables tested by $S$ on input $x$ until $f(x)$ is determined. We say $x \sim p$ if $\Pr(x) = \prod_{i:x_i=1} p_i \prod_{i:x_i=0}(1-p_i)$. Then $\mathrm{cost}_{c,p}(f,S) := \mathbb{E}_{x \sim p}[\mathrm{cost}_c(f,x,S)]$ is the expected evaluation cost of strategy $S$ when $x$ is drawn according to the product distribution induced by $p$.

For fixed $n$, let $\mathcal{A}$ be the set of adaptive strategies on $n$ variables and $\mathcal{N}$ be the set of non-adaptive strategies on $n$ variables. We are interested in the quantities

$$\mathrm{OPT}_{\mathcal{A}}(f,c,p) := \min_{S \in \mathcal{A}} \mathrm{cost}_{c,p}(f,S) \text{ and } \mathrm{OPT}_{\mathcal{N}}(f,c,p) := \min_{S \in \mathcal{N}} \mathrm{cost}_{c,p}(f,S).$$

We will omit $c$ and $p$ from the notation when the costs and probabilities are clear from context. A *(Boolean) read-once formula* is a tree, each of whose internal nodes are labeled either $\vee$ or $\wedge$. The internal nodes have two or more children. Each leaf is labeled with a Boolean variable $x_i \in \{x_1, \ldots, x_n\}$. The formula computes a Boolean function in the usual way.[2] A *(Boolean) DNF formula* is a formula of the form $T_1 \vee T_2 \vee \ldots T_m$ for some $m \geq 1$, such that each *term* $T_i$ is the conjunction $(\wedge)$ of literals. A literal is a variable $x_i$ or a negated variable $\neg x_i$. The DNF formula is read-once if distinct terms contain disjoint sets of variables, without negations. Read-once DNF formulas whose terms all contain the same number $w$ of literals are sometimes known as tribes formulas of width $w$ (cf. [29]).

## 2  Warm Up: Adaptivity Gaps for Read-Once DNFs

Let $f : \{0,1\}^n \to \{0,1\}$ be a read-once DNF formula. Boros and Ünyülurt [3] showed that the following approach gives an optimal adaptive strategy for evaluating $f$ (this has been rediscovered in later papers [18, 26, 27]). Let $f = T_1 \vee T_2 \ldots \vee T_k$ be a DNF formula with $k$ terms. For each term $T_j$, let $\ell(j)$ be the number of variables in term $T_j$. Order the variables of $T_j$ as $x_{j_1}, x_{j_2}, \ldots x_{j_{\ell(j)}}$ in non-decreasing order of the ratio $c_i/(1 - p_i)$, i.e., so that $c_{j_1}/(1 - p_{j_1}) \leq c_{j_2}/(1 - p_{j_2}) \leq \ldots \leq c_{j_{\ell(j)}}/(1 - p_{j_{\ell(j)}})$. For evaluating the single term $T_j$, an optimal strategy tests the variables in $T_j$ sequentially, in the order $x_{j_1}, x_{j_2}, \ldots x_{j_{\ell(j)}}$, until a variable is found to be false, or until all variables are tested and found to be true.

Denote the probability of the term evaluating to true as $P(T_j) = \prod_{i=1}^{\ell(j)} p_{j_i}$ and the expected cost of this evaluation of the term as

$$C(T_j) = \sum_{i=1}^{\ell(j)} \left( \sum_{k=1}^{i} c_{j_k} \prod_{r=1}^{i-1} p_{j_r} \right).$$

An optimal algorithm for evaluating $f$ applies the above strategy sequentially to the terms $T$ of $f$, in non-decreasing order of the ratio $C(T)/P(T)$, until either some term is found to be satisfied by $x$, so $f(x) = 1$, or all terms have been evaluated and found to be falsified by $x$, so $f(x) = 0$. We will use this optimal adaptive strategy in the remainder of the section.

In what follows, we will frequently describe non-adaptive strategies as performing the $n$ possible tests in a particular order. We mean by this that the permutation representing the strategy lists the tests in this order. The testing stops when the value of $f$ can be determined.

---

[2] Some definitions of a read-once formula allow negations in the internal nodes of the formula. By DeMorgan's laws, these negations can be "pushed" into the leaves of the formula, resulting in a formula whose internal nodes are $\vee$ and $\wedge$, such that each variable $x_i$ appears in at most one leaf.

---

**Algorithm 1:** Evaluating a read-once DNF where each variable has unit cost and uniform distribution.

---

**Input**  : $n > 0$, read-once DNF $f : \{0,1\}^n \to \{0,1\}$ with $m$ terms
**Output:** $\pi$          // $O(\log n)$-approximation non-adaptive strategy for $f$
$\pi \leftarrow []$                                                                        // empty list
**for** $i = 1$ **to** $m$ **do**

    **if** $|T_i| \leq 2\log n$ **then**                         // $T_i$ is $i$th shortest term in $f$
    |  $\pi \leftarrow \pi + $ all variables in $T_i$
    **else**
    |  $\pi \leftarrow \pi + $ first $2\log n$ variables in $T_i$
    **end**

**end**
$\pi \leftarrow \pi + $ remaining variables not in $\pi$

---

## 2.1   Unit Costs and the Uniform Distribution

We begin by showing that the adaptivity gap for read-once DNFs, in the case of unit costs and the uniform distribution, is at most $O(\log n)$.

**Theorem 1.** *Let $f : \{0,1\}^n \to \{0,1\}$ be a read-once DNF formula. For unit costs and the uniform distribution, there is a non-adaptive strategy $S$ such that* $\text{cost}(f, S) \leq O(\log n) \cdot \text{OPT}_{\mathcal{A}}(f)$.

*Proof (Proof Sketch).* Using the characterization of the optimal adaptive strategy due to Boros and Ünyülurt [3], we show that Algorithm 1 gives a non-adaptive strategy that has expected cost at most $O(\log n)$ times the optimal adaptive strategy. The algorithm crucially relies on the observation that the optimal adaptive algorithm tests terms in non-decreasing order of length for unit costs and the uniform distribution. To see this, observe $C(T)/P(T)$ is non-decreasing when terms are ordered by length in this setting. For terms with length at most $2\log n$, we can test every variable without paying more than $O(\log n)$ times the optimal adaptive strategy. For terms with length greater than $2\log n$, we can test $2\log n$ variables and only need to continue testing with probability $1/n^2$.

We complement Theorem 1 with a matching lower bound. We prove the theorem by exhibiting a read-once DNF with $\sqrt{n}$ identical terms. We upper bound the optimal adaptive strategy and argue any non-adaptive strategy has to make $\log n$ tests per term to verify $f(x) = 0$ which occurs with constant probability.

**Theorem 2.** *Let $f : \{0,1\}^n \to \{0,1\}$ be a read-once DNF formula. For unit costs and the uniform distribution, $\text{OPT}_{\mathcal{N}}(f) \geq \Omega(\log n) \cdot \text{OPT}_{\mathcal{A}}(f)$.*

## 2.2   Unit Costs and Arbitrary Probabilities

We give an upper bound of the adaptivity gap for read-once DNF formulas with unit costs and arbitrary probabilities in the special case where all terms have the same number of variables. This is known as a tribes formula [29]. Let the number of terms be $m$. We now describe two non-adaptive strategies which yield a $n/m$-approximation and a $m$-approximation, respectively. Then, by choosing the non-adaptive strategy based on the number of terms $m$, we are guaranteed a $\min\{n/m, m\} \leq O(\sqrt{n})$-approximation.

**Lemma 1.** *Consider a read-once DNF $f : \{0,1\}^n \to \{0,1\}$ where each term has the same number of variables. For unit costs and arbitrary probabilities, there is a non-adaptive strategy $S \in \mathcal{N}$ such that $\mathrm{cost}(f, S) \leq n/m \cdot \mathsf{OPT}_\mathcal{A}(f)$.*

*Proof (Proof of Lemma 1).* Consider a random input $x$ and the optimal adaptive strategy described at the start of this section. If $f(x) = 0$, the optimal adaptive strategy must certify that each term is 0 which requires at least $m$ tests. Since any non-adaptive strategy will make at most $n$ tests, the ratio between the cost incurred on $x$ by a non-adaptive strategy, and by the optimal adaptive strategy, is at most $n/m$. Otherwise, if $f(x) = 1$, the optimal adaptive strategy will certify that a term is true after testing some number of false terms. Now consider the non-adaptive version of this optimal adaptive strategy which tests terms in the same fixed order but must test all variables in a term before proceeding to the next term. For each false term that the optimal adaptive strategy tests, the non-adaptive strategy will test every variable for a total of $n/m$ tests. Since the optimal adaptive strategy must make at least one test per false term, the ratio between the cost incurred on $x$ by the non-adaptive strategy, and the cost incurred by the optimal strategy, is at most $n/m$. Since the ratio $n/m$ holds for all $x$, the lemma follows.

**Lemma 2.** *Consider a read-once DNF $f : \{0,1\}^n \to \{0,1\}$ where each term has the same number of variables. For unit costs and arbitrary probabilities, there is a non-adaptive strategy $S \in \mathcal{N}$ with expected cost $\mathrm{cost}(f, S) \leq m \cdot \mathsf{OPT}_\mathcal{A}(f)$.*

*Proof (Proof of Lemma 2).* Fix a random input $x$. If $f(x) = 0$, the optimal adaptive strategy certifies that every term is false. Let $C_i$ be the number of tests it makes until finding a false variable on the $i$th term. Consider the non-adaptive "round-robin" strategy which progresses in rounds, making one test in each term per round. Within a term, the non-adaptive strategy tests variables in the same fixed order as the optimal adaptive strategy. Then the cost of the non-adaptive strategy is $m \cdot \max_i C_i$ whereas the cost of the optimal adaptive strategy is $\sum_{i=1}^m C_i$. It follows that the adaptivity gap is at most $m$. Otherwise, if $f(x) = 1$, the optimal adaptive strategy must certify that a term is true by making at least $n/m$ tests. Any non-adaptive strategy will make at most $n$ tests so the adaptivity gap is at most $m$.

Together, the $O(n/m)$- and $O(m)$-approximations imply the following result.

**Theorem 3.** *Let $f : \{0,1\}^n \to \{0,1\}$ be a read-once DNF formula where each term has the same number of variables. For unit costs and arbitrary probabilities, there is a non-adaptive strategy $S \in \mathcal{N}$ with $\mathrm{cost}(f, S) \leq O(\sqrt{n}) \cdot \mathsf{OPT}_\mathcal{A}(f)$.*

We complement Theorem 3 with a matching lower bound. We prove the theorem by exhibiting a read-once DNF with $2\sqrt{n}$ identical terms. By making one special variable in each term have a low probability of being true and arguing it must always be tested first, the non-adaptive strategy has to search at random for which special variable is true when every other special variable is false which happens with constant probability.

**Theorem 4.** *Let $f : \{0,1\}^n \rightarrow \{0,1\}$ be a read-once DNF formula. For unit costs and arbitrary probabilities, $\mathsf{OPT}_\mathcal{N}(f) \geq \Omega(\sqrt{n}) \cdot \mathsf{OPT}_\mathcal{A}(f)$.*

### 2.3   Arbitrary Costs and the Uniform Distribution

We prove Theorem 5 by exhibiting a read-once DNF with $2^\ell$ terms each of length $\ell$. Within each term, the cost of each variable increases geometrically with a ratio of 2. The challenge is choosing $\ell$ so that $2^\ell \ell = n$. We accomplish this by using a modified Lambert W function [6] which is how we calculate $n_\epsilon$.

**Theorem 5.** *For all $\epsilon > 0$, there exists $n_\epsilon > 0$ such that the following holds for all read-once DNF formulas $f : \{0,1\}^n \rightarrow \{0,1\}$ where $n > n_\epsilon$: There exists a cost assignment such that for the uniform distribution, $\mathsf{OPT}_\mathcal{N}(f) \geq \Omega(n^{1-\epsilon}/\log n) \cdot \mathsf{OPT}_\mathcal{A}(f)$.*

## 3   Main Result: Read-Once Formulas

**Theorem 6.** *Fix $\epsilon > 0$. There is a read-once formula $f : \{0,1\}^n \rightarrow \{0,1\}$, such that for unit costs and $p_i = \frac{1+\epsilon}{2}$ for all $i \in [n]$, $\mathsf{OPT}_\mathcal{N}(f) \geq \Omega\left(\epsilon^3 n^{1-2\epsilon/\log 2}\right) \cdot \mathsf{OPT}_\mathcal{A}(f)$.*

Before we prove Theorem 6, we describe the read-once formula $f$ and present the technical lemmas we use in the proof. Without loss of generality, assume $n = 2 \cdot 2^d - 2$ for some positive integer $d$. We define the function $f(x)$ on inputs $x \in \{0,1\}^n$ in terms of a binary tree with depth $d$. The edges of the tree are numbered 1 through $n$, and variable $x_i$ corresponds to edge $i$. Each variable $x_i$ has a $\frac{1+\epsilon}{2}$ probability of being true. Say that a leaf of the tree is "alive" if $x_i = 1$ for all edges $i$ on the path from the root to the leaf. We define $f(x) = 1$ if and only if at least one leaf of the tree is alive. A strategy for evaluating $f$ will continue testing until it can certify that there is at least one alive leaf, or that no alive leaf exists (Fig. 1).

Now consider the multi-graph that is produced from the tree by merging all leaves into a single node. The function $f$ is the $st$-connectivity function of this multi-graph. It is easy to show, by induction on the depth of the tree, that the multi-graph is two-terminal series-parallel, with $s$ the root, and $t$ the node produced by merging the leaves of the tree. Thus $f$ is computed by a read-once formula.

We refer to the edges of the tree that join a leaf to its parent as leaf edges, and the other edges as internal edges. We say that a non-adaptive strategy $S$

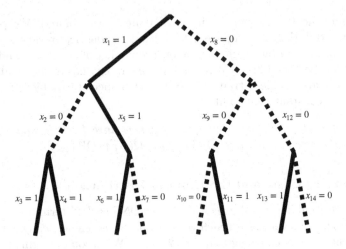

**Fig. 1.** The binary tree corresponding to the read-once formula we construct when $n = 14$. In particular, $f(x) = (x_1 \wedge ((x_2 \wedge (x_3 \vee x_4)) \vee (x_5 \wedge (x_6 \vee x_7)))) \vee (x_8 \wedge ((x_9 \wedge (x_{10} \vee x_{11})) \vee (x_{12} \wedge (x_{13} \vee x_{14}))))$. Notice that $f(x) = 1$ for this $x$ because the third leaf from the left is alive (all its ancestors are true).

is leaf-last if it first tests all non-leaf edges of the tree, and then tests the leaf edges.

In the proof of Theorem 6, we consider an alternative cost assignment where we pay unit costs for the tests on leaf edges, as usual, but tests on internal edges are free. The expected cost of a strategy under the usual unit cost assignment is clearly lower bounded by its expected cost when internal edges are free. Note that when internal edges are free, there is no disadvantage in performing all the tests on internal edges first, so there is an optimal non-adaptive strategy which is leaf-last in the sense that all the leaf edges appear last. Our first technical lemma describes a property of a leaf-last strategy. We defer the proof of this lemma, and of the ones that follow, to the end of this section. In all of the lemma statements, we assume $f$ is as just described, and expected costs are with respect to unit costs and test probabilities $p_i = \frac{1+\epsilon}{2}$. We use $L$ to denote the number of leaves in the tree.

**Lemma 3.** *There exists a leaf-last non-adaptive strategy $S$ for evaluating $f$ which, conditioned on the event that there is at least one alive leaf, has minimum expected cost when internal edges are free relative to all non-adaptive strategies. Further, for any such $S$ and any $\ell \in [L-1]$, conditioned on the existence of at least one alive leaf, the probability that $S$ first finds an alive leaf on the $\ell$th leaf test is at least the probability $S$ first finds an alive leaf on the $(\ell+1)$st leaf test.*

The next lemma gives us an inequality that we will use to lower bound the cost of the optimal non-adaptive strategy.

**Lemma 4.** *Let $L$ be a positive integer and $p_1 \geq p_2 \geq \ldots \geq p_L$ be non-negative real numbers. Now let $p \geq p_1$ and define $L' = \lfloor \sum_{\ell=1}^{L} p_\ell / p \rfloor$. Then $\sum_{\ell=1}^{L} \ell p_\ell \geq \sum_{\ell=1}^{L'} \ell p$.*

Our analysis depends on there being at least constant probability that $f(x) = 1$, or equivalently, that there is at least one alive leaf. The next lemma assures us that this is indeed the case. The proof of the lemma depends on our choice of having each $p_i$ be slightly larger than $1/2$; it would not hold otherwise.

**Lemma 5.** *With probability at least $\epsilon$, there is at least one alive leaf in the binary tree representing $f$.*

With these key lemmas in hand, we prove Theorem 6.

*Proof (Proof of Theorem 6).* We will show that the adaptivity gap is large. Intuitively, we rely on the fact that if there is at least one alive leaf, then an adaptive strategy can find an alive leaf cheaply, by beginning at the root of the tree and moving downward only along edges that are alive. In contrast, a non-adaptive strategy cannot stop searching along "dead" branches. However, it is not immediately clear that the cost of the non-adaptive strategy is high because there are conditional dependencies between the probabilities that two leaves with the same ancestor(s) are alive. To prove the desired result, we need to show that, despite these dependencies, the optimal non-adaptive strategy must have high expected cost.

We begin by showing that the expected cost of any non-adaptive strategy is at least $\frac{\epsilon^2}{16} n^{1 - \frac{\epsilon}{\log 2}}$. We want to lower bound the expected cost of the optimal strategy $\mathsf{OPT}_{\mathcal{N}}(f)$:

$$\min_{S \in \mathcal{N}} \mathbb{E}_x[\mathrm{cost}(f, x, S)] \geq \min_{S} \mathbb{E}[\mathrm{cost}^L(f, x, S)] = \min_{S'} \mathbb{E}[\mathrm{cost}^L(f, x, S')] \quad (1)$$

where $\mathrm{cost}^L(f, x, S)$ is the number of leaf tests $S$ makes on $x$ until $f(x)$ is determined, and $S'$ is a leaf-last strategy. Then

$$(1) = \min_{S'} \left( \sum_{x : f(x) = 1} \Pr(x) \cdot \mathrm{cost}^L(f, x, S') + \sum_{x : f(x) = 0} \Pr(x) \cdot \mathrm{cost}^L(f, x, S') \right)$$

$$\geq \min_{S'} \sum_{x : f(x) = 1} \Pr(x) \cdot \mathrm{cost}^L(f, x, S')$$

$$= \min_{S'} \sum_{\ell=1}^{L} \ell \Pr(S' \text{ first finds alive leaf on } \ell\text{th leaf test})$$

where $L = 2^d$ is the number of leaves in the binary tree. Initially, all leaves have a $\left(\frac{1+\epsilon}{2}\right)^d$ probability of being alive where $d = \log_2((n+2)/2)$. By Lemma 3, the probability that the next leaf is alive cannot increase as the optimal non-adaptive strategy $S^*$ performs its test. Set

$$p = \left(\frac{1+\epsilon}{2}\right)^d$$

and $p_\ell = \Pr(S^*$ first finds alive leaf on $\ell$th test$)$. Therefore, Lemma 4 with $p$ and $p_\ell$ tells us that

$$\sum_{\ell=1}^{L} \ell p_\ell \geq \sum_{\ell=1}^{L'} \ell \left(\frac{1+\epsilon}{2}\right)^d \geq \left(\frac{1+\epsilon}{2}\right)^d \frac{\epsilon^2}{8\left(\frac{1+\epsilon}{2}\right)^{2d}} \geq \frac{\epsilon^2}{8\left(\frac{n+2}{2}\right)^{\log_2\left(\frac{1+\epsilon}{2}\right)}} \geq \frac{\epsilon^2}{16} n^{1-\frac{\epsilon}{\log 2}}$$

(2)

where we use the inequality that $L' \geq \epsilon/(2\left(\frac{1+\epsilon}{2}\right)^d)$. To see this, recall that $\sum_{\ell=1}^{L} p_\ell/p \geq L'$ and, since the right-hand side is greater than 1, $2L' \geq \sum_{\ell=1}^{L} p_\ell/p$. By Lemma 5, $\sum_{\ell=1}^{L} p_\ell \geq \epsilon$ so $L' \geq \epsilon/(2\left(\frac{1+\epsilon}{2}\right)^d)$. The last inequality in Eq. (2) follows from $\log_2(\frac{1+\epsilon}{2}) \leq \frac{\epsilon}{\log 2} - 1$ which can be shown by comparing the $y$-intercepts and derivatives for $\epsilon > 0$.

Next, we show that the expected cost of the adaptive strategy is at most $(n+1)^{\frac{\epsilon}{\log 2}}/\epsilon$. Consider an adaptive strategy which starts by querying the two edges of the root and recurses as follows: if an edge is alive it queries its two child edges and otherwise stops. Observe that this simple depth-first search adaptive strategy will make at most two tests for every alive edge in the binary tree. Therefore the expected number of tests an adaptive strategy must make is at most twice the expected number of alive edges. By the branching process analysis in the proof of Lemma 5, twice the expected number of alive edges is

$$2\sum_{i=0}^{d}(1+\epsilon)^i \leq 2\sum_{i=0}^{\log_2 n}(1+\epsilon)^i = 2\frac{(1+\epsilon)^{\log_2(n)+1}-1}{(1+\epsilon)-1} \leq 4\frac{n^{\log_2(1+\epsilon)}}{\epsilon} \leq 4\frac{n^{\frac{\epsilon}{\log 2}}}{\epsilon}$$

(3)

where the last inequality follows from $\log_2(1+\epsilon) \leq \frac{\epsilon}{\log 2}$ which we can see by comparing the $y$-intercepts and slopes for $\epsilon > 0$. Then Theorem 6 follows from Eqs. (2) and (3).

*Proof (Proof of Lemma 3).* A leaf-last non-adaptive strategy $S$ satisfying the given conditional optimality property clearly exists, because any non-adaptive strategy can be made leaf-last by moving the leaf tests to the end without affecting its cost when internal edges are free. Define $p_\ell(S)$ as the probability that $S$ finds an alive leaf for the first time on leaf test $\ell$. We may write the expected number of leaf tests of $S$ as $\sum_{\ell=1}^{L} \ell p_\ell(S)$.

Now suppose for contradiction that there is some $\ell'$ such that $p_{\ell'}(S) < p_{\ell'+1}(S)$. Let $S'$ be $S$ but with the $\ell'$th and $(\ell'+1)$th tests swapped. We will show that the expected number of leaf tests made by $S'$ is strictly lower than the expected number of leaf tests made by $S$. Observe that

$$\sum_{\ell=1}^{L} \ell p_\ell(S) - \sum_{\ell=1}^{L} \ell p_\ell(S') = \ell'(p_{\ell'}(S) - p_{\ell'}(S')) + (\ell'+1)(p_{\ell'+1}(S) - p_{\ell'+1}(S')).$$

Notice that $p_{\ell'}(S) < p_{\ell'+1}(S) \leq p_{\ell'}(S')$ where the first inequality follows by assumption and the second inequality follows because moving a test on a particular leaf edge to appear earlier in the permutation can only increase the probability that its leaf is the first alive leaf found. In addition, since the combined probability we first find an alive leaf in either the $\ell'$th or $(\ell' + 1)$th test is the same in either order of tests, $p_{\ell'}(S) + p_{\ell'+1}(S) = p_{\ell'}(S') + p_{\ell'+1}(S')$. Together, we have that $-(p_{\ell'}(S) - p_{\ell'}(S')) = p_{\ell'+1}(S) - p_{\ell'+1}(S') > 0$. Therefore $\sum_{\ell=1}^{L} \ell p_\ell(S) - \sum_{\ell=1}^{L} \ell p_\ell(S') = p_{\ell'+1}(S) - p_{\ell'+1}(S') > 0$ and $S'$ makes fewer leaf tests in expectation even though $S$ was optimal by assumption. A contradiction!

*Proof (Proof of Lemma 4).* Define $\delta_\ell = p - p_\ell \geq 0$ for $\ell \in [L']$. Observe that $L' \leq L$ since $p \geq p_\ell$ for all $\ell \in [L']$. Then

$$\sum_{\ell=1}^{L} \ell p_\ell = \sum_{\ell=1}^{L'} \ell(p - \delta_\ell) + \sum_{\ell=L'+1}^{L} \ell p_\ell = \sum_{\ell=1}^{L'} \ell p - \sum_{\ell=1}^{L'} \ell\delta_\ell + \sum_{\ell=L'+1}^{L} \ell p_\ell. \quad (4)$$

Now all that remains to be shown is that the sum of the last two terms in Eq. (4) is non-negative. Notice that

$$\sum_{\ell=1}^{L'} p_\ell + \sum_{\ell=L'+1}^{L} p_\ell \geq \sum_{\ell=1}^{L'} p \implies \sum_{\ell=L'+1}^{L} p_\ell \geq \sum_{\ell=1}^{L'}(p - p_\ell) = \sum_{\ell=1}^{L'} \delta_\ell.$$

Then

$$\sum_{\ell=1}^{L'} \ell\delta_\ell \leq L'\sum_{\ell=1}^{L'} \delta_\ell \leq L' \sum_{\ell=L'+1}^{L} p_\ell \leq \sum_{\ell=L'+1}^{L} \ell p_\ell.$$

*Proof (Proof of Lemma 5).* Let $Z_i$ denote the number of alive edges at level $i$. Then the statement of Lemma 5 becomes $\Pr(0 < Z_d) \geq \epsilon$. Using standard results from the study of branching processes, we know that

$$\mathbb{E}[Z_d] = \mu^d \quad \text{and} \quad \text{Var}(Z_d) = (\mu^{2d} - \mu^d)\frac{\sigma^2}{\mu(\mu - 1)}$$

where $\mu$ is the expectation and $\sigma$ is the variance of the number of alive "children" from a single alive edge (see e.g., p. 6 in Harris [22]). In our construction,

$$\mu = 2\left(\frac{1+\epsilon}{2}\right) = (1 + \epsilon) \quad \text{and} \quad \sigma^2 = 2\left(\frac{1+\epsilon}{2}\right)\left(1 - \frac{1+\epsilon}{2}\right) = \frac{1 - \epsilon^2}{2}$$

since the children of one edge follow the binomial distribution. Then $\mathbb{E}[Z_d] = (1 + \epsilon)^d$ and

$$\text{Var}(Z_d) = ((1 + \epsilon)^{2d} - (1 + \epsilon)^d)\frac{1 - \epsilon^2}{2(1 + \epsilon)\epsilon} \leq \frac{1}{2\epsilon}((1 + \epsilon)^d)^2 = \frac{1}{2\epsilon}\mathbb{E}[Z_d]^2.$$

We will now use Cantelli's inequality (see page 46 in Boucheron et al. [4]) to show that $\Pr(Z_d > 0) \geq \epsilon$. Cantelli's tells us that $\Pr(X - \mathbb{E}[X] \geq \lambda) \leq \frac{\text{Var}(X)}{\text{Var}(X) + \lambda^2}$

for any real-valued random variable $X$ and $\lambda > 0$. Choose $X = -Z_d$ and $\lambda = \mathbb{E}[Z_d]$. Then

$$\Pr(Z_d \leq \mathbb{E}[Z_d] - \mathbb{E}[Z_d]) \leq \frac{\mathrm{Var}(Z_d)}{\mathrm{Var}(Z_d) + \mathbb{E}[Z_d]^2}$$

and, by taking the complement,

$$\Pr(Z_d > 0) \geq \frac{\mathbb{E}[Z_d]^2}{\mathrm{Var}(Z_d) + \mathbb{E}[Z_d]^2} \geq \frac{\mathbb{E}[Z_d]^2}{\frac{1}{2\epsilon}\mathbb{E}[Z_d]^2 + \mathbb{E}[Z_d]^2} = \frac{2\epsilon}{1 + 2\epsilon} \geq \epsilon \quad (5)$$

for $0 < \epsilon \leq 1/2$. Then Lemma 5 follows from Eq. (5).

## 4   DNF Formulas

We will show near-linear and linear in $n$ lower bounds for DNF formulas under the uniform distribution with unit and arbitrary costs, respectively. Since the function we exhibit has linear terms, the lower bounds also apply to the class of linear-size DNF formulas.

**Theorem 7.** *Let $f : \{0,1\}^n \to \{0,1\}$ be a DNF formula. For unit costs and the uniform distribution, $\mathsf{OPT}_{\mathcal{N}}(f) \geq \Omega(n/\log n) \cdot \mathsf{OPT}_{\mathcal{A}}(f)$.*

*Proof.* Without loss of generality, assume $n = 2^d + d$ for some positive integer $d$. Consider the address function $f$ with $2^d$ terms which each consist of $d$ shared variables appearing in all terms, and a single dedicated variable appearing only in that term. We may write $f = T_0 \vee T_1 \vee \cdots \vee T_{2^d-1}$ where $T_i$ consists of the shared variables negated according to the binary representation of $i$ and the single dedicated variable.

By testing the $d$ shared variables, the optimal adaptive strategy can learn which single term is unresolved and test the corresponding dedicated variable in a total of $d+1$ tests. In contrast, any non-adaptive strategy has to search for the unresolved dedicated test at random which gives expected $2^d/2$ cost. (We can ensure the non-adaptive strategy tests the shared variables first by making them free which can only decrease the expected cost.) It follows that the adaptivity gap is $\Omega(2^d/d) = \Omega(n/\log n)$.

We can easily modify the address function in the proof of Theorem 7 to prove an $\Omega(n)$ lower bound for DNF formulas under the uniform distribution and arbitrary costs. In particular, make the cost of each shared variable $1/d$. Then the adaptive strategy pays $d \cdot (1/d)+1 = 2$ while the non-adaptive strategy still pays $\Omega(n)$. The $O(n)$ upper bound comes from the increasing cost strategy and analysis in [25].

**Theorem 8.** *Let $f : \{0,1\}^n \to \{0,1\}$ be a DNF formula. For the uniform distribution, $\mathsf{OPT}_{\mathcal{N}}(f) \geq \Theta(n) \cdot \mathsf{OPT}_{\mathcal{A}}(f)$.*

## 5   Conclusion and Open Problems

We have shown bounds on the adaptivity gaps for the SBFE problem for well-studied classes of Boolean formulas. Our proof of the lower bound for read-once formulas depended on having $p_i$'s that are slightly larger than $1/2$ but we conjecture that a similar or better lower bound holds for the uniform distribution. We note that our lower bound for read-once formulas also applies to (linear-size) monotone DNF formulas, since the given read-once formula based on the binary tree has a DNF formula with one term per leaf. Another open question is to prove a lower bound for monotone DNF formulas that matches our lower bound for general DNF formulas.

A long-standing open problem is whether the SBFE problem for read-once formulas has a polynomial-time algorithm (cf. [18,30]). The original problem only considered adaptive strategies, and it is also open whether there is a polynomial-time (or pseudo polynomial-time) constant or $\log n$ approximation algorithm for such strategies. Happach et al. [21] gave a pseudo polynomial-time approximation algorithm for the non-adaptive version of the problem, which outputs a non-adaptive strategy with expected cost within a constant factor of the optimal non-adaptive strategy. Because of the large adaptivity gap for read-once formulas, as shown in this paper, the result of Happach et al. does not have any implications for the open question of approximating the adaptive version of the SBFE problem for read-once formulas.

## A   Additional Proofs

*Proof (Proof of Theorem 1).* Suppose $f$ is a read-once DNF formula. We will prove that for unit costs and the uniform distribution, there is a non-adaptive strategy $S$ such that $\text{cost}(f, S) \leq O(\log n) \cdot \text{OPT}_{\mathcal{A}}(f)$.

Let $m$ be the number of terms in $f$. Because each variable $x_i$ appears in at most one term, we have that $m \leq n$. As a warm-up, we begin by proving adaptivity gaps for two special cases of $f$.

*Case 1: All Terms have at Most $2 \log n$ Variables.* Under the uniform distribution and with unit costs, the $p_i$ are all equal, and the $c_i$ are all equal. Thus in this case, the optimal adaptive strategy described previously tests terms in increasing order of length. The adaptive strategy *skips* in the sense that if it finds a variable in a term that is false, it moves to the next term without testing the remaining variables in the term. Suppose we eliminate skipping from the optimal adaptive strategy, making the strategy non-adaptive. Since all terms have at most $2 \log n$ variables, this increases the testing cost for any given $x$ by a factor of at most $2 \log n$. Thus the cost of evaluating $f(x)$ for a fixed $x$ increases by a factor of at most $2 \log n$ from an optimal adaptive strategy to a non-adaptive strategy, leading to an adaptivity gap of at most $2 \log n$.

*Case 2: All Terms have More than $2 \log n$ Variables* Consider the following non-adaptive strategy that operates in two phases. In Phase 1, the strategy tests a fixed subset of $2 \log n$ variables from each term, where the terms are taken in increasing length order. In Phase 2, it tests the remaining untested variables in fixed arbitrary order. Since each term has more than $2 \log n$ variables, the value $f$ can only be determined in Phase 1 if a false variable is found in each term during that phase.

Say that an assignment $x$ is *bad* if the value of $f$ cannot be determined in Phase 1, meaning that a false variable is not found in every term during the phase. The probability that a random $x$ satisfies all the tested $2 \log n$ variables of a particular term is $1/n^2$. Then, by the union bound, the probability that $x$ is bad is at most $m/n^2 \le n/n^2 = 1/n$.

Now let us focus on the *good* (not bad) assignments $x$. For each good $x$, our strategy must find a false variable in each term of $f$, which requires at least one test per term for any adaptive or non-adaptive strategy. The cost incurred by our non-adaptive strategy on a good $x$ is at most $2m \log n$, since the strategy certifies that $f(x) = 0$ by the end of Phase 1. Therefore, the expected cost incurred by our non-adaptive strategy $S$ is

$$
\begin{aligned}
\mathrm{cost}(f, S) &\le \Pr(x \text{ good}) \cdot \mathbb{E}[\mathrm{cost}(f, x, S)|x \text{ good}] \\
&+ \Pr(x \text{ bad}) \cdot \mathbb{E}[\mathrm{cost}(f, x, S)|x \text{ bad}] \\
&\le 1 \cdot 2m \log n + \frac{1}{n} \cdot n \le 3m \log n
\end{aligned}
$$

using the fact that $\mathbb{E}[\mathrm{cost}(f, x, S)|x \text{ bad}] \le n$, since there are only $n$ tests, with unit costs.

The expected cost of any strategy, including the optimal adaptive strategy, is at least

$$
\mathrm{OPT}_{\mathcal{A}}(f) \ge \min_{S \in \mathcal{A}} \Pr(f(x) = 0) \cdot \mathbb{E}[\mathrm{cost}(f, x, S)|f(x) = 0] \ge P(f(x) = 0) \cdot m
$$

$$
= (1 - \Pr(f(x) = 1)) \cdot m \ge (1 - \Pr(x \text{ bad})) \cdot m \ge \left(1 - \frac{1}{n}\right) \cdot m \ge \frac{m}{2}
$$

for $n \ge 2$. It follows that the adaptivity gap is at most $6 \log n$.

*Case 3: Everything Else.* We now generalize the ideas in the above two cases. Let $f$ be a read-once DNF that does not fall into Case 1 or Case 2. We can break this DNF into two smaller DNFs, $f = f_1 \vee f_2$ where $f_1$ contains the terms of $f$ of length at most $2 \log n$ and $f_2$ contains the terms of $f$ of length greater than $2 \log n$.

Let $S$ be the non-adaptive strategy that first applies the strategy in Case 1 to $f_1$ and then, if $f_1(x) = 0$, the strategy in Case 2 to $f_2$. Since $S$ cannot stop testing until it determines the value of $f$, in the case that $f_1(x) = 0$, it will test all variables in $f_1$ and then proceed to test variables $f_2$.

Let $S^*$ be the optimal adaptive strategy for evaluating read-once DNFs, described above. We know $S^*$ will test terms in non-decreasing order of length since all tests are equivalent. So, like $S$, $S^*$ tests $f_1$ first and then, if $f_1(x) = 0$, it continues to $f_2$. It follows that we can write the expected cost of $S$ on $f$ as

$$\mathbb{E}[\text{cost}(f, x, S)] = \mathbb{E}[\text{cost}(f_1, x, S_1)] + \Pr(f_1(x) = 0) \cdot \mathbb{E}[\text{cost}(f_2, x, S_2)|f_1(x) = 0]$$

where $S_1$ is the first stage of $S$, where $f_1$ is evaluated, and $S_2$ is the second stage of $S$, where $f_2$ is evaluated. Notice that, by the independence of variables, $\mathbb{E}[\text{cost}(f_2, x, S_2)|f_1(x) = 0] = \mathbb{E}[\text{cost}(f_2, x, S_2)]$. We can similarly write the expected cost of $S^*$ on $f$. Then the adaptivity gap is

$$\frac{\text{OPT}_{\mathcal{N}}(f)}{\text{OPT}_{\mathcal{A}}(f)} \leq \frac{\mathbb{E}[\text{cost}(f_1, x, S_1)] + \Pr(f_1(x) = 0) \cdot \mathbb{E}[\text{cost}(f_2, x, S_2)]}{\mathbb{E}[\text{cost}(f_1, x, S_1^*)] + \Pr(f_1(x) = 0) \cdot \mathbb{E}[\text{cost}(f_2, x, S_2^*)]} \quad (6)$$

where $S_1^*$ is $S^*$ applied to $f_1$ and $S_2^*$ is $S^*$ beginning from the point when it starts evaluating $f_2$.

Using the observation that $(a + b)/(c + d) \leq \max\{a/c, b/d\}$ for positive real numbers $a, b, c, d$, we know that

$$(6) \leq \max\left\{\frac{\mathbb{E}[\text{cost}(f_1, x, S_1)]}{\mathbb{E}[\text{cost}(f_1, x, S_1^*)]}, \frac{\mathbb{E}[\text{cost}(f_2, x, S_2)]}{\mathbb{E}[\text{cost}(f_2, x, S_2^*)]}\right\} = O(\log n)$$

where the upper bound follows from the analysis of Cases 1 and 2.

*Proof (Proof of Theorem 2).* Suppose $f$ is a read-once DNF formula. For unit costs and the uniform distribution, we will show that $\text{OPT}_{\mathcal{N}}(f) \geq \Omega(\log n) \cdot \text{OPT}_{\mathcal{A}}(f)$.

For ease of notation, assume $\sqrt{n}$ is an integer. Consider a read-once DNF $f$ with $\sqrt{n}$ terms where each term has $\sqrt{n}$ variables. By examining the number of tests in each term, we can write the optimal adaptive cost as

$$\text{OPT}_{\mathcal{A}}(f) \leq \sqrt{n} \sum_{i=1}^{\sqrt{n}} \frac{i}{2^i} \leq \sqrt{n} \sum_{i=1}^{\infty} \frac{i}{2^i} = 2\sqrt{n}.$$

The key observation is that, within a term, the adaptive strategy queries variables in any order since each variable is equivalent to any other. Then the probability that the strategy queries exactly $i \leq \sqrt{n}$ variables is $1/2^i$.

Next, we will lower bound the expected cost of the optimal non-adaptive strategy

$$\text{OPT}_{\mathcal{N}}(f) = \min_{S \in \mathcal{N}} \mathbb{E}_{x \sim \{0,1\}^n}[\text{cost}(f, x, S)]$$

$$\geq \min_{S \in \mathcal{N}} \Pr(f(x) = 0)\mathbb{E}[\text{cost}(f, x, S)|f(x) = 0]$$

where $x \sim \{0, 1\}^n$ indicates $x$ is drawn from the uniform distribution. First, we know $\Pr(f(x) = 0) \geq .5$. To see this, consider a random input $x \sim \{0, 1\}^n$. The probability that a particular term is true is $1/2^{\sqrt{n}}$ so the probability that all terms are false (i.e., $f(x) = 0$) is

$$\left(1 - \frac{1}{2^{\sqrt{n}}}\right)^{\sqrt{n}} = \left(\left(1 - \frac{1}{2^{\sqrt{n}}}\right)^{2^{\sqrt{n}}}\right)^{\sqrt{n}/2^{\sqrt{n}}} \geq \left(\frac{1}{2e}\right)^{\sqrt{n}/2^{\sqrt{n}}} \geq .5$$

where the first inequality follows from the loose lower bound that $(1 - 1/x)^x \geq 1/(2e)$ when $x \geq 2$ and the second inequality follows when $n \geq 8$. Second, we know

$$\mathbb{E}[\text{cost}(f, x, S) | f(x) = 0]$$
$$\geq \Pr(\text{one term needs}\Omega(\log n)\text{tests} | f(x) = 0) \cdot \frac{\log_4 n}{2} \cdot \frac{\sqrt{n}}{2}$$

where we used the symmetry of the terms to conclude that if any term needs $\Omega(\log n)$ tests to evaluate it then any non-adaptive strategy will have to spend $\Omega(\log n)$ on half the terms in expectation.

All that remains is to lower bound the probability one term requires $\Omega(\log n)$ tests given $f(x) = 0$. Observe that this probability is

$$1 - (1 - \Pr(\text{a particular term needs}\Omega(\log n)\text{tests} | f(x) = 0))^{\sqrt{n}}$$

$$\geq 1 - \left(1 - \frac{1}{\sqrt{n}}\right)^{\sqrt{n}} \geq 1 - \frac{1}{e} \geq .63$$

where we will now show the first inequality. We can write the probability that a particular term needs $\log_4(n)/2$ tests given $f(x) = 0$ as

$$\Pr\left(x_1 = 1 | f(x) = 0\right) \cdots \Pr\left(x_{\log_4(n)/2} = 1 | f(x) = 0, x_1 = \cdots = x_{\log_4(n)/2-1} = 1\right)$$
$$= \frac{2^{\sqrt{n}-1} - 1}{2^{\sqrt{n}} - 1} \cdots \frac{2^{\sqrt{n}-1-\log_4(n)/2} - 1}{2^{\sqrt{n}-\log_4(n)/2} - 1} \geq \left(\frac{2^{\sqrt{n}-1-\log_4(n)/2} - 1}{2^{\sqrt{n}-\log_4(n)/2} - 1}\right)^{\log_4(n)/2}$$
$$\geq \left(\frac{1}{4}\right)^{\log_4(n)/2} = \frac{1}{\sqrt{n}}.$$

For the first equality, we use the observation that conditioning on $f(x) = 0$ eliminates the possibility every variable is true so the probability of observing a true variable is slightly smaller. For the first inequality, notice that $(2^{i-1} - 1)/(2^i - 1)$ is monotone increasing in $i$. For the second, observe that $i \geq \sqrt{n} - \log_4(n)/2$ for our purposes and so $(2^{i-1} - 1)/(2^i - 1) \geq 1/4$ when $n \geq 16$.

*Proof (Proof of Theorem 4).* Suppose $f$ is a read-once DNF. For unit costs and arbitrary probabilities, we prove $\text{OPT}_\mathcal{N}(f) \geq \Omega(\sqrt{n}) \cdot \text{OPT}_\mathcal{A}(f)$.

Consider the read-once DNF with $m = 2\sqrt{n}$ identical terms where each term has $\ell = \sqrt{n}/2$ variables. In each term, let one variable have $1/\ell$ probability of being true and the remaining variables have a $(\ell/m)^{1/(\ell-1)}$ probability of being true. Within a term, the optimal adaptive strategy will test the variable with the lowest probability of being true first. Using this observation, we can write

$$\text{OPT}_\mathcal{A}(f) \leq [\Pr(x_1 = 0) \cdot 1 + \Pr(x_1 = 1) \cdot \ell] \cdot m$$
$$\leq [(1 - 1/\ell) \cdot 1 + (1/\ell) \cdot \ell] \cdot m \leq 4\sqrt{n}$$

where $x_1$ is the first variable tested in each term. The first inequality follows by charging the optimal adaptive strategy for all $\ell$ tests in the term if the first one is true. The second inequality follows since the variable with probability $1/\ell$ of being true is tested first for $n \geq 18$ (i.e., $1/\ell < (\ell/m)^{1/(\ell-1)}$ for such $n$).

In order to lower bound the cost of the optimal non-adaptive strategy, we will argue that there is a constant probability of an event where the non-adaptive strategy has to test $\Omega(n)$ variables. In particular,

$$\mathsf{OPT}_{\mathcal{N}}(f) \geq \min_{S \in \mathcal{N}} \Pr(\text{exactly one term is true})$$

$$\cdot \mathbb{E}[\text{cost}(f, x, S)| \text{ exactly one term is true}].$$

By the symmetry of the terms, observe that

$$\mathbb{E}[\text{cost}(f, x, S)| \text{ exactly one term is true}] \geq \sqrt{n}/2 \cdot \sqrt{n} = n/2.$$

That is, the optimal non-adaptive strategy has to search blindly for the single true term among all $2\sqrt{n}$ terms, making $\sqrt{n}/2$ tests each for half the terms in expectation.

All that remains is to show there is a constant probability exactly one term is true. The probability a particular term is true is $(1/\ell)((\ell/m)^{1/(\ell-1)})^{(\ell-1)} = 1/m$. Since all variables are independent, the probability that exactly one of the $m$ terms is true is

$$m \cdot \Pr(\text{a term is true}) \cdot \Pr(\text{a term is false})^{m-1}$$

$$= m \cdot \frac{1}{m} \cdot \left(1 - \frac{1}{m}\right)^{m-1} \geq \frac{1}{2e}^{(m-1)/m} \geq \frac{1}{2e}.$$

It follows that $\mathsf{OPT}_{\mathcal{N}}(f) \geq \frac{1}{2e} \cdot \frac{n}{2} = \Omega(n)$ so the adaptivity gap is $\Omega(\sqrt{n})$.

*Proof (Proof of Theorem 5).* Suppose $f$ is a read-once formula. For arbitrary costs and the uniform distribution, $\mathsf{OPT}_{\mathcal{N}}(f) \geq \Omega(n^{1-\epsilon}/\log n) \cdot \mathsf{OPT}_{\mathcal{A}}(f)$.

Define $W(w) := w^{1-\epsilon} \log_2(w^{1-\epsilon})$ for positive real numbers $w$.[3] We will choose $n_\epsilon$ in terms of the function $W$ so that $W(n) < n$ for $n \geq n_\epsilon$. First, consider the first and second derivatives of $W$:

$$W'(w) = \frac{1 - \epsilon}{w^\epsilon}\left(\log_2(w^{1-\epsilon}) + \frac{1}{\log 2}\right)$$

$$W''(w) = \frac{1 - \epsilon}{w^{1+\epsilon}}\left[-\epsilon\left(\log_2(w^{1-\epsilon}) + \frac{1}{\log 2}\right) + \frac{1 - \epsilon}{\log 2}\right].$$

For fixed $\epsilon > 0$, observe that as $w$ goes to infinity, $W(w) < w$, $W'(w) < 1$, and $W''(w) < 0$. Therefore there is some point $n_\epsilon$ so that for all $n \geq n_\epsilon$, the slope of $W$ is decreasing, the slope of $W$ is less than the slope of $n$, and $W(n)$ is less than $n$. Equivalently, $n \geq W(n) = n^{1-\epsilon} \log_2(n^{1-\epsilon})$. We will use this inequality when lower bounding the asymptotic behavior of the adaptivity gap.

---

[3] Notice that $W$ is similar to a Lambert W function $e^y y$, after changing the base of the logarithm and substituting $y = \log(w^{1-\epsilon})$ [6].

For $n \geq n_\epsilon$ we construct the $n$-variable read-once DNF formula $f$ as follows. First, let $r_n$ be a real number such that $n = n^{1-r_n} \log_2(n^{1-r_n})$. We know that $r_n$ exists for all $n \geq 4$ by continuity since $n^{1-0} \log_2(n^{1-0}) \geq n \geq n^{1-1} \log_2(n^{1-1})$. Let $f$ be the read-once DNF formula with $m$ terms of length $\ell$, where $\ell = \log_2(n^{1-r_n})$ and $m = 2^\ell$. Thus the total number of variables in $f$ is $m\ell = n^{1-r_n} \log_2(n^{1-r_n}) = n$ as desired. We assume for simplicity that $\ell$ is an integer. The bound holds by a similar proof without this assumption.

To obtain our lower bound on evaluating this formula, we consider expected evaluation cost with respect to the uniform distribution and the following cost assignment: in each term, choose an arbitrary ordering of the variables and set the cost of testing the $i$th variable in the term to be $2^{i-1}$.

Consider a particular term. Recall the optimal adaptive strategy for evaluating a read-once DNF formula presented at the start of Sect. 2. Within a term, this optimal strategy tests the variables in non-decreasing cost order, since each variable has the same probability of being true. Since it performs tests within a term until finding a false variable or certifying the term is true, we can upper bound the expected cost of this optimal adaptive strategy in evaluating $f$ as follows:

$$\mathsf{OPT}_\mathcal{A}(f) \leq m \cdot \left[ \frac{1}{2} \cdot (1) + \frac{1}{4} \cdot (1+2) + \ldots + \frac{1}{2^\ell} \cdot (1 + \ldots + 2^{\ell-1}) \right] \leq m \cdot \ell.$$

In contrast, the optimal non-adaptive strategy does not have the advantage of stopping tests in a term when it finds a false variable. We will lower bound the expected cost of the optimal non-adaptive strategy in the case that exactly one term is true. By symmetry, any non-adaptive strategy will have to randomly search for the term and so pay $2^\ell$ for half the terms in expectation.

All that remains is to show there is a constant probability exactly one term is true. The probability that a particular term is true is $1/2^\ell$ and so the probability that exactly one term is true is

$$m \cdot \frac{1}{2^\ell} \cdot \left( 1 - \frac{1}{2^\ell} \right)^{m-1} \geq \frac{m}{2^\ell} \cdot \left( \frac{1}{2e} \right)^{(m-1)/2^\ell} \geq \frac{1}{2e}$$

where the last inequality follows since $m = 2^\ell$. Then the expected cost $\mathsf{OPT}_\mathcal{N}(f)$ of the optimal non-adaptive strategy is at least

$$\Pr(\text{exactly one term is true}) \cdot 2^\ell \cdot \frac{m}{2} = \Omega(m \cdot 2^\ell) = \Omega(m \cdot n^{1-r_n}) \geq \Omega(m \cdot n^{1-\epsilon})$$

where we used that $2^\ell = n^{1-r_n}$ and $n^{1-r_n} \log_2(n^{1-r_n}) = n \geq n^{1-\epsilon} \log_2(n^{1-\epsilon})$ since $n \geq n_\epsilon$. It follows that the adaptivity gap is $\Omega(n^{1-\epsilon}/\log n)$.

## References

1. Agarwal, A., Assadi, S., Khanna, S.: Stochastic submodular cover with limited adaptivity. In: Chan, T.M. (ed.) Proceedings of the Thirtieth Annual ACM-SIAM Symposium on Discrete Algorithms, SODA, pp. 323–342. SIAM (2019)

2. Asadpour, A., Nazerzadeh, H., Saberi, A.: Stochastic submodular maximization. In: Papadimitriou, C., Zhang, S. (eds.) WINE 2008. LNCS, vol. 5385, pp. 477–489. Springer, Heidelberg (2008). https://doi.org/10.1007/978-3-540-92185-1_53
3. Boros, E., Unyulurt, T.: Sequential testing of series-parallel systems of small depth. In: Laguna, M., Velarde, J.L.G. (eds.) Computing Tools for Modeling, Optimization and Simulation: Interfaces in Computer Science and Operations Research, pp. 39–73. Springer, Boston (2000). https://doi.org/10.1007/978-1-4615-4567-5_3
4. Boucheron, S., Lugosi, G., Massart, P.: Concentration Inequalities: A Nonasymptotic Theory of Independence. Oxford University Press, Oxford (2013)
5. Bradac, D., Singla, S., Zuzic, G.: (Near) optimal adaptivity gaps for stochastic multi-value probing. In: Achlioptas, D., Végh, L.A. (eds.) Approximation, Randomization, and Combinatorial Optimization. Algorithms and Techniques, APPROX/RANDOM 2019, 20–22 September 2019, Massachusetts Institute of Technology, Cambridge, MA, USA. LIPIcs, vol. 145, pp. 49:1–49:21. Schloss Dagstuhl - Leibniz-Zentrum für Informatik (2019)
6. Bronstein, M., Corless, R.M., Davenport, J.H., Jeffrey, D.J.: Algebraic properties of the Lambert W function from a result of Rosenlicht and of Liouville. Integral Transform. Spec. Funct. **19**(10), 709–712 (2008)
7. Dean, B.C., Goemans, M.X., Vondrák, J.: Approximating the stochastic knapsack problem: the benefit of adaptivity. In: Proceedings of 45th Symposium on Foundations of Computer Science (FOCS 2004), 17–19 October 2004, Rome, Italy, pp. 208–217. IEEE Computer Society (2004)
8. Dean, B.C., Goemans, M.X., Vondrák, J.: Adaptivity and approximation for stochastic packing problems. In: Proceedings of the Sixteenth Annual ACM-SIAM Symposium on Discrete Algorithms, SODA 2005, Vancouver, British Columbia, Canada, 23–25 January 2005, pp. 395–404. SIAM (2005)
9. Dean, B.C., Goemans, M.X., Vondrák, J.: Approximating the stochastic knapsack problem: the benefit of adaptivity. Math. Oper. Res. **33**(4), 945–964 (2008)
10. Deshpande, A., Hellerstein, L., Kletenik, D.: Approximation algorithms for stochastic submodular set cover with applications to boolean function evaluation and min-knapsack. ACM Trans. Algorithms **12**(3), 42:1–42:28 (2016)
11. El-Neweihi, E., Proschan, F., Sethuraman, J.: Optimal allocation of components in parallel-series and series-parallel systems. J. Appl. Probab. **23**(3), 770–777 (1986)
12. Eppstein, D.: Parallel recognition of series-parallel graphs. Inf. Comput. **98**(1), 41–55 (1992)
13. Fu, L., Fu, X., Xu, Z., Peng, Q., Wang, X., Lu, S.: Determining source-destination connectivity in uncertain networks: modeling and solutions. IEEE/ACM Trans. Netw. **25**(6), 3237–3252 (2017)
14. Ghuge, R., Gupta, A., Nagarajan, V.: Non-adaptive stochastic score classification and explainable halfspace evaluation. CoRR abs/2111.05687 (2021)
15. Ghuge, R., Gupta, A., Nagarajan, V.: The power of adaptivity for stochastic submodular cover. In: Proceedings of the 38th International Conference on Machine Learning, ICML (2021)
16. Gkenosis, D., Grammel, N., Hellerstein, L., Kletenik, D.: The stochastic score classification problem. In: 26th Annual European Symposium on Algorithms, ESA 2018, 20–22 August 2018, Helsinki, Finland, pp. 36:1–36:14 (2018)
17. Goemans, M., Vondrák, J.: Stochastic covering and adaptivity. In: Correa, J.R., Hevia, A., Kiwi, M. (eds.) LATIN 2006. LNCS, vol. 3887, pp. 532–543. Springer, Heidelberg (2006). https://doi.org/10.1007/11682462_50
18. Greiner, R., Hayward, R., Jankowska, M., Molloy, M.: Finding optimal satisficing strategies for and-or trees. Artif. Intell. **170**(1), 19–58 (2006)

19. Gupta, A., Nagarajan, V., Singla, S.: Algorithms and adaptivity gaps for stochastic probing. In: Krauthgamer, R. (ed.) Proceedings of the Twenty-Seventh Annual ACM-SIAM Symposium on Discrete Algorithms, SODA 2016, Arlington, VA, USA, 10–12 January 2016, pp. 1731–1747. SIAM (2016)

20. Gupta, A., Nagarajan, V., Singla, S.: Adaptivity gaps for stochastic probing: submodular and XOS functions. In: Klein, P.N. (ed.) Proceedings of the Twenty-Eighth Annual ACM-SIAM Symposium on Discrete Algorithms, SODA 2017, Barcelona, Spain, Hotel Porta Fira, 16–19 January 2017, pp. 1688–1702. SIAM (2017)

21. Happach, F., Hellerstein, L., Lidbetter, T.: A general framework for approximating min sum ordering problems. INFORMS J. Comput. **34**(3), 1437–1452. https://doi.org/10.1287/ijoc.2021.1124

22. Harris, T.E., et al.: The Theory of Branching Processes, vol. 6. Springer, Berlin (1963)

23. Harvey, N.J., Patrascu, M., Wen, Y., Yekhanin, S., Chan, V.W.: Non-adaptive fault diagnosis for all-optical networks via combinatorial group testing on graphs. In: IEEE INFOCOM 2007–26th IEEE International Conference on Computer Communications, pp. 697–705. IEEE (2007)

24. Hellerstein, L., Kletenik, D., Lin, P.: Discrete stochastic submodular maximization: adaptive vs. non-adaptive vs. offline. In: Proceedings of the 9th International Conference on Algorithms and Complexity (CIAC) (2015)

25. Kaplan, H., Kushilevitz, E., Mansour, Y.: Learning with attribute costs. In: Gabow, H.N., Fagin, R. (eds.) Proceedings of the 37th Annual ACM Symposium on Theory of Computing, Baltimore, MD, USA, 22–24 May 2005, pp. 356–365. ACM (2005)

26. Kaplan, H., Kushilevitz, E., Mansour, Y.: Learning with attribute costs. In: Proceedings of the 37th Annual ACM Symposium on Theory of Computing, (STOC), pp. 356–365 (2005)

27. Kowshik, H.J.: Information aggregation in sensor networks. University of Illinois at Urbana-Champaign (2011)

28. Liva, G., Paolini, E., Chiani, M.: Optimum detection of defective elements in non-adaptive group testing. In: 2021 55th Annual Conference on Information Sciences and Systems (CISS), pp. 1–6. IEEE (2021)

29. O'Donnell, R.: Analysis of Boolean Functions. Cambridge University Press, Cambridge (2014)

30. Ünlüyurt, T.: Sequential testing of complex systems: a review. Discret. Appl. Math. **142**(1–3), 189–205 (2004)

31. Wikipedia: Series and parallel circuits – Wikipedia, the free encyclopedia (2022). Accessed 8 Feb 2022

# On Streaming Algorithms for Geometric Independent Set and Clique

Sujoy Bhore[1], Fabian Klute[2], and Jelle J. Oostveen[2]([✉])

[1] Indian Institute of Science Education and Research, Bhopal, India
[2] Utrecht University, Utrecht, The Netherlands
{f.m.klute,j.j.oostveen}@uu.nl

**Abstract.** We study the maximum geometric independent set and clique problems in the streaming model. Given a collection of geometric objects arriving in an insertion only stream, the aim is to find a subset such that all objects in the subset are pairwise disjoint or intersect respectively.

We show that no constant factor approximation algorithm exists to find a maximum set of independent segments or 2-intervals without using a linear number of bits. Interestingly, our proof only requires a set of segments whose intersection graph is also an interval graph. This reveals an interesting discrepancy between segments and intervals as there does exist a 2-approximation for finding an independent set of intervals that uses only $O(\alpha(\mathcal{I}) \log |\mathcal{I}|)$ bits of memory for a set of intervals $\mathcal{I}$ with $\alpha(\mathcal{I})$ being the size of the largest independent set of $\mathcal{I}$. On the flipside we show that for the geometric clique problem there is no constant-factor approximation algorithm using less than a linear number of bits even for unit intervals. On the positive side we show that the maximum geometric independent set in a set of axis-aligned unit-height rectangles can be 4-approximated using only $O(\alpha(\mathcal{R}) \log |\mathcal{R}|)$ bits.

**Keywords:** Geometric independent set · Streaming algorithms · Geometric intersection graphs · Communication lower bounds

## 1 Introduction

The independent set problem is one of the fundamental combinatorial optimization problems in theoretical computer science, with a wide range of applications. Given a graph $G = (V, E)$, a set of vertices $M \subset V$ is *independent* if no two vertices in $M$ are adjacent in $G$. A *maximum independent set* is a maximum cardinality independent set. Maximum independent set is one of the most well-studied algorithmic problems and is one of Karp's 21 classic NP-complete problems [33]. Moreover, it is well-known to be hard to approximate: no polynomial time algorithm can achieve an approximation factor $n^{1-\epsilon}$, for any constant $\epsilon > 0$,

Fabian Klute supported by the Austrian Science Foundation (FWF) grant J4510. Jelle Oostveen is partially supported by the NWO grant OCENW.KLEIN.114 (PACAN).

P. Chalermsook and B. Laekhanukit (Eds.): WAOA 2022, LNCS 13538, pp. 211–224, 2022.
https://doi.org/10.1007/978-3-031-18367-6_11

unless P = NP [28,39]. Maximum independent set serves as a natural model for many real-life optimization problems, including map labeling, computer vision, information retrieval, and scheduling; see [1,6,9].

*Geometric Independent Set.* In the geometric setting we are given a set of geometric objects $\mathcal{S}$, and we say a subset $\mathcal{S}' \subseteq \mathcal{S}$ is *independent* if no two objects in $\mathcal{S}'$ intersect and we say $\mathcal{S}'$ is a *clique* if every two objects pairwise intersect. Let $\alpha(\mathcal{S})$ be the cardinality of the largest subset $\mathcal{S}' \subseteq \mathcal{S}$ such that $\mathcal{S}'$ is an independent set and $\omega(\mathcal{S})$ the cardinality of the largest subset of $\mathcal{S}' \subseteq \mathcal{S}$ such that $\mathcal{S}'$ is a clique. The *geometric maximum independent set* and *geometric maximum clique* problem ask for a set $S \subseteq \mathcal{S}$ of independent objects (that induce a clique) such that $S = \alpha(\mathcal{S})$ ($S = \omega(\mathcal{S})$).

Given a set $\mathcal{S}$ of geometric objects, we define the *geometric intersection graph* $\mathcal{G}(S)$ as the simple graph in which each vertex corresponds to an object in $\mathcal{S}$ and two vertices are connected by an edge if their corresponding objects intersect.

Stronger results are known for the geometric maximum independent set problem is known in comparison to general graphs. A fundamental problem is the 1-dimensional case, where all objects are intervals. This problem is also known as *interval selection* problem which has applications to scheduling and resource allocation [7]. For intervals geometric maximum independent set can be solved in $O(n \log n)$ time, by a simple greedy algorithm that sweeps the line from left to right and at each step picks the interval with the leftmost right endpoint, see e.g. [34]. In contrast, the geometric maximum independent set problem is already NP-hard for sets of segments in the plane using only two directions [35], or 2-intervals [8]. For some restricted classes of segment intersection graphs, such as permutation [31] and circle graphs [25], the geometric maximum independent set problem can be solved in polynomial time. Efficient approximation algorithms exist for example for unit square intersection graphs [21] and more generally for pseudo disks [15], as well as segments [3,22]. In a recent breakthrough work [37], it was shown that there exists a constant-factor approximation scheme for maximum independent set for a set of axis-aligned rectangles. Very recently, this factor was improved to 3 [24]. Also, the geometric maximum independent set problem has been extensively studied for dynamic geometric objects, i.e., objects can be inserted and deleted [10,12,17,26,29].

*Streaming Algorithms.* In this paper, we study the geometric maximum independent set and geometric maximum clique problems for *insertion only streams* of geometric objects. In the streaming model we consider data that is too large to fit at once into the working memory. Instead the data is dealt with in a data stream in which we receive the elements of the input one after another in no specific order and have access to only a limited amount of memory. More specifically, in this model, we have bounds on the amount of available memory. As the data arrives sequentially, and we are not allowed to look at input data of the past, unless the data was stored in our limited memory. This is effectively equivalent to assuming that we can only make one or a few passes over the input data. We refer to [36, 38] and the lecture notes of Chakrabati [13] for an overview on the general topic of streaming algorithms.

For maximum independent set Halldórsson et al. [27] studied the problem for graphs and hypergraps in linear space in the semi-streaming model: Their model work in poly-logarithmic space, like in the case of the classical streaming model, but they can access and update the output buffer, treating it as an extra piece of memory. Kane et al. [32] gave the first optimal algorithm for estimating the number of distinct elements in a data stream.

Streaming algorithms for geometric data have seen a flurry of results in recent years; see [2, 16, 19, 23, 30]. Note that one can also view a stream of geometric objects $S$ as a vertex stream, also called *implicit vertex stream*, of its associated geometric intersection graph $\mathcal{G}(S)$ [18]. Finding an independent, i.e. disjoint, set of geometric objects in a data stream has been among the most studied problems in this geometric direction. Emek et al. [20] studied the interval selection problem, where the input is a set of intervals $\mathcal{I}$ with real endpoints, and the objective is to find an independent subset of largest cardinality. They studied the interval selection problem using $O(\alpha(\mathcal{I}))$ space. They presented a 2-approximation algorithm for the case of arbitrary intervals and a $(3/2)$-approximation for the case of unit intervals, i.e., when all intervals have the same length. These bounds are also known to be the best possible [20]. Cabello et al. [11] studied the question of estimating $\alpha(\mathcal{I})$ for a set $\mathcal{I}$ of intervals and gave simpler proofs of the algorithms presented by Emek et al. [20].

Cormode et al. [18] considered unit balls in the $L^1$ and $L^\infty$ norms, i.e. squares in $\mathbb{R}^2$. For a set of such unit balls $\mathcal{B}$ they obtained a 3-approximation using $O(\alpha(\mathcal{B}))$ space and show that there is no $\frac{5}{2} - \varepsilon$ approximation using $o(|\mathcal{B}|)$ space. Finally, Bakshi et al. [5] considered Turnstile streams, i.e., deletion is also allowed, of (weighted) unit intervals and disks.

## 1.1   Our Results

In this paper we investigate several geometric objects that have not been studied in the context of streaming algorithms. We show in Sect. 2 that there is no constant-factor approximation in the streaming model for finding an independent set of $n$ segments using $o\left(\frac{n}{p}\right)$ bits of memory for any constant number $p$ of passes and this bound holds even if the endpoints of the segments are on two parallel lines. In other words, our bound holds even when the geometric intersection graph of the segments is a permutation graph.

Our construction leads to an interesting consequence. Namely, the intersection graph created in our reduction is not only a permutationa graph, but also an interval graph and the cardinality of its maximum independent set is not dependent on the input size. Since there exists a 2-approximation algorithm for geometric independent set of a set of intervals $\mathcal{I}$ in the streaming model that uses only $O(\alpha(\mathcal{I}))$ space this implies that there is a difference between an interval graph being streamed as a set of intervals or as a set of segments. We discuss this implication in Sect. 3. In Sect. 4 we show that for streams of 2-intervals there is no one-pass algorithm that achieves a constant-factor approximation using less than $o(n)$ bits. On the positive side we show in Sect. 5 that for $n$ axis-aligned unit height rectangles there exists a one pass streaming algorithm achieving a 4-approximation of the largest set of disjoint rectangles using $O(\alpha(\mathcal{R}) \log n)$ bits.

Finally, we show in Sect. 6 that the distinction between segments and intervals observed for the geometric independent set problem does not occur for the same objects in the geometric clique problem by showing that there does not exist a $p$-pass algorithm using less than $o\left(\frac{n}{p}\right)$ bits of memory and achieving a constant-factor approximation of the geometric clique problem in streams of $n$ unit intervals. We complement this hardness result by showing how to obtain an exact solution for the geometric clique problem in streams of $n$ intervals using only $O(n \log \omega(\mathcal{I}))$ bits of memory.

## 2    Independent Sets in Streams of Segments

In this section we establish our lower bound for the memory necessary to approximate the maximum independent set problem to any constant factor on streams of segments. We employ a lower bound reduction technique that uses multi-party set disjointness, which gives us space bounds not only for single-pass algorithms, but also for multi-pass algorithms. The following problem was first studied by Alon, Matias, and Szegedy [4].

**Definition 1 (Multi-Party Set Disjointness).** *There are $t$ players $P_1, \ldots, P_t$. Each player $P_i$ has a size $n$ bit string $x^i$. The players want to find out if there is an index $j \in [n]$ where $x_j^i = 1$ for all $i$. So, $\text{DISJ}_{n,t}(x^1, \ldots, x^t) = \bigvee_{j=1}^{n} \bigwedge_{i=1}^{t} x_j^i$.*

In our proof we are going to make use of the following result on the communication complexity of MULTI-PARTY SET DISJOINTNESS.

**Theorem 1 (Chakrabarti et al. [14]).** *For an error probability $0 < \delta < 1/4$, to decide $\text{DISJ}_{n,t}$ the players need $\Omega(\frac{n}{t \log t})$ bits of communication, even for a family of instances $(x^1, \ldots, x^t)$ satisfying the following properties*

$$|\{j : x_j^i = 1\}| = n/2t \qquad \forall i \in [t] \qquad (1)$$

$$|\{i : x_j^i = 1\}| \in \{0, 1, t\} \qquad \forall j \in [n] \qquad (2)$$

$$|\{j : |\{i : x_j^i = 1\}| = t\}| \leq 1 \qquad (3)$$

We can use Theorem 1 by having $t$ players use a streaming algorithm to answer $\text{DISJ}_{n,t}$. The players construct the stream by creating some part of the stream and giving it to the algorithm, and then passing the memory state of the algorithm to the next player who does the same. This way, the space used by the algorithm must abide to the lower bound on the communication between the players. We can use the $t$ players (rather than just 2) to create a bigger gap between the yes and no answer, excluding the possibility for any constant factor approximation algorithms.

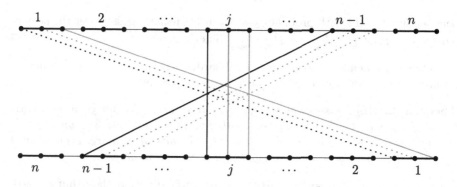

**Fig. 1.** Lower bound for Independent Set in permutation graphs, with $t = 3$ players. Here the independent set has size $t$, by the $j$-th group, where $x_j^i = 1$ for all $i \in [t]$.

**Lemma 1.** *For any $p \geq 1, t \geq 2$, any algorithm for geometric maximum independent set that can distinguish between instances with an independent set of size 1 and $t$ and succeeds with probability at least 3/4 on segment streams using $p$ passes must use at least $\Omega(\frac{n}{p \cdot t \log t})$ bits of memory, even when the segment endpoints lie on two parallel lines.*

*Proof.* Let $(x^1, \ldots, x^t)$ be an instance of DISJOINTNESS with $t$ players, each with $n$ bits. We construct a permutation graph depending on the input to DISJOINT-NESS, as illustrated in Fig. 1 for $t = 3$. Let the permutation graph have $n$ groups of $t$ points both on the top, labelled $1, \ldots, n$ from left to right. On the bottom do the same, but we label from $n$ to 1 from left to right. For $i \in [t], j \in [n]$, player $i$ creates a segment from the $i$-th point in group $j$ at the top to the $i$-th point in group $j$ at the bottom when $x_j^i = 1$. This creates a permutation graph with $n' = n/2$ vertices by Property 1 of Theorem 1.

The players construct the segment stream for some algorithm for MAX-CLIQUE as follows, starting from player 1, player $i$ inputs all their $n/2t$ segments, then passes the memory state of the algorithm to player $i + 1$, and this continues until all players have input their segments.

We claim that the graph will contain an independent set of size $t$ if exactly $\text{DISJ}_{n,t}(x^1, \ldots, x^t) = 1$, and otherwise the maximum independent set size is 1.

First notice that any segment inserted to a group $j \in [n]$ intersects all other segments in the graph, except for segments inserted to group $j$, as these segments are parallel to it. Hence, any independent set can only contain vertices that correspond to segments inserted to the same group $j$, for some $j \in [n]$, and the size of the independent set is the number of 1's present over all players at index $j$. Now by Property 2 of Theorem 1, any independent set can have only size 1 or $t$ in the graph. And indeed, an independent set of size $t$ implies that all $t$ players inserted a segment for some group $j \in [n]$, and hence all have a 1 for index $j$.

Now it follows from Theorem 1 that any algorithm for maximum independent set on a permutation graph with $n'$ vertices that can discern between independent set size 1 and $t$ with probability at least $3/4$ using $p$ passes over the stream must use at least $\Omega(\frac{n}{p \cdot t \log t}) = \Omega(\frac{n'}{p \cdot t \log t})$ bits of memory.                    □

We now use Lemma 1 to give a general hardness statement for approximation geometric maximum independent set in segment streams.

**Theorem 2.** *Any constant-factor approximation algorithm for geometric maximum independent set that succeeds with probability at least $3/4$ on segment streams using $p$ passes must use at least $\Omega(n/p)$ bits of memory, even when it is known that the segments correspond to a permutation graph.*

*Proof.* Let us be given some constant-factor approximation algorithm for geometric maximum independent set that succeeds with probability at least $3/4$ on segment streams of permutation graphs in $p$ passes. Then there exists some $c \geq 2$ such that the algorithm can distinguish between an independent set of size 1 or $c$ in a given graph. But then we can apply Lemma 1 to get that this algorithm must use at least $\Omega(\frac{n}{p \cdot c \log c}) = \Omega(n/p)$ bits of memory.                    □

## 3    Intervals and Segments are Different

When we consider the intersection graph $G$ of the set of segments constructed in the proof of Theorem 2 one can see that it has a straight-forward representation as a set of intervals whose intersection graph is isomorphic to $G$. Also, notice that the size of the independent set of the construction is not dependent on the length of the bit strings, but only on the number of players. For example, we can rule out the existence of a 2-approximation streaming algorithm using any constant number $p$ of passes and $o(n/p)$ bits of memory, already when $t/2 \geq 2 \iff t \geq 4$ players are used in the construction presented in Lemma 1.

At the same time there exists a 2-approximation one pass streaming algorithm for independent sets of intervals that uses only $O(\alpha(\mathcal{I}) \log |\mathcal{I}|)$ bits of memory where $\mathcal{I}$ is the set of input intervals [11,20]. Hence, these algorithms find a 2-approximation of the independent set of the intersection graph constructed in the proof of Lemma 1 using only $O(\log n)$ bits of memory if the graph was given as a set of intervals. This leads to the following corollary.

**Corollary 1.** *Given a stream of segments $\mathcal{S}$ whose intersection graph $\mathcal{G}(\mathcal{S})$ is in the intersection of permutation and interval graphs, there is no algorithm that uses $o(n/p)$ bits of memory and $p \geq 1$ passes and computes a stream of intervals $\mathcal{I}$ such that $\mathcal{G}(\mathcal{I})$ is isomorphic to $\mathcal{G}(\mathcal{S})$.*

## 4    Independent Sets in Streams of $c$-Intervals

A *$c$-interval* is a set of non-overlapping intervals $\{I_1, \ldots, I_c\}$ on the real line. We say two $c$-intervals intersect if at least two of their intervals have one point in common. We call a family of $c$-intervals *separated* if the intervals can be split into independent groups of intervals, each containing at most one interval from each

$c$-interval, without changing which $c$-intervals intersect. Since we only consider 2-intervals we can talk of left and right intervals. Formally, let $T = \{L, R\}$ be a 2-interval such that the startpoint of $L$ is left of the startpoint of $R$, then we denote $L$ as the *left* interval of $T$ and $R$ as the *right* interval of $T$.

For the reduction we use the $\text{CHAIN}_t$ communication problem which was introduced by Cormode et al. [18].

**Definition 2 (Cormode et al. [18]).** *The $t$-party chained index problem* $\text{CHAIN}_t$ *consists of $t - 1$ $n$-bit binary vectors $\{x^i\}_{i=1}^{t-1}$, along with corresponding indices $\{\sigma_i\}_{i=1}^{t-1}$ from the range $[n]$. We have the promise that the entries $\{x^i_{\sigma_i}\}_{i=1}^{t-1}$ are all equal to the desired bit $z \in \{0, 1\}$. The input is initially allocated as follows:*

- *The first party $P_1$ knows $x^1$*
- *Each intermediate party $P_p$ for $1 < p < k$ knows $x^p$ and $\sigma_{p-1}$*
- *The final party $P_t$ knows just $\sigma_{t-1}$*

*Communication proceeds as follows: $P_1$ sends a single message to $P_2$, then $P_2$ communicates to $P_3$, and so on, with each party sending exactly one message to its immediate successor. After all messages are sent, $P_k$ must correctly output $z$, succeeding with probability at least $2/3$. If the promise condition is violated, any output is considered correct.*

Cormode et al. [18] showed the following result in the same paper.

**Theorem 3 (Cormode et al. [18]).** *Any communication scheme $\mathcal{B}$ which solves $\text{CHAIN}_t$ must communicate at least $\Omega(\frac{n}{t^2})$ bits.*

In the following we use the $\text{CHAIN}_t$ to show that there is no one-pass streaming algorithm that computes a constant-factor approximation of the maximum independent set for a family of separated 2-intervals using $o(n)$ bits of memory.

*Remark 1.* For $\text{CHAIN}_t$ we may assume that party $i > 1$ knows all indices before $\sigma_{i-1}$. To realize this, just assume that every party $i$ appends all $i - 1$ previous indices to its message. This uses only $O(t \log n)$ bits in each such message and hence $O(t^2 \log n)$ bits in total. As this is a lower order term with respect to the bound in Theorem 3 we retain the linear communication bound of $\Omega(\frac{n}{t^2})$.

We define an *interval stack* as a set of intervals $I_1, \ldots, I_n$ on the real line where first all startpoints appear in order of the indices and then all endpoints, again in order of the interval indices. We denote as *left-gap* the space between the startpoint of $I_i$ and $I_{i+1}$ for $i = 1, \ldots, n-1$ and the startpoint of $I_n$ and the endpoint of $I_1$. Observe that any interval containing a point of the left gap of $I_i$ intersects all intervals $I_j$ with $1 \leq j \leq i$. As for independent sets of segments we first show a technical lemma.

**Lemma 2.** *For any $t \geq 2$, any algorithm for geometric maximum independent set that can distinguish between instances admitting an independent set of size 1 and $t$ and succeeds with probability at least $2/3$ on streams of 2-intervals must use at least $\Omega(\frac{n}{t^3})$ bits of memory, even if the union of the 2-intervals is separated.*

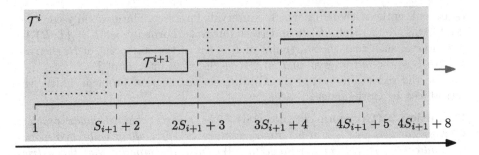

**Fig. 2.** Left intervals created for party $i$ in the construction of Lemma 2. The intervals in the grey box are for party $i$, the intervals outside are of party $i-1$, and the boxes mark the left gap spaces for party $i+1$. The red interval is the interval corresponding to a 1 bit at $\sigma_i$. Dotted intervals and boxes are not actually inserted by the parties. The shown coordinates are the local coordinates for the interval stack inserted by party $i$. (Color figure online)

*Proof.* Given an instance of CHAIN$_t$ with $\frac{n}{t}$-length bit strings $x^i$ and indices $\sigma_j$. Let $N = \frac{n}{t}$ and will assume for simplicity that $\frac{n}{t}$ is a whole number. We create one 2-interval $T_j^i = (L_j^i, R_j^i)$ for each 1 bit at index $j$ of bit string $i$ and one additional 2-interval for party $t$. In the following we first describe the construction of the left intervals. See Fig. 2 for an illustration.

Create an interval stack $\mathcal{L}^i = \{L_j^i \mid \forall j \in [N] : x^i \text{ has a 1 bit at index } j\}$. To simplify the presentation we assume that all $\frac{n}{t}$ intervals are present in $T^i$. When actually constructing the intervals in a stream party $i$ simply does not add an interval when the $j$th bit is set to 0, but still shifts the coordinates accordingly. We initially place the intervals of $\mathcal{L}^1$ and then place $\mathcal{L}^i$ for $i > 1$ in the left gap of $L_{\sigma_{i-1}}^{i-1}$. For player $t$ we add one interval $L_1^t$ in the left gap of $L_{\sigma_{t-1}}^{t-1}$. Let $\mathcal{L}$ be the union of all $\mathcal{L}^i$.

To complete the construction we create the same construction using the reversed bit strings for each party. This creates the interval stacks $\mathcal{R}^i$, $i = 1, \ldots, n$ and with $R_j^i$ we denote the right interval inserted by the $i$th party for the 1 bit at index $j$ in the non-reversed bit string $x^i$. Let $\mathcal{R}$ be the union of all $\mathcal{R}^i$. Finally, we create a set of 2-intervals as $\mathcal{T} = \{(L_j^i, R_j^i) \mid L_j^i \in \mathcal{L}^i \text{ and } R_j^i \in \mathcal{R}^i\}$.

Consider a 2-interval $T_j^i = (L, R)$ inserted for bit string $x^i$ such that $j \neq \sigma_i$ for any $i \in \{1, \ldots, t-1\}$. Then, $L$ is contained in all intervals $L_b^a$ with $a < i$ and $b < \sigma_a$. Moreover, $L$ contains every $L_b^a$ with $a > i$ and $b < \sigma_i$. Similarly, $R$ is contained in all intervals $R_b^a$ with $a > i$ and $b > \sigma_j$ and contains every $R_b^a$ with $a > i$ and $b > \sigma_i$. Consequently if $T_j^i$ is part of an independent set $\mathcal{I} \subseteq \mathcal{T}$ we can only add 2-intervals $T_j^a$ to $\mathcal{I}$ with $a < i$ and $j = \sigma_a$.

If the answer bit is 0 then no 2-interval corresponding to some $\sigma_i$ index exists and hence by the above argumentation the largest independent set has size one. If the answer bit is 1 then there is a 2-interval $T_{\sigma_i}^i$ for every $\sigma_i$ with $i = 1, \ldots, t$ and all $t$ of them are independent. Hence, the largest independent set in this case has size $t$.

It remains to describe the precise coordinates of the intervals and argue that we only need $O(\log n)$ bits to represent the construction for every fixed $t$. For each party $i = 1, \ldots, t$ let $S_i$ be the length of the left interval stack. Since for party $t$ we insert only one 2-interval we set $S_t = 2$. Let $\mathcal{T}$ be the set of 2-intervals created as above. In the following we consider only the left intervals of every $T_j^i \in \mathcal{T}$. The calculation and placement works analogously for the right intervals after reversing every $x^i$. Let $L_j^i$ be a left interval for party $i$ and index $j$. We put the startpoint of $L_j^i$ at position $1 + (j-1) \cdot (S_{i+1} + 1)$ and its endpoint at $j + N \cdot (S_{i+1} + 1)$. Hence, for party $i$ its left interval stack has length at most

$$S_i = N + N \cdot (S_{i+1} + 1) = N \cdot (S_{i+1} + 2).$$

This can be written as a closed formula

$$S_i = 4N^{t-i} \cdot 2 \sum_{j=1}^{t-(i+1)} N^j = 4N^{t-i} \cdot 2 \left( \frac{N^{t-i} - N}{N - 1} \right).$$

Now, party $i$ places its left stack at

$$P_i = 1 + \sum_{j=1}^{i-1} \left( (\sigma_j - 1) \cdot (S_{j+1} + 1) + 1 \right).$$

The last party places an interval of length two at position $P_t$. Since every left interval placed by a party $i > 1$ is nested by the intervals inserted into the stream by the first party we can conclude that $S_1 \in O\left(N^{t-1}\right)$. This number can be represented using $O(t \log N) = O(t \log \frac{n}{t})$ bits. Since $t$ can be treated as a constant we get that we only require $O(t \log \frac{n}{t}) = O(\log n)$ bits.     □

We conclude Theorem 4 from Lemma 2 in the same way as for Theorem 2.

**Theorem 4.** *Any constant-factor approximation algorithm for geometric maximum independent set that succeeds with probability at least 2/3 on streams of 2-intervals requires at least $\Omega(n)$ bits of memory, even if the 2-intervals are separated.*

## 5   Independent Sets in Streams of Unit-Height Rectangles

In this section, we study the independent set problem for a stream of unit height arbitrary width rectangles. To conform with previous work we assume in this section that one cell of memory can store one rectangle, i.e., one cell of memory has $\Theta(\log n)$ bits where all coordinates of the rectangles are assumed to be in $O(n)$. Cabello and Pérez-Lantero [11] studied the independent set problem for streams of intervals on the real line and achieved the following result.

**Theorem 5 (Theorem 5 [11]).** *Let $\mathcal{I}$ be a set of intervals in the real line that arrive in a data stream. There is a data stream algorithm to compute a 2-approximation to the largest independent subset of $\mathcal{I}$ that uses $O(\alpha(\mathcal{I}))$ space and handles each interval of the stream in $O(\log \alpha(\mathcal{I}))$ time.*

Using Theorem 5 we obtain a constant-factor approximation for finding the largest independent set of rectangles in a stream of axis-aligned unit height rectangles in one pass using $O(\alpha(\mathcal{R}))$ space. The below notation is similar to the one used by Cabello and Pérez-Lantero [11].

We divide the $y$-axis into size two intervals. Similar to [11] we define windows $W_\ell = [\ell, \ell + 2)$. Then, we form two partitions $\mathcal{W}_0$ and $\mathcal{W}_1$ of the $y$-axis as $\mathcal{W}_z = \{W_{z+2i} \mid i \in \mathbb{Z}\}$ for $z \in \{0, 1\}$. We denote with $\mathcal{R}_z \subseteq \mathcal{R}$ and $z \in \{0, 1\}$ the set of rectangles that is contained in any window of $\mathcal{W}_z$. Observe, that every rectangle is fully contained in only one of the two partitions.

Computing an independent set for the rectangles $\mathcal{R}_z$ now amasses to computing independent sets for each set of rectangles lying in one window $w_\ell$ of $\mathcal{W}_z$. By only considering windows that contain at least one interval and using Theorem 5 we can compute for every $\mathcal{W}_z$ and $z \in \{0, 1\}$ a 2-approximation of its largest independent set using $\alpha(\mathcal{R}_z)$ space in one pass. Let $\alpha'(\mathcal{R}_z)$ be such a 2-approximation, $\alpha(\mathcal{R}_z)$ the size of an optimal independent set of $\mathcal{R}_z$, and $\mathcal{R}_I \subseteq \mathcal{R}$ an optimal independent set of $\mathcal{R}$, then it holds that

$$2\max\{\alpha'(\mathcal{R}_0), \alpha'(\mathcal{R}_1)\} \geq \alpha'(\mathcal{R}_0) + \alpha'(\mathcal{R}_1) \geq \frac{1}{2}(\alpha(\mathcal{R}_0) + \alpha(\mathcal{R}_1))$$

$$\geq \frac{1}{2}(|\mathcal{R}_I \cap \mathcal{R}_0| + |\mathcal{R}_I \cap \mathcal{R}_1|) \geq \frac{1}{2}|\mathcal{R}_I| \geq \frac{1}{2}\alpha(\mathcal{R}).$$

From this it follows that $\max\{\alpha'(\mathcal{R}_0), \alpha'(\mathcal{R}_1)\} \geq \frac{1}{4}\alpha(\mathcal{R})$.

**Theorem 6.** *Let $\mathcal{R}$ be a set of axis-aligned unit height rectangles that arrive in a data stream, there is an algorithm that compute a 4-approximation to the maximum independent set of $\mathcal{R}$, uses $O(\alpha(\mathcal{R}))$ space, and handles each rectangle in polylog time.*

Note, that this algorithm restricted to axis-aligned squares matches the approximation factor of three due to Cormode et al. [18] since for unit intervals we can use the $\frac{3}{2}$-approximation algorithm from Cabello et al. [11].

## 6    Clique in Streams of Intervals and Segments

We can make an identical statement as Theorem 2 for maximum clique instead of maximum independent set by observing the complement graph of the construction in Lemma 1.

**Theorem 7.** *Any constant-factor approximation algorithm for geometric maximum clique that succeeds with probability at least 3/4 on segment streams using $p$ passes must use at least $\Omega(n/p)$ bits of memory, even when the endpoints of the segments lie on two lines.*

*Proof.* The complement graph of the construction of Lemma 1 is also a permutation graph and admits the property that it contains either a clique of size 1 or $t$. It is given by reversing the permutation (exactly mirroring the bottom) of the

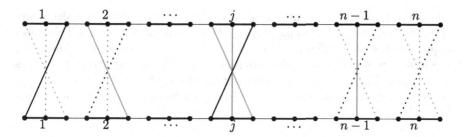

**Fig. 3.** Lower bound for Clique in permutation graphs, with $t = 3$ players. In this example, $x_j^i = 1$ for all players $i \in [t]$.

construction in Lemma 1. This construction is illustrated in Fig. 3. It follows that Lemma 1 also holds for geometric maximum clique instead of geometric maximum independent set. The theorem now follows from the proof of Theorem 2, using maximum clique instead of maximum independent set.     □

For streams of intervals, we show a simple upper bound, using that there are at most $2n$ different endpoints of intervals.

**Theorem 8.** *Let $\mathcal{I}$ be a set of intervals in the real line that arrive in a data stream. There is an algorithm to compute the largest clique size, $\omega(\mathcal{I})$, in 1 pass using $O(n\log(\omega(\mathcal{I}))$ bits of memory, using time $O(n^2)$ total. In a second pass, the intervals that make up the clique can be recovered, which can be streamed without extra memory use, or stored using $O(\omega(\mathcal{I})\log n)$ bits of memory.*

*Proof.* We keep a counter for every possible endpoint of an interval, which are $2n$ counters total. We keep the order of counters fixed, but need no labels for a counter, because of the assumption that the range of endpoints is $1, \ldots, 2n$. When an interval appears in the stream, we increment all counters that are contained in the interval, including its endpoints. At the end of the stream, $\omega(\mathcal{I})$ is given by the largest counter, as this coordinate is a witness to $\omega(\mathcal{I})$ intervals co-intersecting. This is the correct maximum, as the number of intersecting intervals can only change at an endpoint of an interval.

In the second pass, we can recover the intervals that make up the clique can be recovered by pushing every interval that overlaps the coordinate of the maximum counter found in the first pass to the output.     □

The result of Theorem 8 is nearly tight, as the construction of Theorem 7 can also be constructed as a stream of unit intervals.

## 7 Conclusion

We studied the geometric independent set and clique problems for a variety of geometric objects. Interestingly, we showed that the type of geometric object used for the implicit stream of a geometric intersection graph can make a substantial difference even for simple objects like segments and intervals. This raises the question if such a difference also exists for other types of objects. Moreover, the complexity of finding an independent set in a stream of arbitrary rectangles remains open. Finally, studying streams of geometric objects in other streaming models, such as turnstile streams, provides an interesting direction for future research.

# References

1. Agarwal, P.K., van Kreveld, M.J., Suri, S.: Label placement by maximum independent set in rectangles. Comput. Geom.: Theory Appl. **11**(3–4), 209–218 (1998). https://doi.org/10.1016/S0925-7721(98)00028-5
2. Agarwal, P.K., Krishnan, S., Mustafa, N.H., Venkatasubramanian, S.: Streaming geometric optimization using graphics hardware. In: Di Battista, G., Zwick, U. (eds.) ESA 2003. LNCS, vol. 2832, pp. 544–555. Springer, Heidelberg (2003). https://doi.org/10.1007/978-3-540-39658-1_50
3. Agarwal, P.K., Mustafa, N.H.: Independent set of intersection graphs of convex objects in 2D. Comput. Geom.: Theory Appl. **34**(2), 83–95 (2006). https://doi.org/10.1016/j.comgeo.2005.12.001
4. Alon, N., Matias, Y., Szegedy, M.: The space complexity of approximating the frequency moments. J. Comput. Syst. Sci. **58**(1), 137–147 (1999). https://doi.org/10.1006/jcss.1997.1545
5. Bakshi, A., Chepurko, N., Woodruff, D.P.: Weighted maximum independent set of geometric objects in turnstile streams. In: Proceedings of the Annual International Conference on Approximation, Randomization, and Combinatorial Optimization. Algorithms and Techniques (APPROX/RANDOM 2020). LIPIcs, vol. 176, pp. 64:1–64:22. Schloss Dagstuhl - Leibniz-Zentrum für Informatik (2020). https://doi.org/10.4230/LIPIcs.APPROX/RANDOM.2020.64
6. Balas, E., Yu, C.S.: Finding a maximum clique in an arbitrary graph. SIAM J. Comput. **15**(4), 1054–1068 (1986). https://doi.org/10.1137/0215075
7. Bar-Noy, A., Bar-Yehuda, R., Freund, A., Naor, J., Schieber, B.: A unified approach to approximating resource allocation and scheduling. J. ACM **48**(5), 1069–1090 (2001). https://doi.org/10.1145/502102.502107
8. Bar-Yehuda, R., Halldórsson, M.M., Naor, J., Shachnai, H., Shapira, I.: Scheduling split intervals. SIAM J. Comput. **36**(1), 1–15 (2006). https://doi.org/10.1137/S0097539703437843
9. van Bevern, R., Mnich, M., Niedermeier, R., Weller, M.: Interval scheduling and colorful independent sets. J. Sched. **18**(5), 449–469 (2014). https://doi.org/10.1007/s10951-014-0398-5
10. Bhore, S., Cardinal, J., Iacono, J., Koumoutsos, G.: Dynamic geometric independent set. CoRR abs/2007.08643 (2020). https://arxiv.org/abs/2007.08643
11. Cabello, S., Pérez-Lantero, P.: Interval selection in the streaming model. Theor. Comput. Sci. **702**, 77–96 (2017). https://doi.org/10.1016/j.tcs.2017.08.015
12. Cardinal, J., Iacono, J., Koumoutsos, G.: Worst-case efficient dynamic geometric independent set. In: Proceedings of the 29th Annual European Symposium on Algorithms (ESA 2021). LIPIcs, vol. 204, pp. 25:1–25:15. Schloss Dagstuhl - Leibniz-Zentrum für Informatik (2021). https://doi.org/10.4230/LIPIcs.ESA.2021. 25

13. Chakrabarti, A.: CS49: data stream algorithms lecture notes (2020). http://www.cs.dartmouth.edu/ac/Teach/data-streams-lecnotes.pdf
14. Chakrabarti, A., Khot, S., Sun, X.: Near-optimal lower bounds on the multi-party communication complexity of set disjointness. In: 18th Annual IEEE Conference on Computational Complexity (Complexity 2003), Aarhus, Denmark, 7–10 July 2003, pp. 107–117. IEEE Computer Society (2003). https://doi.org/10.1109/CCC.2003.1214414
15. Chan, T.M., Har-Peled, S.: Approximation algorithms for maximum independent set of pseudo-disks. Discret. Comput. Geom. **48**(2), 373–392 (2012). https://doi.org/10.1007/s00454-012-9417-5
16. Chen, X., Jayaram, R., Levi, A., Waingarten, E.: New streaming algorithms for high dimensional EMD and MST. In: Proceedings of the 54th Annual ACM SIGACT Symposium on Theory of Computing (STOC 2022), pp. 222–233. ACM (2022). https://doi.org/10.1145/3519935.3519979
17. Compton, S., Mitrovic, S., Rubinfeld, R.: New partitioning techniques and faster algorithms for approximate interval scheduling. CoRR abs/2012.15002 (2020). https://arxiv.org/abs/2012.15002
18. Cormode, G., Dark, J., Konrad, C.: Independent sets in vertex-arrival streams. In: Proceedings of the 46th International Colloquium on Automata, Languages, and Programming, (ICALP 2019). LIPIcs, vol. 132, pp. 45:1–45:14. Schloss Dagstuhl - Leibniz-Zentrum für Informatik (2019). https://doi.org/10.4230/LIPIcs.ICALP.2019.45
19. Czumaj, A., Jiang, S.H., Krauthgamer, R., Veselý, P.: Streaming algorithms for geometric steiner forest. CoRR abs/2011.04324 (2020). https://arxiv.org/abs/2011.04324
20. Emek, Y., Halldórsson, M.M., Rosén, A.: Space-constrained interval selection. ACM Trans. Algorithms **12**(4), 51:1–51:32 (2016). https://doi.org/10.1145/2886102
21. Erlebach, T., Jansen, K., Seidel, E.: Polynomial-time approximation schemes for geometric intersection graphs. SIAM J. Comput. **34**(6), 1302–1323 (2005). https://doi.org/10.1137/S0097539702402676
22. Fox, J., Pach, J.: Computing the independence number of intersection graphs. In: Proceedings of the 22nd Annual ACM-SIAM Symposium on Discrete Algorithms (SODA 2011), pp. 1161–1165. SIAM (2011)
23. Frahling, G., Sohler, C.: Coresets in dynamic geometric data streams. In: Proceedings of the 37th Annual ACM Symposium on Theory of Computing (STOC 2005), pp. 209–217. ACM (2005). https://doi.org/10.1145/1060590.1060622
24. Gálvez, W., Khan, A., Mari, M., Mömke, T., Pittu, M.R., Wiese, A.: A 3-approximation algorithm for maximum independent set of rectangles. In: Proceedings of the 2022 ACM-SIAM Symposium on Discrete Algorithms, (SODA 2022), pp. 894–905. SIAM (2022). https://doi.org/10.1137/1.9781611977073.38
25. Gavril, F.: Algorithms for a maximum clique and a maximum independent set of a circle graph. Networks **3**(3), 261–273 (1973). https://doi.org/10.1002/net.3230030305
26. Gavruskin, A., Khoussainov, B., Kokho, M., Liu, J.: Dynamic algorithms for monotonic interval scheduling problem. Theor. Comput. Sci. **562**, 227–242 (2015). https://doi.org/10.1016/j.tcs.2014.09.046
27. Halldórsson, B.V., Halldórsson, M.M., Losievskaja, E., Szegedy, M.: Streaming algorithms for independent sets. In: Abramsky, S., Gavoille, C., Kirchner, C., Meyer auf der Heide, F., Spirakis, P.G. (eds.) ICALP 2010. LNCS, vol. 6198, pp. 641–652. Springer, Heidelberg (2010). https://doi.org/10.1007/978-3-642-14165-2_54

28. Håstad, J.: Clique is hard to approximate within $n^{1-\epsilon}$. In: Proceedings of the 37th Annual Symposium on Foundations of Computer Science (FOCS 1996), pp. 627–636. IEEE Computer Society (1996). https://doi.org/10.1109/SFCS.1996.548522

29. Henzinger, M., Neumann, S., Wiese, A.: Dynamic approximate maximum independent set of intervals, hypercubes and hyperrectangles. In: Proceedings of the 36th International Symposium on Computational Geometry (SoCG 2020). LIPIcs, vol. 164, pp. 51:1–51:14. Schloss Dagstuhl - Leibniz-Zentrum für Informatik (2020). https://doi.org/10.4230/LIPIcs.SoCG.2020.51

30. Indyk, P.: Streaming algorithms for geometric problems. In: Lodaya, K., Mahajan, M. (eds.) FSTTCS 2004. LNCS, vol. 3328, pp. 32–34. Springer, Heidelberg (2004). https://doi.org/10.1007/978-3-540-30538-5_3

31. Trotter, L.E., Jr.: Algorithmic graph theory and perfect graphs, by Martin C. Golumbic, academic, New York, 284 pp. price: $34.00. Networks **13**(2), 304–305 (1983). https://doi.org/10.1002/net.3230130214

32. Kane, D.M., Nelson, J., Woodruff, D.P.: An optimal algorithm for the distinct elements problem. In: Proceedings of the 29th ACM SIGMOD-SIGACT-SIGART Symposium on Principles of Database Systems (PODS 2010), pp. 41–52. ACM (2010). https://doi.org/10.1145/1807085.1807094

33. Karp, R.M.: Reducibility among combinatorial problems. In: Proceedings of a Symposium on the Complexity of Computer Computations. The IBM Research Symposia Series, pp. 85–103. Plenum Press, New York (1972). https://doi.org/10.1007/978-1-4684-2001-2_9

34. Kleinberg, J.M., Tardos, É.: Algorithm Design. Addison-Wesley, Boston (2006)

35. Kratochvíl, J., Nešetřil, J.: Independent set and clique problems in intersection-defined classes of graphs. Comment. Math. Univ. Carol. **31**(1), 85–93 (1990)

36. McGregor, A.: Graph stream algorithms: a survey. SIGMOD Rec. **43**(1), 9–20 (2014). https://doi.org/10.1145/2627692.2627694

37. Mitchell, J.S.B.: Approximating maximum independent set for rectangles in the plane. In: Proceedings of the 62nd IEEE Annual Symposium on Foundations of Computer Science, FOCS 2021, pp. 339–350. IEEE (2021). https://doi.org/10.1109/FOCS52979.2021.00042

38. Muthukrishnan, S., et al.: Data streams: algorithms and applications. Found. Trends® Theor. Comput. Sci. **1**(2), 117–236 (2005)

39. Zuckerman, D.: Linear degree extractors and the inapproximability of max clique and chromatic number. Theory Comput. **3**(1), 103–128 (2007). https://doi.org/10.4086/toc.2007.v003a006

# Approximating Length-Restricted Means Under Dynamic Time Warping

Maike Buchin[1] , Anne Driemel[2,3] , Koen van Greevenbroek[4] ,
Ioannis Psarros[5] , and Dennis Rohde[1(✉)]

[1] Faculty of Computer Science, Ruhr-University Bochum, Bochum, Germany
{maike.buchin,dennis.rohde-t1b}@rub.de
[2] Hausdorff Center for Mathematics, Bonn, Germany
[3] University of Bonn, Bonn, Germany
driemel@cs.uni-bonn.de
[4] Department of Computer Science, UiT The Arctic University of Norway,
Tromsø, Norway
koen.v.greevenbroek@uit.no
[5] Athena Research Center, Marousi, Greece
ipsarros@di.uoa.gr

**Abstract.** We study variants of the mean problem under the $p$-Dynamic Time Warping ($p$-DTW) distance, a popular and robust distance measure for sequential data. In our setting we are given a set of finite point sequences over an arbitrary metric space and we want to compute a mean point sequence of given length that minimizes the sum of $p$-DTW distances, each raised to the $q^{\text{th}}$ power, between the input sequences and the mean sequence. In general, the problem is NP-hard and known not to be fixed-parameter tractable in the number of sequences. On the positive side, we show that restricting the length of the mean sequence significantly reduces the hardness of the problem. We give an exact algorithm running in polynomial time for constant-length means. We explore various approximation algorithms that provide a trade-off between the approximation factor and the running time. Our approximation algorithms have a running time with only linear dependency on the number of input sequences. In addition, we use our mean algorithms to obtain clustering algorithms with theoretical guarantees.

## 1 Introduction

The $p$-Dynamic Time Warping distance – in short $p$-DTW – is a popular distance measure for temporal data sequences, which has been studied and applied extensively over the past decades. While it was first applied to speech recognition [30], it also showed effective for other kinds of sequential data and now there is a

I. Psarros—The author was partially supported by the European Union's Horizon 2020 Research and Innovation programme, under the grant agreement No. 957345: "MORE". Part of this work was done while the author was a postdoctoral researcher at the Hausdorff Center for Mathematics and the University of Bonn, Germany.

P. Chalermsook and B. Laekhanukit (Eds.): WAOA 2022, LNCS 13538, pp. 225–253, 2022.
https://doi.org/10.1007/978-3-031-18367-6_12

broad variety of applications in numerous domains, cf. [1,4,13,21,25,26,37]. Its particular strength is the ability to handle differences in the length and in the temporal properties (e.g. phase or sampling rates) of the data. Furthermore, it less sensitive to outliers in the sequences, e.g. from measurement errors or noise, because it is realized as a sum of point-to-point distances as compared to other distance measures which are correlated with the maximum point-to-point distance. It is based on monotonic alignments of the sequences, i.e., every element of the first sequence is paired to an element of the second sequence in a monotonic fashion (along the temporal axis). To compensate differences in length, samplings rates or in phase, several elements of the first sequence can be paired to a single element of the second sequence and vice versa. Such an alignment is called a warping and the $p$-DTW is the $p^{\text{th}}$ root of the sum of distances, each raised to the $p^{\text{th}}$ power, between all pairs of elements determined by an optimal warping, i.e., a warping that minimizes said quantity. The distances between elements are determined by an underlying space, which itself is determined by the application at hand. The $p$-DTW can be computed by a dynamic program with running time quadratic in the lengths of the given sequences, and it cannot be computed in strongly subquadratic time unless the Exponential Time Hypothesis is false [2,8]. Apart from its apparent benefits, $p$-DTW has the drawback that it is not a metric since it does not fulfill the identity of indiscernibles nor the triangle inequality. This rules out a wealth of techniques that were developed for proper metrics.

In this work, we consider the problem of computing a mean under $p$-DTW. Here, we are given a finite set of $n$ point sequences over an arbitrary metric space, each of complexity, i.e., the number of elements of the sequence, bounded by a number $m$ and we want to compute a point sequence (the mean) that minimizes the sum of $p$-DTW distances, each raised to the $q^{\text{th}}$ power, between the given sequences and the mean.[1] We call this problem the *unrestricted* $(p,q)$-mean problem. It is known to be NP-hard [10,12] and all known algorithms that solve it either suffer from exponential running time, only work for binary alphabets, or are of heuristic nature [6,28,32].

We show that when we restrict the complexity of the mean to be bounded by some constant $\ell$ – we call this the *restricted* $(p,q)$-mean problem – the problem becomes tractable, i.e., there exist polynomial time approximation algorithms. This restriction also comes with a practical motivation, i.e., to suppress overfitting, see also the discussion in [5].

## 1.1    Related Work

Among many practical approaches for the problem of computing a mean, one very influential heuristic is the DTW Barycentric Average (DBA) method, as formalized by Petitjean, Ketterlin and Gançarski [28]. The core idea behind DBA is a Lloyd's style ($k$-means) iterative strategy, which has been rediscovered many

---

[1] For $q = 1$ this is an adaption of the Euclidean median and for $q = 2$ an adaption of the Euclidean mean.

times for this problem in the past (see e.g. [3,18,29]). DBA iteratively improves the solution as follows: given a candidate average sequence $c = (c_1, \ldots, c_\ell)$, it first computes the warpings between $c$ and all input sequences, and then given each set of input vertices $S_i$ matched with the same vertex $c_i$, it substitutes $c_i$ with the mean of $S_i$. DBA has inspired many recent solutions that are successful in practice [15,20,24,27,33]. However, it does not give any guarantees. Just like the $k$-means algorithm, it may even converge to a local optimum that is arbitrarily far from the global optimum in terms of the $(p,q)$-mean target function.

There are few results in the literature with formal guarantees on the running time or the quality of the solution. Brill et al. [6,7] presented an algorithm for solving the unrestricted $(2,2)$-mean problem defined over $\mathbb{Q}$ with the Euclidean distance, with an asymptotic bound on the time complexity. Their algorithm is based on dynamic programming, and computes the (unrestricted) $(2,2)$-mean, in time $O(m^{2n+1}2^n n)$. The algorithm can be slightly modified, to compute a restricted $(2,2)$-mean. Brill et al. [6,7] also show that the unrestricted $(2,2)$-mean problem defined over $\{0,1\}$ with the Euclidean distance can be solved in $O(nm^3)$ time. This was later improved by Schaar, Froese and Niedermeier [32] to $O(nm^{1.87})$ time.

All previous hardness results concern the exact computation of the $(p,q)$-mean. Bulteau, Froese and Niedermeier [12] proved that the $(2,2)$-mean problem defined over $\mathbb{Q}$ with the Euclidean distance is NP-hard and W[1]-hard with the number of input sequences $n$ as the parameter. Moreover, they show that the problem cannot be solved in time $O(f(n)) \cdot m^{o(n)}$ for any computable function $f$ unless the Exponential Time Hypothesis (ETH) fails. Buchin, Driemel and Struijs [10] presented an alternative proof of the above statements, which more generally applies to the unrestricted $(p,q)$-mean problem for any $p,q \in \mathbb{N}$.

## 1.2   Overview of Results

In this section we give an overview of our results.[2] Since, we mainly study the case $p = q$, we shorthand call the $(p,p)$-mean, the $p$-mean problem. In Sect. 2.1, we present an exact algorithm with polynomial running time for the problem of computing a restricted $(2,2)$-mean in Euclidean space. Our approach is based on a decomposition of the solution space by an arrangement of polynomial surfaces such that each cell corresponds to a set of means with uniquely defined optimal warpings to all input sequences. The algorithm has running time $O(n^{2d\ell}m^{8d\ell^2})$, where the input are $n$ sequences of $m$ points in $\mathbb{R}^d$. Note that the running time is exponential in $\ell$ and $d$. In the remainder of the paper, our goal is to improve upon this with the help of approximation and randomization techniques. We will show that linear dependency in $n$ is possible.

In Sect. 2.2, we present a randomized constant-factor approximation algorithm for the restricted $p$-mean problem that works for sequences from any fixed

---

[2] An earlier version of this manuscript claimed hardness of approximation for the problem of computing the mean under dynamic time warping. However, the proof turned out to be flawed. We leave it as an open problem to show hardness of approximation for this problem.

metric space. As such, this result is applicable to the classical median problem under the 1-DTW distance and the classical mean problem under the 2-DTW distance. The main idea is to uniformly sample from the union of points of all given sequences, then enumerate all sequences of complexity $\ell$ from the sampled vertices and return the sequence with the lowest cost. We also show how to derandomize the algorithm.

In Sect. 2.4 we present a $(1 + \varepsilon)$-approximation algorithm for the restricted $(p, 1)$-mean problem for point sequences defined over the Euclidean space. This algorithm is based on an exhaustive search over a carefully constructed set of candidate means. The crucial ingredients for this are presented in Sect. 2.3: an efficient approximation algorithm for simplification under $p$-DTW and a weak triangle inequality for $p$-DTW. The result holds for the important case of sequences that stem from a Euclidean space. A nice property of this result is that it provides a complete trade-off between the approximation factor and the running time.

Finally, in Sect. 3, we briefly discuss an application of the newly developed techniques to the problem of clustering for $p$-DTW distances. In particular, we can use the random sampling techniques developed in Sects. 2.2 and 2.3, in combination with a known algorithmic scheme that reduces the computation of $k$-medians to a problem of computing 1-median candidates for an arbitrary subset of the input set.

## 1.3 Preliminaries

In the following $d \in \mathbb{N}$ is an arbitrary constant. For $n \in \mathbb{N}$ we define $[n] = \{1, \ldots, n\}$. Let $\mathcal{X} = (X, \rho)$ be a metric space. For $x \in X$ and $r \in \mathbb{R}_{\geq 0}$ we denote by $B(x, r) = \{y \in X \mid \rho(x, y) \leq r\}$ the ball of radius $r$ centered at $x$. We define sequences of points over $\mathcal{X}$.

**Definition 1.** *A point sequence $\sigma$ over $\mathcal{X}$ is a tuple $(\sigma_1, \ldots, \sigma_m) \in X^m$, where $m \in \mathbb{N}_{>1}$ is called its complexity, denoted by $|\sigma|$, and $\sigma_1, \ldots, \sigma_m$ are called its vertices.*

By $X^* = \bigcup_{i=1}^{\infty} X^i$ we define the set of all point sequences over $\mathcal{X}$ and by $X^{\leq m} = \bigcup_{i=1}^{m} X^i$ we define the subset of point sequences of complexity at most $m$. The *concatenation* of a point sequence $\pi = (\pi_1, \ldots, \pi_m)$ with a sequence $\tau = (\tau_1, \ldots, \tau_m)$ is denoted by $\pi \oplus \tau$ and is defined as the point sequence $(\pi_1, \ldots, \pi_m, \tau_1, \ldots, \tau_m)$. We define the $p$-Dynamic Time Warping distance.

**Definition 2.** *For $m_1, m_2 \in \mathbb{N}_{>1}$, let $\mathcal{W}_{m_1, m_2}$ denote the set of all $(m_1, m_2)$-warpings, that is, the set of all sequences $(i_1, j_1), \ldots, (i_n, j_n)$ with*

- *$i_1 = j_1 = 1$, $i_n = m_1$, $j_n = m_2$ and*
- *$(i_k - i_{k-1}, j_k - j_{k-1}) \in \{(0,1), (1,0), (1,1)\}$ for each $k \in \{2, \ldots, n\}$.*

*For $p \in [1, \infty)$ and $\sigma = (\sigma_1, \ldots, \sigma_{m_1}) \in X^{m_1}$, $\tau = (\tau_1, \ldots, \tau_{m_2}) \in X^{m_2}$ the p-Dynamic Time Warping distance is defined as*

$$d_{\text{DTW}p}(\sigma, \tau) = \min_{W \in \mathcal{W}_{m_1, m_2}} \left( \sum_{(i,j) \in W} \rho(\sigma_i, \tau_j)^p \right)^{1/p}.$$

Here we assume that $\rho(\cdot, \cdot)$ can be evaluated in constant time.[3]

Let $\sigma = (\sigma_1, \ldots, \sigma_{m_1})$, $\tau = (\tau_1, \ldots, \tau_{m_2})$ be point sequences over $\mathcal{X}$ of complexity $m_1$ and $m_2$. We call a warping $W \in \underset{W \in \mathcal{W}_{m_1, m_2}}{\arg\min} \left( \sum_{(i,j) \in W} \rho(\sigma_i, \tau_j)^p \right)^{1/p}$ an optimal $p$-warping between $\sigma$ and $\tau$.

**Definition 3.** *The restricted $(p, q)$-mean problem is defined as follows, where $\ell \in \mathbb{N}_{>1}$ and $p, q \in [1, \infty)$ are fixed (constant) parameters of the problem: given a set $T = \{\tau_1, \ldots, \tau_n\} \subseteq X^{\leq m}$ of point sequences, compute a point sequence $c \in X^{\leq \ell}$, such that $\text{cost}_p^q(T, c) = \sum_{i=1}^n d_{\text{DTW}p}(c, \tau_i)^q$ is minimal.*

If $p$ is clear from the context, we drop it from our notation. By the *unrestricted* $(p, q)$-mean problem we define the problem that is similar to the restricted $(p, q)$-mean problem with the only difference that we compute a mean sequence $c \in X^*$. We mainly study the case that $p = q$. Therefore, we shorthand call the restricted, respectively unrestricted, $(p, p)$-mean problem the restricted, respectively unrestricted, $p$-mean problem. We emphasize that these problems are prevalent in the literature.

## 2 Tractability of the Restricted Mean Problem

We study exact and approximation algorithms for the restricted $p$-mean and $(p, 1)$-mean problems.

### 2.1 Exact Computation of a Restricted 2-Mean in Euclidean Space

In the following, we make use of a simplified structure of the solution space, which holds in the case $p = q$. This is captured in the notion of sections, which we define as follows.

**Definition 4 (sections).** *Let $T = \{\tau_1, \ldots, \tau_n\} \subseteq X^{\leq m}$ be a set of point sequences and $c = (c_1, \ldots, c_\ell) \in X^\ell$ be a point sequence. For $i \in [n]$ and $p \in [1, \infty)$, let $W_i$ be an optimal $p$-warping between $c$ and $\tau_i$. For $j \in [\ell]$ we define the $j^{th}$ section of $c$ with respect to $T$ (and $W_1, \ldots, W_n$): $S_j(c, T, W_1, \ldots, W_n) = \{\tau_{i,k} \mid i \in [n], (j, k) \in W_i\}$, where $\tau_{i,k}$ is the $k^{th}$ vertex of $\tau_i$.*

---

[3] This restriction is only for the sake of simplicity of presentation. Our results can be easily extended to metric spaces that do not have a constant-time distance oracle.

If $T$ is clear from the context, we omit it from the notation. Also, we will always omit $W_1, \ldots, W_n$ from the notation, because the specific choice of optimal $p$-warpings is not of interest. We will then write $S_j^p(T, c)$ to clarify that the sections are defined with respect to optimal $p$-warpings. An immediate consequence of this definition is the following identity:

$$\text{cost}_p^p(T, c) = \sum_{j=1}^{\ell} \sum_{v \in S_j^p(c, T)} \rho(c_j, v)^p,$$

where $\ell$ denotes the complexity of $c$.

A central observation is that the vertices of an optimal restricted $p$-mean $c = (c_1, \ldots, c_{\ell'})$ must minimize the sum of distances, each raised to the $p^{\text{th}}$ power, to the vertices in their section, i.e., for all $j \in [\ell']$: $c_j \in \arg\min_{w \in X} \sum_{v \in S_j^p(c, T)} \rho(w, v)^p$.

Using this, we obtain the following result.

**Theorem 1.** *There exists an algorithm that, given a set $T \subset (\mathbb{Q}^d)^{\leq m}$ of $n$ point sequences, computes an optimal restricted 2-mean (defined over the Euclidean distance) in time $O(n^{2d\ell} m^{8d\ell^2})$.*

The strategy is as follows. For the sake of simplicity, suppose that we want to compute the best mean of complexity exactly $\ell' \in [\ell]$. To compute the optimal restricted 2-mean, it suffices to find the best mean for every $\ell' \in [\ell]$. For a fixed $\ell' \in [\ell]$, the main idea is to compute for any two warpings between a point sequence of complexity $\ell'$ and an input point sequence, a polynomial function whose sign indicates which of the warpings yields a smaller distance between the sequences. These functions are then used to define an arrangement that partitions the space $(\mathbb{R}^d)^{\ell'}$. The trick is that while there is an infinite number of point sequences in $(\mathbb{R}^d)^{\ell'}$, to each input point sequence there are only $O(m^{2\ell'})$ warpings and in each face of the arrangement the point sequences have the same optimal warpings to the input point sequences. Therefore, for an arbitrary point sequence from each face of the arrangement, we can compute the optimal warpings to the input sequences and then use the resulting sections to compute an optimal point sequence for these warpings, obtaining the optimal mean of complexity exactly $\ell'$ when we eventually hit the face containing it. We use a tool of computational algebraic geometry, namely an algorithm to compute a stratification of $(\mathbb{R}^d)^{\ell'}$ (which is a refinement of the arrangement), to compute the arrangement and obtain an element of each face.

Before we now prove the theorem we note that the employed stratification algorithm only works for polynomials with real algebraic coefficients. Therefore, here we assume that the input point sequences have rational coordinates. This is indeed a realistic assumption since physical computers are not capable of storing arbitrary real numbers.

*Proof (of Theorem 1).*

To simplify our exposition, we restrict ourselves to means of complexity exactly $\ell' \in [\ell]$, i.e. in the rest of the proof we describe an algorithm for computing an optimal mean of complexity exactly $\ell'$. The complete algorithm consists of iteratively computing the optimal mean of complexity $\ell'$, for each $\ell' \in [\ell]$.

For each $\tau = (\tau_1, \ldots, \tau_{|\tau|}) \in T$ and all $W_1, W_2 \in \mathcal{W}_{\ell',|\tau|}$ we define for $c = (c_1, \ldots, c_{\ell'}) \in (\mathbb{R}^d)^{\ell'}$ the polynomial function

$$P_{\tau,W_1,W_2}(c) = \left( \sum_{(i,j)\in W_1} \|c_i - \tau_j\|^2 \right) - \left( \sum_{(i,j)\in W_2} \|c_i - \tau_j\|^2 \right).$$

Clearly, iff $W_1$ yields a smaller distance between $c$ and $\tau$ than $W_2$, then

$$P_{\tau,W_1,W_2}(c) < 0$$

and iff $W_2$ yields a smaller distance between $c$ and $\tau$ than $W_2$, then

$$P_{\tau,W_1,W_2}(c) > 0.$$

Iff $P_{\tau,W_1,W_2}(c) = 0$, both yield the same distance.

Let $F = \{P_{\tau,W_1,W_2} \mid \tau \in T, W_1, W_2 \in \mathcal{W}_{\ell',|\tau|}\}$ be the set of these polynomials. The central observation is that if all functions in $F$ have the same sign for any $c_1, c_2 \in (\mathbb{R}^d)^{\ell'}$, then $c_1$ and $c_2$ have the same optimal 2-warpings to the point sequences in $T$. To see this, for each $\tau \in T$ let $W_\tau \in \mathcal{W}_{\ell',|\tau|}$ be an optimal 2-warping between $c_1$ and $\tau$. Clearly, $P_{\tau,W_\tau,W}(c_1) \leq 0$ for all $\tau \in T$ and $W \in \mathcal{W}_{\ell',|\tau|}$. Now, if all functions in $F$ have the same sign for $c_1$ and $c_2$ it must be that $P_{\tau,W_\tau,W}(c_2) \leq 0$ for all $\tau \in T$ and $W \in \mathcal{W}_{\ell',|\tau|}$. Thus, $W_\tau$ is an optimal 2-warping between $c_2$ and $\tau$ for each $\tau \in T$.

Now, we compute an arrangement of the zero sets of the polynomials in $F$ (cf. [22]), i.e., a partition of $(\mathbb{R}^d)^{\ell'}$ into regions where all functions in $F$ have the same sign. To be precise, we compute a refinement of this arrangement, namely a sign-invariant stratification of $(\mathbb{R}^d)^{\ell'}$ consisting of $O(|F|^{2d\ell'-2})$ cylindrical Tarski cells, which can be done in time $O(|F|^{2d\ell'-1} \log|F|)$ by [14, Theorem 3.4]. This computation includes obtaining a sample point from each Tarski cell and we discard all following sample points from a face of the arrangement after obtaining the first one. For each sample $c$ from some face of the arrangement we first compute the optimal 2-warpings between $c$ and the input point sequences $\tau \in T$ in time $O(nm)$. Second we compute all sections $S_j(c)$ of $c$ and store the point sequence $c' = (c'_1, \ldots, c'_{\ell'})$ consisting of the optimal means $c'_j = \frac{1}{|S_j(c)|} \sum_{v \in S_j(c)} v$, for $j \in [\ell']$. This takes time $O(nm)$.

At some point, we obtain a sample from the face containing the optimal restricted 2-mean $c^* = (c^*_1, \ldots, c^*_{\ell'})$ (where $c^*_j$ is the mean of $S_j(c^*)$ for each $j \in [\ell']$), which we return when we finally return the point sequence $c'$ that minimizes the objective function. This takes time $O(nmA)$, where $A$ is the number of cells in the arrangement.

To conclude the proof, note that for each $\tau \in T$ we have that $|W_{\ell',|\tau|}| \leq m^{2\ell'}$, thus $|F| \leq nm^{4\ell'}$. Hence, $A \leq \left(\frac{100nm^{4\ell'}}{\ell'd}\right)^{\ell'd}$ by [22, Theorem 6.2.1].

As we have already mentioned, we iteratively run the above algorithm to compute means of complexity $\ell'$, for each $\ell' \in [\ell]$, in order to find an optimal restricted 2-mean. Each iteration runs in time $O(n^{2d\ell'}m^{8d\ell'^2})$. Since $\ell$ is constant, the running time is in $O(n^{2d\ell}m^{8d\ell^2})$.

## 2.2    Constant-Factor Approximation of the Restricted $p$-Mean

We start by describing a simple approximation algorithm that reveals the basic idea underlying the following algorithms. The algorithm relies on the following observation. If $p$-DTW is defined over a metric, then the triangle inequality holds for the point-to-point-distances in the sum that defines the $p$-DTW distance (albeit not for $p$-DTW distance itself). Assume for simplicity that $p = 1$. In this case, there always exists a 2-approximate mean that is formed by points from the input sequences. Enumerating all possible such sequences, then, if the input consists of $n$ point sequences of length $m$, leads to an algorithm with running time in $O((nm)^{\ell+1})$, where $\ell$ denotes the largest allowed complexity of the mean. This approach also extends to other variants of the mean problem for different choices of $p$ and $q$ (with varying approximation factors). One obvious disadvantage of this simple algorithm is the high running time. In the following, we use similar observations as above and show that the dependency on the number of input sequences $n$ can be improved to linear while still achieving approximation factors close to 2.

**Randomized Algorithm.** We present a randomized constant-factor approximation algorithm for the restricted $p$-mean problem. The approximation factor of the algorithm depends on $p$, and the best it can achieve is $2 + \varepsilon$ for $p = 1$ and $4 + \varepsilon$ for $p = 2$, which resemble the famous Euclidean median and mean problems. The idea of the algorithm is to obtain for each $j \in [\ell']$ from the corresponding section $S_j^p(c, T)$ of an optimal restricted $p$-mean $c = (c_1, \ldots, c_{\ell'})$ one of the closest input vertices to $c_j$. The obtained vertices in the corresponding order form an approximate restricted $p$-mean. We formalize the idea in the following lemma.

**Lemma 1.** Let $T = \{\tau_1 = (\tau_{1,1}, \ldots, \tau_{1,|\tau_1|}), \ldots, \tau_n = (\tau_{n,1}, \ldots, \tau_{n,|\tau_n|})\} \subseteq X^{\leq m}$ be a set of point sequences and let $P = \bigcup_{i=1}^{n} \bigcup_{j=1}^{|\tau_i|} \{\tau_{i,j}\}$. For any $p \in [1, \infty)$, $\ell \in \mathbb{N}_{>1}$ and $\varepsilon \in (0, \infty)$ there exists an $\ell' \leq \ell$ and balls $B_1, \ldots, B_{\ell'} \subseteq P$, of cardinality at least $\frac{\varepsilon n}{2^{p-1}+\varepsilon}$ each, such that every point sequence $c' = (c'_1, \ldots, c'_{\ell'})$, with $c'_i \in B_i$ for each $i \in [\ell']$, is a $(2^p + \varepsilon)$-approximate restricted $p$-mean for $T$.

*Proof.* Without loss of generality we assume that $\tau_{1,1}, \ldots, \tau_{1,|\tau_1|}, \ldots, \tau_{n,1}, \ldots, \tau_{|\tau_n|}$ are distinct points. Let $c = (c_1, \ldots, c_{\ell'}) \in X^{\ell'}$ be an optimal restricted $p$-mean for $T$ and for $j \in [\ell']$ let $S_j = \{s_{j,1}, \ldots, s_{j,n_j}\} = S_j^p(c)$ for brevity. Define

$\Delta(S_j) = \sum\limits_{v \in S_j} \rho(c_j, v)^p$. We immediately have $\text{cost}_p^p(T, c) = \sum_{j=1}^{\ell'} \Delta(S_j)$. Now, for $j \in [\ell']$, let $\pi_j$ be a permutation of the index set $[n_j]$, such that

$$\rho(c_j, s_{j,\pi_j^{-1}(1)})^p \leq \cdots \leq \rho(c_j, s_{j,\pi_j^{-1}(n_j)})^p.$$

Let $\varepsilon' = \frac{\varepsilon}{2^{p-1}+\varepsilon}$. For the sake of simplicity, we assume that $\varepsilon'n$ is integral. Further, for $j \in [\ell']$, we define $C_j = \{s_{j,\pi_j^{-1}(1)}, \ldots, s_{j,\pi_j^{-1}(\varepsilon'n)}\}$. We have that $\rho(c_j, s_{j,\pi_j^{-1}(\varepsilon'n)})^p \leq \frac{\Delta(S_j)}{|S_j|-(\varepsilon'n-1)}$ by the fact that $\rho(c_j, s_{j,\pi_j^{-1}(\varepsilon'n)})^p$ is of maximal value, iff $\rho(c_j, s')^p = 0$ for each $s' \in C_j \setminus \{s_{j,\pi_j^{-1}(\varepsilon'n)}\}$ and $\rho(c_j, s')^p = \rho(c_j, s_{j,\pi_j^{-1}(\varepsilon'n)})^p$ for each $s' \in S_j \setminus C_j$. For $j \in [\ell']$, we now define

$$B_j = \{x \in P \mid \rho(c_j, x)^p \leq \rho(c_j, s_{j,\pi_j^{-1}(\varepsilon'n)})^p\}$$

and by definition we have $\rho(c_j, x)^p \leq \frac{\Delta(S_j)}{|S_j|-\varepsilon'n+1} \leq \frac{\Delta(S_j)}{|S_j|-\varepsilon'n}$ for each $x \in B_j$ and $j \in [\ell']$. Then let $c' = (c'_1, \ldots, c'_{\ell'})$ be a point sequence with $c'_j \in B_j$ for each $j \in [\ell']$. We bound its cost:

$$\text{cost}_p^p(T, c') = \sum_{j=1}^{\ell'} \sum_{v \in S_j} \rho(c'_j, v)^p \leq \sum_{j=1}^{\ell'} \sum_{v \in S_j} (\rho(c_j, v) + \rho(c_j, c'_j))^p$$

$$\leq \sum_{j=1}^{\ell'} \sum_{v \in S_j} 2^{p-1}(\rho(c_j, v)^p + \rho(c_j, c'_j)^p)$$

$$\leq 2^{p-1} \sum_{j=1}^{\ell'} \sum_{v \in S_j} \left( \rho(c_j, v)^p + \frac{\Delta(S_j)}{|S_j| - \varepsilon'n} \right)$$

$$\leq 2^{p-1} \text{cost}_p^p(T, c) + 2^{p-1} \sum_{j=1}^{\ell'} \sum_{v \in S_j} \frac{\Delta(S_j)}{(1 - \varepsilon')|S_j|}$$

$$= \left( 2^{p-1} + \frac{2^{p-1}}{1 - \varepsilon'} \right) \text{cost}_p^p(T, c) = (2^p + \varepsilon) \text{cost}_p^p(T, c).$$

The first inequality follows from the triangle-inequality and the last inequality holds, because a vertex from each $\tau_i \in T$ must be warped to each $c_j \in c$, thus $|S_j| \geq n$ for each $j \in [\ell']$.

Now we present the first algorithm. The idea is to uniformly sample from the set of all vertices of all point sequences, to obtain at least one vertex from each ball guaranteed by the previous lemma, with high probability. After the sampling, the algorithm enumerates all point sequences of at most $\ell$ elements from the sample and returns a point sequence with lowest cost.

---

**Algorithm 1.** Restricted $p$-Mean Constant-Factor Approximation

---

1: **procedure** MEAN-C$(T = \{(\tau_{1,1}, \ldots, \tau_{1,|\tau_1|}), \ldots, (\tau_{n,1}, \ldots, \tau_{n,|\tau_n|})\}, \delta, \varepsilon, p)$

2:     $P \leftarrow \bigcup_{i=1}^{n} \bigcup_{j=1}^{|\tau_i|} \{\tau_{i,j}\}$

3:     $S \leftarrow$ sample $\left\lceil \frac{m(\ln(\ell)+\ln(1/\delta))}{\varepsilon/(2^{p-1}+\varepsilon)} \right\rceil$ points from $P$
        uniformly and independently at random with replacement

4:     $C \leftarrow S^{\leq \ell}$

5:     **return** an arbitrary element from $\arg\min_{c \in C} \text{cost}_p^p(T, c)$

---

The correctness of Algorithm 1 follows by an application of Lemma 1.

**Theorem 2.** *Given a set $T = \{\tau_1, \ldots, \tau_n\} \subseteq X^{\leq m}$ of point sequences (defined over any metric), three parameters $\delta \in (0,1)$, $\varepsilon \in (0,\infty)$ and $p \in [1,\infty)$, Algorithm 1 returns with probability at least $1 - \delta$ a $(2^p + \varepsilon)$-approximate restricted $p$-mean for $T$, in time $O\left(nm^{\ell+1}\ln(1/\delta)^\ell \left(1 + \frac{2^{p-1}}{\varepsilon}\right)^\ell\right)$.*

*Proof.* For the given $\varepsilon$, let $\varepsilon' = \frac{\varepsilon}{2^{p-1}+\varepsilon}$ and let $B_1, \ldots, B_{\ell'}$, $\ell' \leq \ell$, be the balls guaranteed by Lemma 1. Recall that each ball has size at least $\varepsilon'n$. For each $i \in [\ell']$ and $s \in S$ we have

$$\Pr[s \notin B_i] \leq (1 - \frac{\varepsilon'n}{|P|}) \leq (1 - \frac{\varepsilon'n}{nm}) = (1 - \frac{\varepsilon'}{m}) \leq \exp(-\varepsilon'/m).$$

By independence, for each $i \in [\ell']$ we have

$$\Pr[B_i \cap S = \emptyset] \leq \exp(-\varepsilon'/m)^{\left\lceil \frac{m(\ln(\ell)-\ln(\delta))}{\varepsilon'} \right\rceil} \leq \delta/\ell.$$

Using a union bound we conclude that with probability at least $1 - \delta$, $S$ contains at least one element of $B_i$, for each $i \in [\ell']$, and thus Algorithm 1 returns a $(2^p + \varepsilon)$-approximate restricted $p$-mean for $T$ with probability at least $1 - \delta$ by Lemma 1.

The running time of the algorithm is dominated by computing the cost of all point sequences of complexity at most $\ell$ over $S$. Since $|S^{\leq \ell}|$ is in $O\left(\frac{\ln(1/\delta)^\ell m^\ell}{(\varepsilon')^\ell}\right) = O\left(\ln(1/\delta)^\ell m^\ell \frac{(2^{p-1}+\varepsilon)^\ell}{\varepsilon^\ell}\right)$ and every distance can be computed in time $O(m)$, this takes time $O\left(\ln(1/\delta)^\ell m^{\ell+1} n \frac{(2^{p-1}+\varepsilon)^\ell}{\varepsilon^\ell}\right)$.

**Derandomization.** In this section, we consider finite metric spaces $(X, \rho)$ for which the set of all metric balls $\mathcal{B} = \{B(x,r) \mid x \in X, r \in \mathbb{R}_{\geq 0}\}$ forms a range space $(X, \mathcal{B})$ with bounded VC dimension $\mathcal{D}$. We present a deterministic algorithm for the restricted $p$-mean problem which is applicable under the additional assumption that there is a subsystem oracle for $(X, \mathcal{B})$. We show that this is the case for the Euclidean metric. Note that our algorithm depends on the existence of a subsystem oracle, which is not always obvious for a given metric.

We formally define range spaces and the associated concepts. A range space is defined as a pair of sets $(X, \mathcal{R})$, where $X$ is the *ground set* and $\mathcal{R} \subseteq 2^X$ is the *range set*. For $Y \subseteq X$, we denote $\mathcal{R}_{|Y} = \{R \cap Y \mid R \in \mathcal{R}\}$ and if $\mathcal{R}_{|Y}$ contains all subsets of $Y$, then $Y$ is *shattered* by $\mathcal{R}$. A measure of the combinatorial complexity of such a range space is the VC dimension.

**Definition 5 (VC dimension).** *The Vapnik-Chervonenkis dimension [31, 34, 35] of $(X, \mathcal{R})$ is the maximum cardinality of a shattered subset of $X$.*

Range spaces need not to be finite and can be discretized by means of $\varepsilon$-nets.

**Definition 6 ($\varepsilon$-net).** *A set $N \subset X$ is an $\varepsilon$-net for $(X, \mathcal{R})$ if for any range $R \in \mathcal{R}$, it holds that $R \cap N \neq \emptyset$ if $|R \cap X| \geq \varepsilon|X|$.*

To compute $\varepsilon$-nets deterministically, we need a subsystem oracle, which we now define.

**Definition 7 (subsystem oracle).** *Let $(X, \mathcal{R})$ be a finite range space. A subsystem oracle is an algorithm which for any $Y \subseteq X$, lists all sets in $\mathcal{R}_{|Y}$ in time $O(|Y|^{\mathcal{D}+1})$, where $\mathcal{D}$ is the VC dimension of $(X, \mathcal{R})$.*

We use the following theorem to obtain $\varepsilon$-nets when provided with a subsystem oracle.

**Theorem 3 ([9, Theorem 2.1]).** *Let $(X, \mathcal{R})$ be a range space with finite ground set and VC dimension $\mathcal{D}$, and $\varepsilon > 0$ be a given parameter. Assume that there is a subsystem oracle for $(X, \mathcal{R})$. Then an $\varepsilon$-net of size $O\left(\frac{\mathcal{D}}{\varepsilon} \log \frac{\mathcal{D}}{\varepsilon}\right)$ can be computed deterministically in time $O(\mathcal{D}^{3\mathcal{D}}) \cdot \left(\frac{1}{\varepsilon} \log \frac{1}{\varepsilon}\right)^{\mathcal{D}} \cdot |X|$.*

The following algorithm is a modification of Algorithm 1. We replace the sampling step with a computation of an $(\varepsilon/m)$-net. Since the balls guaranteed by Lemma 1 are of appropriate size, the $(\varepsilon/m)$-net stabs all of them and by enumeration of all point sequences of at most $\ell$ points from the $(\varepsilon/m)$-net, we again find an approximate restricted $p$-mean.

---

**Algorithm 2.** Restricted $p$-Mean Constant-Factor Approximation

1: **procedure** MEAN-C-D$(T = \{(\tau_{1,1}, \ldots, \tau_{1,|\tau_1|}), \ldots, (\tau_{n,1}, \ldots, \tau_{n,|\tau_n|})\}, \varepsilon, p)$
2:     $\varepsilon' \leftarrow \frac{\varepsilon}{2^{p-1}+\varepsilon}$, $P \leftarrow \bigcup_{i=1}^{n} \bigcup_{j=1}^{|\tau_i|} \{\tau_{i,j}\}$
3:     $S \leftarrow$ compute an $(\varepsilon'/m)$-net of $(P, \mathcal{B})$     ▷ $\mathcal{B} = \{B(x,r) \mid x \in X, r \in \mathbb{R}_{\geq 0}\}$
4:     $C \leftarrow S^{\leq \ell}$
5:     **return** an arbitrary element from $\arg\min_{c \in C} \text{cost}_p^p(T, c)$

---

The correctness of Algorithm 2 follows from Definition 6.

**Theorem 4.** *Given a set $T \subseteq X^{\leq m}$ of $n$ point sequences, parameters $\varepsilon \in (0, \infty)$, and $p \in [1, \infty)$, Algorithm 2 returns a $(2^p + \varepsilon)$-approximate restricted $p$-mean for $T$.*

*Proof.* By Lemma 1, for any $\varepsilon \in (0, \infty)$ there exist balls $B_1, \ldots, B_{\ell'} \subseteq P$, $\ell' \leq \ell$ of cardinality at least $\varepsilon'n$ each, such that any point sequence $c' = (c'_1, \ldots, c'_{\ell'})$, with $c'_i \in B_i$ for each $i \in [\ell']$, is a $(2^p + \varepsilon)$-approximate restricted $p$-mean for $T$, where $\varepsilon' = \frac{\varepsilon}{2^p - 1 + \varepsilon}$. Since we compute an $(\varepsilon'/m)$-net of $P$ and $|P| \leq nm$, $S$ contains at least one point from each of $B_1, \ldots, B_{\ell'}$ by Definition 6. Hence, $S^{\leq \ell}$ contains a $(2^p + \varepsilon)$-approximate restricted $p$-mean for $T$.

We now turn to the Euclidean setting. First, we prove that there exists a subsystem oracle for $(X, \mathcal{B})$, when $X \subset \mathbb{R}^d$ is a finite subset of the $d$-dimensional Euclidean space.

**Lemma 2.** *There is a subsystem oracle for the range space $(X, \mathcal{B})$, where $X$ is a finite subset of $\mathbb{R}^d$.*

*Proof.* The VC dimension of $(P, \mathcal{B})$ is bounded by $d+1$, see [17]. For any $Y \subseteq P$, we need to compute the set $\mathcal{B}_{|Y}$ explicitly in time $O(|Y|^{d+2})$. We first apply the standard lifting $\phi: (x_1, \ldots, x_d) \mapsto \left(x_1, \ldots, x_d, \sum_{i=1}^{d} x_i^2\right)$. A point $p \in Y$ belongs to some ball $B \in \mathcal{B}$, with center $c = (c_1, \ldots, c_d) \in \mathbb{R}^d$ and radius $r > 0$, if and only if $\phi(p)$ lies below the hyperplane $h_B$, where $h_B$ is the hyperplane defined by the equation $\langle a_B, x \rangle = b_B$, where $a_B = (2c_1, 2c_2, \ldots 2c_d, 1)$ and $b_B = r^2 - \sum_{i=1}^{d} c_i^2$. Notice that $h_B$ is nonvertical by definition. Then we dualize: for any point $\phi(p) = (y_1, \ldots, y_{d+1})$, $D(\phi(p)) = \{(x_1, \ldots, x_{d+1}) \in \mathbb{R}^{d+1} \mid x_{d+1} = \sum_{i=1}^{d} x_i y_i - y_{d+1}\}$ is a nonvertical hyperplane in $\mathbb{R}^{d+1}$ and for any nonvertical hyperplane $h_B$, $D^{-1}(h_B)$ is a point in $\mathbb{R}^{d+1}$. A standard fact about duality is that a point $\phi(p)$ lies below a hyperplane $h_B$ if and only if the hyperplane $D(\phi(p))$ lies above point $D^{-1}(h_B)$. Finally we construct the arrangement of hyperplanes in the dual space in time $O(|Y|^{d+1})$, using the algorithm in [16]. For each of the at most $O(|Y|^{d+1})$ cells, we return a subset $X \subseteq Y$ corresponding to the hyperplanes lying above. The overall running time is $O(|Y|^{d+2})$.

Then we can analyze the running time of Algorithm 2 in the Euclidean setting.

**Theorem 5.** *Given a set $T \subset (\mathbb{R}^d)^{\leq m}$ of $n$ point sequences (defined over the Euclidean space), parameters $\varepsilon \in (0, \infty)$, and $p \in [1, \infty)$, Algorithm 2 can be implemented to run in $O\left(nm\left(\left(\frac{m}{\varepsilon'} \log \frac{m}{\varepsilon'}\right)^{d+1} + \left(\frac{m}{\varepsilon'} \log \frac{m}{\varepsilon'}\right)^{\ell}\right)\right)$ deterministic time, where $\varepsilon' = \frac{\varepsilon}{2^p - 1 + \varepsilon}$.*

*Proof.* The VC dimension of the range space $(P, \mathcal{B})$ is bounded by $d+1$, see [17]. By Lemma 2, we can use Theorem 3 to compute an $(\varepsilon'/m)$-net $S$ of $(P, \mathcal{B})$, with size $|S| = O\left(\frac{m}{\varepsilon'} \log \left(\frac{m}{\varepsilon'}\right)\right)$, in time $O\left(nm \left(\frac{m}{\varepsilon'} \log \left(\frac{m}{\varepsilon'}\right)\right)^{d+1}\right)$. We then compute the $d_{\mathrm{DTW}_p}$ distance of any of the $|S^{\leq \ell}|$ candidates with the $n$ input point sequences in time $O\left(|S|^{\ell} \cdot nm\right)$.

## 2.3   Simplifications and the Triangle Inequality

In this section, we give an efficient approximation algorithm for simplification under $p$-DTW and show that a weak triangle inequality for $p$-DTW holds, which improves upon previous similar statements [19] when one bounds the DTW distance of two short point sequences.

    These results are then used in Sect. 2.4, where we provide an algorithm for the restricted $(p, 1)$-mean problem, which achieves an approximation factor of $(1 + \varepsilon)$, for any $\varepsilon \in (0, \infty)$. The result holds for the important case of sequences lying in the Euclidean space. In particular, we will bound the expected cost of a mean obtained by randomly sampling an input point sequence and then computing its approximate minimum-error simplification.

**Minimum-Error Simplification.** We first define the notion of simplification of a point sequence under the $p$-DTW distance.

**Definition 8.** *Let $\pi \in X^*$. An $(\alpha, \ell)$-simplification of $\pi$, under the $\mathrm{d}_{\mathrm{DTW}p}$ distance, is a point sequence $\tilde{\pi} \in X^{\leq \ell}$ such that*

$$\forall \pi' \in X^{\leq \ell} : \mathrm{d}_{\mathrm{DTW}p}(\pi, \tilde{\pi}) \leq \alpha \cdot \mathrm{d}_{\mathrm{DTW}p}(\pi, \pi').$$

    We present a dynamic programming solution for the problem of computing an approximate simplification. Each subproblem is parameterized by the length of a prefix of the input point sequence and the maximum length of a simplification of that prefix. Our algorithm can be seen as a special case of the result of Brill et al. [6,7] for computing a mean of restricted complexity, but since our statement is different, we include a proof for completeness.

---

**Algorithm 3.** 2-Approximate Simplification

---

1: **procedure** 2-APPROXIMATE-SIMPLIFICATION($\pi = (x_1, \ldots, x_m), \ell, p$)
2:      Initialize $m \times \ell$ table $D$ with elements in $\mathbb{R}$
3:      Initialize $m \times \ell$ table $C$ with elements in $\mathbb{R}^{\leq \ell}$
4:      $P \leftarrow \{x_1, \ldots, x_m\}$
5:      **for each** $i = 1, \ldots, m$ **do**
6:          **for each** $j = 1, \ldots, \ell$ **do**
7:              **if** $j = 1$ **then**
8:                  $x^* \leftarrow \arg\min_{x \in P} \sum_{k=1}^{i} \rho(x_k, x)^p$
9:                  $D(i, j) \leftarrow \sum_{k=1}^{i} \rho(x_k, x^*)^p$
10:                $C(i, j) \leftarrow (x^*)$
11:              **else**
12:                  $i' \leftarrow \arg\min_{i' \leq i} \left( D(i', j-1) + \min_{x \in P} \sum_{k=i'}^{i} \rho(x_k, x)^p \right)$
13:                  $x^* \leftarrow \arg\min_{x \in P} \sum_{k=i'}^{i} \rho(x_k, x)^p$
14:                  $D(i, j) \leftarrow D(i', j-1) + \sum_{k=i'}^{i} \rho(x_k, x^*)^p$
15:                  $C(i, j) \leftarrow C(i', j-1) \oplus (x^*)$
16:      $j^* \leftarrow \arg\min_{j \in [\ell]} D(m, j)$
17:      **return** $C(m, j^*)$

---

**Lemma 3.** *Let $\mathcal{X} = (X, \rho)$ be a metric space. Given as input a point sequence $\pi = (x_1, \ldots, x_m) \in X^m$, Algorithm 3 returns a point sequence from $\{x_1, \ldots, x_m\}^{\leq \ell}$, which minimizes the $\mathrm{d_{DTW}}_p$ distance to $\pi$, among all point sequences in $\{x_1, \ldots, x_m\}^{\leq \ell}$.*

*Proof.* We show that $C(m, j^*)$ satisfies

$$\mathrm{d_{DTW}}_p(\pi, C(m, j^*)) = \min_{\pi' \in P^{\leq \ell}} \mathrm{d_{DTW}}_p(\pi, \pi').$$

We claim that there is a point sequence $\tilde{\pi} \in P^{\leq \ell}$ such that

$$\mathrm{d_{DTW}}_p(\pi, \tilde{\pi}) = \min_{\pi' \in P^{\leq \ell}} \mathrm{d_{DTW}}_p(\pi, \pi'),$$

and such that the optimal warping between $\pi$ and $\tilde{\pi}$ does not match two vertices of $\tilde{\pi}$ with the same vertex of $\pi$. To see this, consider an optimal warping $W$ between $\pi$ and some point sequence $\pi' = (p'_1, \ldots, p'_j) \in P^{\leq \ell}$. Let $(t_1, t_2) \in W$ and $(t_1, t_2 + 1) \in W$. If $(t_1 - 1, t_2) \in W$ then removing $(t_1, t_2)$ yields a new warping with a cost at most equal to the cost of $W$. Similarly, if $(t_1 + 1, t_2 + 1) \in W$ then removing $(t_1, t_2 + 1)$ from $W$ yields a new warping with a cost at most equal to the cost of $W$. If $(t_1 - 1, t_2) \notin W$, then we can remove $p'_{t_2}$ from $\pi'$. If $(t_1 + 1, t_2 + 1) \notin W$, then we can remove $p'_{t_2+1}$ from $\pi'$. We conclude that there exists a point sequence $\pi'' \in P^{\leq \ell}$ such that $\mathrm{d_{DTW}}_p(\pi, \pi'') \leq \mathrm{d_{DTW}}_p(\pi, \pi')$, and an optimal warping $W$ between $\pi$ and $\pi''$ for which there are no $t_1 \in [m], t_2 \in [\ell]$ such that both $(t_1, t_2) \in W$ and $(t_1, t_2 + 1) \in W$.

For each $i \in [m]$, let $\pi_{|i} = (x_1, \ldots, x_i)$. By construction, each $D(i, j)$ stores the minimum distance between $\pi_{|i}$ and any point sequence $x$ from $P^j$, where the distance is attained by a warping that does not match two vertices of $x$ to the same vertex of $\pi$. Hence, $D(m, j^*)$ stores the minimum distance between $\pi$ and any point sequence in $P^{\leq \ell}$, and $C(m, j^*)$ stores a point sequence from $P^{\leq \ell}$ with distance $D(m, j^*)$ from $\pi$. $\qquad \square$

**Lemma 4.** *Let $\mathcal{X} = (X, \rho)$ be a metric space. Given as input a point sequence $\pi \in X^m$, Algorithm 3 returns a $(2, \ell)$-simplification under the $\mathrm{d_{DTW}}_p$ distance.*

*Proof.* Let $P = \{x_1, \ldots, x_m\}$. By Lemma 3, $C(m, j^*)$ is a point sequence in $P^{\leq \ell}$ that minimizes the distance to $\pi$, among all point sequences in $P^{\leq \ell}$. We show that $C(m, j^*)$ is a $(2, \ell)$-simplification. Let $\pi^* = (x_1^*, \ldots, x_{\ell'}^*)$ be a $(1, \ell)$-simplification of $\pi$, and let $\tilde{\pi}^* = (\tilde{x}_1^*, \ldots, \tilde{x}_{\ell'}^*)$, where for each $i \in [\ell']$, $\tilde{x}_i^* := \arg\min_{x \in P} \rho(x, x_i^*)$. Let $W^* \in \mathcal{W}_{m, \ell'}$ be an optimal warping of $\pi$ and $\pi^*$. Then,

$$\mathrm{d_{DTW}}_p^1(\pi, C(m, j^*)) \leq \mathrm{d_{DTW}}_p^1(\pi, \tilde{\pi}^*)$$

$$= \min_{W \in \mathcal{W}_{m\ell'}} \left( \sum_{(i,j) \in W} \rho(x_i, \tilde{x}_j^*)^p \right)^{1/p}$$

$$\leq \left( \sum_{(i,j) \in W^*} \rho(x_i, \tilde{x}_j^*)^p \right)^{1/p}.$$

$$\leq \left( \sum_{(i,j)\in W^*} (\rho(x_i, x_j^*) + \rho(x_j^*, \tilde{x}_j^*))^p \right)^{1/p} \tag{1}$$

$$\leq \left( \sum_{(i,j)\in W^*} 2^p \left( \rho(x_i, x_j^*) \right)^p \right)^{1/p}$$

$$= 2 \cdot d_{\mathrm{DTW}p}(\pi, \pi^*),$$

where in Step (1) we applied the triangle inequality.

**Theorem 6.** *There is an algorithm that given as input $\pi \in X^m$, computes a $(2, \ell)$-simplification of $\pi$ under the $d_{\mathrm{DTW}p}$, in time $O(m^4\ell)$.*

*Proof.* Correctness of Algorithm 3 follows from Lemma 4. It remains to bound the running time of the algorithm. To do so, we consider the operations taking place in the body of the nested loop. For each $i, j$, we iterate over $O(m)$ values for $i'$ and for each value of $i'$ we compute $\min_{x\in P} \sum_{k=i'}^{i} \rho(x_k, x)^p$ in time $O((i - i') \cdot m) = O(m^2)$. Hence, the total running time is $O(m^4\ell)$. ∎

**Weak Triangle Inequality.** While DTW is not a metric and it is known that the triangle inequality fails for certain instances, there is a weak version of the triangle inequality that is satisfied. In particular, Lemire [19] shows that given $x, y, z \in X^m$, and $p \in [1, \infty)$, we have

$$d_{\mathrm{DTW}p}(x, z) \leq m^{1/p} \cdot \left( d_{\mathrm{DTW}p}(x, y) + d_{\mathrm{DTW}p}(y, z) \right).$$

We slightly generalize the above inequality in a way that implies a better bound for the distance between two short point sequences using the distances to a potentially longer point sequence.

**Lemma 5.** *For any $m_1, m_2 \in \mathbb{N}$, let $x, z \in X^{\leq m_1}$, $y \in X^{m_2}$, and $p \in [1, \infty)$. Then,*

$$d_{\mathrm{DTW}p}(x, z) \leq m_1^{1/p} \cdot \left( d_{\mathrm{DTW}p}(x, y) + d_{\mathrm{DTW}p}(y, z) \right).$$

*Proof.* Let $W_{xz} \in \mathcal{W}_{|x|,|z|}$ be the optimal warping between $x$ and $z$. Let $W_{xy} \in \mathcal{W}_{|x|,|y|}$ be the optimal warping between $x$ and $y$ and $W_{yz} \in \mathcal{W}_{|y|,|z|}$ be the optimal warping between $y$ and $z$. Let $S_{xz} = \{(i, k, j) \in [|x|] \times [|y|] \times [|z|] \mid (i, k) \in W_{xy} \text{ and } (k, j) \in W_{yz}\}$ and $W'_{xz} = \{(i, j) \in [|x|] \times [|z|] \mid \exists k \ (i, k, j) \in S_{xz}\}$. Then,

$$d_{\mathrm{DTW}p}(x, z) = \left( \sum_{(i,j)\in W_{xz}} \rho(x_i, z_j)^p \right)^{1/p}$$

$$\leq \left( \sum_{(i,j)\in W'_{xz}} \rho(x_i, z_j)^p \right)^{1/p}$$

$$\leq \left( \sum_{(i,k,j) \in S_{xz}} (\rho(x_i, y_k) + \rho(y_k, z_j))^p \right)^{1/p}$$

$$\leq \left( \sum_{(i,k,j) \in S_{xz}} \rho(x_i, y_k)^p \right)^{1/p} + \left( \sum_{(i,k,j) \in S_{xz}} \rho(y_k, z_j)^p \right)^{1/p}$$

$$\leq m_1^{1/p} \cdot \mathrm{d}_{\mathrm{DTW}p}(x, y) + m_1^{1/p} \cdot \mathrm{d}_{\mathrm{DTW}p}(y, z),$$

where the second inequality holds by the triangle inequality and the third inequality holds by Minkowski's inequality.

The following theorem uses the weak triangle inequality and provides an upper bound on the expected cost of the restricted $(p, 1)$-mean obtained by first sampling an input point sequence uniformly at random and then computing an $(\alpha, \ell)$-simplification of this point sequence. This theorem will be useful in the next section, where we design an approximation scheme for the mean problem that relies on a first rough estimation of the cost.

**Theorem 7.** *Let $T = \{\tau_1, \ldots, \tau_n\} \subseteq X^{\leq m}$ be a set of point sequences and let $p \in [1, \infty)$. Let $\pi$ be a point sequence picked uniformly at random from $T$, and let $\tilde{\pi}$ be an $(\alpha, \ell)$-simplification of $\pi$ under $\mathrm{d}_{\mathrm{DTW}p}$, where $\ell \leq m$. Then,*

$$\mathrm{E}_\pi \left[ \mathrm{cost}_p^1(T, \tilde{\pi}) \right] \leq (2 + \alpha) m^{1/p} \ell^{1/p} \cdot \mathrm{OPT}_\ell,$$

*where $\mathrm{OPT}_\ell$ denotes the cost of the optimal restricted $(p, 1)$-mean of $T$.*

*Proof.* Let $c$ be an optimal $(p, 1)$-mean of $T$ with cost $\mathrm{OPT}_\ell$. Then,

$$\mathrm{E}_\pi[\mathrm{cost}(T, \tilde{\pi})] = \mathrm{E}_\pi \left[ \sum_{i=1}^n \mathrm{d}_{\mathrm{DTW}p}(\tau_i, \tilde{\pi}) \right]$$

$$\leq \mathrm{E}_\pi \left[ m^{\frac{1}{p}} \sum_{i=1}^n \left( \mathrm{d}_{\mathrm{DTW}p}(\tau_i, c) + \mathrm{d}_{\mathrm{DTW}p}(c, \tilde{\pi}) \right) \right] \qquad (2)$$

$$= m^{1/p} \cdot (\mathrm{OPT}_\ell + n \cdot \mathrm{E}_\pi[\mathrm{d}_{\mathrm{DTW}p}(c, \tilde{\pi})])$$

$$\leq m^{1/p} \cdot (\mathrm{OPT}_\ell + n \cdot \ell^{1/p} \cdot \mathrm{E}_\pi[\mathrm{d}_{\mathrm{DTW}p}(c, \pi) + \mathrm{d}_{\mathrm{DTW}p}(\pi, \tilde{\pi})]) \qquad (3)$$

$$\leq m^{1/p} \cdot (\mathrm{OPT}_\ell + (1 + \alpha) \cdot n \cdot \ell^{1/p} \cdot \mathrm{E}_\pi[\mathrm{d}_{\mathrm{DTW}p}(c, \pi)])$$

$$= m^{1/p} \cdot \left( \mathrm{OPT}_\ell + (1 + \alpha) \cdot n \cdot \ell^{1/p} \cdot \sum_{\pi \in T} \mathrm{d}_{\mathrm{DTW}p}(c, \pi) \cdot \frac{1}{n} \right)$$

$$= m^{1/p} \cdot (\mathrm{OPT}_\ell + (1 + \alpha) \cdot \ell^{1/p} \cdot \mathrm{OPT}_\ell)$$

$$\leq (2 + \alpha) m^{1/p} \ell^{1/p} \cdot \mathrm{OPT}_\ell,$$

where in Step (2) and Step (3), we applied Lemma 5.

## 2.4   Approximation Scheme for Point Sequences in Euclidean Spaces

Now we study the restricted $(p, 1)$-mean problem (which is to compute one median point sequence of complexity at most $\ell$, under the $p$-DTW distance) for point sequences in the Euclidean space. Formally, input point sequences belong to $(\mathbb{R}^d)^{\leq m}$ and we compute a median point sequence in $(\mathbb{R}^d)^{\leq \ell}$. The distance between any two points $x, y \in \mathbb{R}^d$ is measured by the Euclidean distance $\|x - y\|$, thus for any $x \in \mathbb{R}^d$ and $r > 0$ we here denote $B(x, r) = \{y \in \mathbb{R}^d \mid \|x - y\| \leq r\}$. We also use Euclidean grids:

**Definition 9 (grid).** *Given* $r \in \mathbb{R}_+$, *for* $x = (x_1, \ldots, x_d) \in \mathbb{R}^d$ *we define by* $G(r, x) = (\lfloor x_1/r \rfloor \cdot r, \ldots, \lfloor x_d/r \rfloor \cdot r)$ *the r-grid-point of x. Let* $P \subseteq \mathbb{R}^d$ *be a subset of* $\mathbb{R}^d$. *The grid of cell width r that covers P is the set* $\mathbb{G}(P, r) = \{G(r, x) \mid x \in P\}$.

A grid partitions $\mathbb{R}^d$ into cubic regions. For any $r \in \mathbb{R}_+$, $x \in P$, we have $\|x - G(r, x)\| \leq r\sqrt{d}$.

**Algorithm.** We build upon ideas developed in Sect. 2.3 and we design a $(1+\varepsilon)$-approximation algorithm for the restricted $(p, 1)$-mean problem. The algorithm is randomized and succeeds with probability $1 - \delta$, where $\delta$ is a user-defined parameter.

---

**Algorithm 4.** Restricted $(p, 1)$-mean $(1 + \varepsilon)$-Approximation

---

1: **procedure**   MED-APPR($T$ = $\{\tau_1$ = $(\tau_{1,1}, \ldots, \tau_{1,|\tau_1|}), \ldots, \tau_n$ = $(\tau_{n,1}, \ldots, \tau_{n,|\tau_n|})\}, \varepsilon, p, \delta$)
2:     $S \leftarrow$ sample $\lceil \log(2/\delta) \rceil$ point sequences from $T$ uniformly and independently at random with replacement
3:     $\mathcal{R} \leftarrow \emptyset, C \leftarrow \emptyset$
4:     **for each** $\tau_i \in S$ **do**
5:         $\tau_i' \leftarrow (2, \ell)$-simplification of $\tau_i$ under $\mathrm{d}_{\mathrm{DTW}_p}$
6:         $\mathcal{R} \leftarrow \mathcal{R} \cup \{\mathrm{cost}_p^1(T, \tau_i')\}$
7:     $R \leftarrow \min \mathcal{R}$, $\beta \leftarrow 2 \cdot \left( \frac{68m^{1/p}}{\varepsilon} + 5 \right)^d$, $I_R \leftarrow \left\{ \frac{R \cdot 2^{-i}}{n} \mid i = 0, \ldots, \lceil 3 + \log(m\ell)/p \rceil \right\}$
8:     **for each** $r \in I_R$ **do**
9:         $\gamma \leftarrow \frac{\varepsilon \cdot r}{(2m)^{1/p}\sqrt{d}}$
10:        **for each** $\tau_i \in S$ **do**
11:            $\mathcal{B}(\tau_i, 4r) \leftarrow \bigcup_{j=1}^{|\tau_i|} B(\tau_{i,j}, 4r)$, $N \leftarrow \mathbb{G}(\mathcal{B}(\tau_i, 4r), \gamma)$
12:            **if** $|N| \leq \ell \cdot \beta$ **then**
13:                $C \leftarrow C \cup N^{\leq \ell}$
14:    **return** an arbitrary element of $\arg\min_{c \in C} \mathrm{cost}_p^1(T, c)$.

---

The high-level idea is the following. Given a set $T$ of $n$ point sequences, we first compute a rough estimate of the optimal cost. To do so, we sub-sample a sufficiently large number of input sequences that we store in a set $S$, and we compute a $(2, \ell)$-simplification for each one of them. We detect a sequence in $S$ whose simplification minimizes the restricted $(p, 1)$-mean cost; this cost is denoted by $R$. The results of Sect. 2.3 imply that with good probability, $R$ is a $O((m\ell)^{1/p})$ approximation of the optimal cost. We can now use $R$ to "guess" a refined estimate for the restricted $(p, 1)$-mean cost which is within a constant factor from the optimal, by enumerating multiples of 2 in the interval $[\Omega(R(m\ell)^{-1/p}), R]$. Assuming that we have such an estimate, we can use it to fine-tune a grid, which is then intersected with balls centered at the points of sequences in $S$. We use the resulting grid points to compute a set of candidate solutions. The idea here is that with good probability one of the point sequences in $S$ is very close to the optimal solution, so one of the candidate solutions will be a good approximation.

**Analysis.** Now we analyze the running time and correctness of Algorithm 4. We begin with a bound on the probability that $R$ is a rough approximation of the optimal median cost.

**Lemma 6.** *Let $c$ be an optimal restricted $(p, 1)$-mean of $T$. With probability at least $1 - \delta/2$,*

$$R \leq 8m^{1/p}\ell^{1/p} \cdot \mathrm{cost}_p^1(T, c).$$

*Proof.* For any point sequence $\tau_i \in T$, let $\tau_i'$ be a $(2, \ell)$-simplification. Let $\tau_j$ be a randomly sampled point sequence from $T$. By Theorem 7,

$$\mathrm{E}\left[\mathrm{cost}_p^1(T, \tau_j')\right] \leq 4m^{1/p}\ell^{1/p} \cdot \mathrm{cost}_p^1(T, c).$$

By Markov's inequality, $\Pr\left[\mathrm{cost}_p^1(T, \tau_j') \geq 8m^{1/p}\ell^{1/p} \cdot \mathrm{cost}_p^1(T, c)\right] \leq \frac{1}{2}$. Hence, the probability that $R \geq 8m^{1/p}\ell^{1/p} \cdot \mathrm{cost}_p^1(T, c)$ is equal to

$$\Pr\left[\forall \tau_i \in S : \mathrm{cost}_p^1(T, \tau_i') \geq 8m^{1/p}\ell^{1/p} \cdot \mathrm{cost}_p^1(T, c)\right] \leq \frac{1}{2^{|S|}} \leq \frac{\delta}{2}.$$

Next, we bound the probability that a point sequence in the sample $S$ is conveniently close to the optimal median.

**Lemma 7.** *Let $c$ be an optimal restricted $(p, 1)$-mean of $T$. With probability at least $1 - \delta/2$, there exists a $\tau_i \in S$ such that $\mathrm{d}_{\mathrm{DTW}p}(\tau_i, c) < (2/n) \cdot \mathrm{cost}_p^1(T, c)$.*

*Proof.* Let $\tau_i$ be a randomly sampled point sequence from $T$:

$$\mathrm{E}_{\tau_i}\left[\mathrm{d}_{\mathrm{DTW}p}(\tau_i, c)\right] = \sum_{i=1}^n \mathrm{d}_{\mathrm{DTW}p}(\tau_i, c) \cdot \frac{1}{n} = \frac{\mathrm{cost}_p^1(T, c)}{n}.$$

By Markov's inequality, $\Pr\left[\mathrm{d}_{\mathrm{DTW}p}(\tau_i, c) > 2 \cdot \frac{\mathrm{cost}_p^1(T,c)}{n}\right] \leq \frac{1}{2}$. Hence,

$$\Pr\left[\forall \tau_i \in S : \mathrm{d}_{\mathrm{DTW}p}(\tau_i, c) > 2 \cdot \frac{\mathrm{cost}_p^1(T, c)}{n}\right] \leq \frac{1}{2^{|S|}} \leq \frac{\delta}{2}.$$

The set $I_R$ contains a value $r$ such that $nr$ is within a factor of 2 from the optimal cost.

**Lemma 8.** *Let $c$ be an optimal restricted $(p, 1)$-mean of $T$. If*

$$R \le 8m^{1/p}\ell^{1/p} \operatorname{cost}_p^1(T, c),$$

*then there exists $r \in I_R$ such that $\operatorname{cost}_p^1(T, c) \in [nr, 2nr]$.*

*Proof.* Since $R$ is the cost of a curve of complexity at most $\ell$, we have that $\operatorname{cost}_p^1(T, c) \le R$. By assumption, $\operatorname{cost}_p^1(T, c) \ge R/(8m^{1/p}\ell^{1/p})$. By the definition of $I_R$, there exists $j \ge 0$ such that

$$2^{-(j+1)} \cdot \frac{R}{n} \le \frac{\operatorname{cost}_p^1(T, c)}{n} \le 2^{-j} \cdot \frac{R}{n}.$$

Hence, the lemma is true for $r = 2^{-(j+1)} \cdot \frac{R}{n}$. ∎

The following is an upper bound on the number of grid cells needed to cover a Euclidean ball. Similar bounds appear often in the literature, but they are typically asymptotic and not sufficient for our needs. Therefore, we prove an exact (non-asymptotic) upper bound.

**Lemma 9.** *Let $x \in \mathbb{R}^d$, $r > 0$, $\gamma > 0$.*

$$|\mathbb{G}\left(B(x, 8r), \gamma\right)| \le 2 \cdot \left(\frac{34r}{\gamma\sqrt{d}} + 5\right)^d.$$

*Proof.* We use Binet's second expression [36] for the Gamma function $\ln \Gamma(z)$:

$$\ln \Gamma(z) = z \ln(z) - z + \frac{1}{2}\ln\left(\frac{2\pi}{z}\right) + \int_0^\infty \frac{2\arctan\left(\frac{t}{z}\right)}{e^{2\pi t} - 1} dt.$$

Since $\arctan(x) \ge 0$ for $x \ge 0$ and $e^{2\pi x} - 1 \ge 0$ for $x \ge 0$, we have the following inequality:

$$\ln \Gamma(z) \ge z \ln(z) - z + \frac{1}{2}\ln\left(\frac{2\pi}{z}\right)$$

$$\Longleftrightarrow \ln \Gamma(z) \ge \ln(z^z) - \ln(e^z) + \ln\left(\sqrt{\frac{2\pi}{z}}\right)$$

$$\Longleftrightarrow \Gamma(z) \ge z^z e^{-z}\sqrt{\frac{2\pi}{z}}$$

$$\Longleftrightarrow \Gamma(z) \ge \sqrt{2\pi}z^{z-\frac{1}{2}}e^{-z}. \tag{4}$$

We apply a standard volumetric argument to upper bound $|\mathbb{G}\left(B(x,8r),\gamma\right)|$.

$$
\begin{aligned}
|\mathbb{G}\left(B(x,8r),\gamma\right)| &\leq \frac{\text{vol}(B(x,8r+\gamma\sqrt{d}))}{\gamma^d} \\
&= \frac{\pi^{d/2}}{\Gamma(\frac{d}{2}+1)} \cdot \frac{(8r+\gamma\sqrt{d})^d}{\gamma^d} \\
&\leq \frac{\pi^{d/2}e^{d/2+1}}{\sqrt{2\pi}\left(\frac{d}{2}+1\right)^{d/2+1/2}} \cdot \frac{(8r+\gamma\sqrt{d})^d}{\gamma^d} \quad\quad (5) \\
&\leq \frac{2^{d/2+1/2}\pi^{d/2}e^{d/2+1}}{\sqrt{2\pi}\cdot d^{d/2+1/2}} \cdot \frac{(8r+\gamma\sqrt{d})^d}{\gamma^d} \\
&\leq \frac{e\cdot(4.2)^d}{\sqrt{\pi}\cdot d^{d/2}} \cdot \frac{(8r+\gamma\sqrt{d})^d}{\gamma^d} \\
&\leq 2\cdot\left(\frac{34r}{\gamma\sqrt{d}}+5\right)^d,
\end{aligned}
$$

where in (5) we used (4).

We now focus on the iteration of the algorithm with $r \in I_R$, $\tau_i \in S$, such that $r$ satisfies the property guaranteed by Lemma 8 and $\tau_i$ satisfies the property guaranteed by Lemma 7. We claim that in that iteration, an $(1+\varepsilon)$-approximate median is inserted to $C$.

**Lemma 10.** *Let $c$ be an optimal restricted $(p,1)$-mean of $T$. Let $r^*$ be such that $\text{cost}_p^1(T,c) \in [nr^*,2nr^*]$ and let $\gamma^* = \frac{\varepsilon r^*}{(2m)^{1/p}\sqrt{d}}$. If $\tau_i \in S$ is such that $d_{\text{DTW}p}(\tau_i,c) \leq (2/n)\cdot\text{cost}_p^1(T,c)$ then*

i) $|\mathbb{G}(\mathcal{B}(\tau_i,4r^*),\gamma^*)| \leq \ell\cdot 2\cdot\left(\frac{34r^*}{\gamma^*\sqrt{d}}+5\right)^d$ *and*

ii) *there exists $c' \in \mathbb{G}(\mathcal{B}(\tau_i,4r^*),\gamma^*)^{\leq\ell}$ such that $\text{cost}_p^1(T,c') \leq (1+\varepsilon)\cdot\text{cost}_p^1(T,c)$.*

*Proof.* Let $c = (c_1,\dots,c_{\ell'})$, where $\ell' \leq \ell$. To prove i), notice that $d_{\text{DTW}p}(\tau_i,c) \leq 4r^*$, which implies that for any vertex $\tau_{i,j}$ of $\tau_i$, there exists a vertex $c_z$ of $c$ such that $\tau_{i,j} \in B(c_z,4r^*)$. By the triangle inequality $B(\tau_{i,j},4r^*) \subseteq B(c_z,8r^*)$. Hence,

$$
\mathcal{B}(\tau_i,4r^*) \subseteq \bigcup_{z=1}^{\ell} B(c_z,8r^*) \implies |\mathbb{G}(\mathcal{B}(\tau_i,4r^*),\gamma^*)| \leq \left|\mathbb{G}\left(\bigcup_{z=1}^{\ell'} B(c_z,8r^*),\gamma^*\right)\right|
$$

$$
\leq \sum_{z=1}^{\ell}|\mathbb{G}\left(B(c_z,8r^*),\gamma^*\right)|.
$$

By Lemma 9, we obtain

$$
|\mathbb{G}(\mathcal{B}(\tau_i,4r^*),\gamma^*)| \leq \ell\cdot 2\cdot\left(\frac{34r^*}{\gamma^*\sqrt{d}}+5\right)^d.
$$

To prove ii), notice that all vertices of $c$ are contained in $\mathcal{B}(\tau_i, 4r^*)$. Hence, for each point $c_z$ there exists a grid point $\tilde{c}_z \in \mathbb{G}(\mathcal{B}(\tau_i, r^*), \gamma^*)$ such that $\|c_z - \tilde{c}_z\| \leq \gamma^* \sqrt{d}$. We will show that the point sequence $\tilde{c} = (\tilde{c}_1, \ldots, \tilde{c}_{\ell'})$ is a $(1+\varepsilon)$-approximation. For each $i \in [n]$, $W_i^*$ denotes the optimal warping of $\tau_i$ with $c$.

$$
\begin{aligned}
\mathrm{cost}_p^1(T, \tilde{c}) &= \sum_{i=1}^n \mathrm{d_{DTW}}_p(\tau_i, \tilde{c}) \\
&= \sum_{i=1}^n \min_{W \in \mathcal{W}_{|\tau_i|, \ell}} \left( \sum_{(k,j) \in W} \|\tau_{i,k} - \tilde{c}_j\|^p \right)^{1/p} \\
&\leq \sum_{i=1}^n \left( \sum_{(k,j) \in W_i^*} \|\tau_{i,k} - \tilde{c}_j\|^p \right)^{1/p} \\
&\leq \sum_{i=1}^n \left( \sum_{(k,j) \in W_i^*} (\|\tau_{i,k} - c_j\| + \|c_j - \tilde{c}_j\|)^p \right)^{1/p} \\
&\leq \sum_{i=1}^n \left( \left( \sum_{(k,j) \in W_i^*} \|\tau_{i,k} - c_j\|^p \right)^{1/p} + \left( \sum_{(k,j) \in W_i^*} \|c_j - \tilde{c}_j\|^p \right)^{1/p} \right) \\
&\leq \sum_{i=1}^n \left( \mathrm{d_{DTW}}_p(\tau_i, c) + |W_i^*|^{1/p} \cdot \gamma^* \sqrt{d} \right) \\
&\leq \sum_{i=1}^n \left( \mathrm{d_{DTW}}_p(\tau_i, c) + \frac{\mathrm{cost}_p^1(T, c) \cdot \varepsilon}{n} \right) \\
&= (1+\varepsilon) \cdot \mathrm{cost}_p^1(T, c),
\end{aligned}
$$

where the second inequality follows from the triangle inequality, and the third inequality follows from Minkowski's inequality. We also make use of the fact that $|W_i^*| \leq 2m$.

The correctness of our algorithm follows by combining the above.

**Lemma 11.** *Given a set $T \subset \left( \mathbb{R}^d \right)^{\leq m}$, $\varepsilon > 0$, $p \in [1, \infty)$, $\delta \in (0, 1)$, Algorithm 4 returns a $(1+\varepsilon)$-approximate restricted $(p, 1)$-mean with probability of success $1 - \delta$.*

*Proof.* Let $c$ be an optimal restricted $(p, 1)$-mean of $T$. Applying a union bound over the events of Lemma 6 and Lemma 7, we conclude that with probability at least $1 - \delta$, we have $R \leq 8m^{1/p} \ell^{1/p} \cdot \mathrm{cost}_p^1(T, c)$, and there exists a $\tau_i \in S$ such that $\mathrm{d_{DTW}}_p(\tau_i, c) < (2/n) \cdot \mathrm{cost}_p^1(T, c)$. We show correctness assuming that the above two events hold. By Lemma 8 we know that there exists an $r^* \in I_R$ such that $\mathrm{cost}_p^1(T, c) \in [nr^*, 2nr^*]$.

We focus on the iteration where $r^*$ is considered. Let $\gamma^*$ be the value of $\gamma$ in that iteration and let $N^*$ be the set $N$ in that iteration. By Lemma 10 i),

$|N^*| \leq \ell\beta$ and all point sequences of complexity at most $\ell$ defined by points in $N^*$ will be considered as possible solutions. By Lemma 10 ii), there is a point sequence in $(N^*)^{\leq \ell}$ which is a $(1 + \varepsilon)$-approximate solution.

Finally, we bound the running time of Algorithm 4.

**Theorem 8.** *Given a set $T \subset (\mathbb{R}^d)^{\leq m}$ of $n$ point sequences (defined over the Euclidean space), $\varepsilon \in (0, m^{1/p}]$, $p \in [1, \infty)$, $\delta \in (0, 1)$, Algorithm 4 returns a $(1 + \varepsilon)$-approximate restricted $(p, 1)$-mean with probability of success $1 - \delta$ and has running time in $O\left( \left( m^4 + nm \cdot \left( \frac{m^{1/p}}{\varepsilon} \right)^{d\ell} \cdot \frac{\log(m)}{p} \right) \cdot \log\left(\frac{1}{\delta}\right) \right)$.*

*Proof.* Correctness follows from Lemma 11. It remains to bound the running time. For each one of the point sequences in $S$, we compute its $(2, \ell)$-simplification in $O(dm^4\ell)$ time using Theorem 6 and its median cost in $O(dnm\ell)$ time. Hence, the total time needed to compute $\mathcal{R}$ and then $R$ is $O((m^4 + nm) \cdot d\ell \log(1/\delta))$. The set $I_R$ has cardinality $|I_R| = O(\log(m\ell)/p)$. For each value $r \in I_R$, we add at most $\sum_{i=1}^{\ell} |N|^i \cdot |S| \leq \ell \left( \ell \cdot 2 \cdot \left( 68\, m^{1/p}\varepsilon^{-1} + 5 \right)^d \right)^\ell \cdot |S|$ candidates. For each candidate point sequence in $C$, we compute the cost in time $O(dnm\ell)$. Since $d$ and $\ell$ are considered constants, the total running time is

$$O\left( \left( m^4 + nm \cdot \left( m^{1/p}\varepsilon^{-1} \right)^{d\ell} \cdot \log(m)/p \right) \cdot \log\left(1/\delta\right) \right).$$

## 3    Application to Clustering

We can apply the results of Sects. 2.2 and 2.3 to the problem of clustering of point sequences, which we define as follows.

**Definition 10 $((k, \ell, p, q)$-clustering).** *The $(k, \ell, p, q)$-clustering problem is defined as follows, where $k \in \mathbb{N}$, $\ell \in \mathbb{N}_{>1}$ and $p, q \in [1, \infty)$ are fixed (constant) parameters of the problem: given a set $T = \{\tau_1, \ldots, \tau_n\} \subseteq X^{\leq m}$ of point sequences, compute a set $C \subseteq X^{\leq \ell}$ of $k$ point sequences, such that $\mathrm{cost}_p^q(T, C) = \sum_{i=1}^n \min_{c \in C} d_{\mathrm{DTW}p}(c, \tau_i)^q$ is minimal.*

Solving an instance of the $(k, \ell, p, q)$-clustering problem is equivalent to solving an instance of the $k$-median problem, where the distance between any center $c$ and any other element $x$ is measured by $d_{\mathrm{DTW}p}(x, c)^q$. To solve the $k$-medians problem, one can apply a general framework represented by the following two theorems, which are proven in [11], and appear slightly rephrased here. The two theorems provide sufficient conditions for a solution to the $k$-medians problem. We specialize the statements to our case of interest, the $d_{\mathrm{DTW}p}$ distances, raised to the power of $q$.

**Theorem 9 (Theorem 7.2 [11]).** *Let $T = \{\tau_1, \ldots, \tau_n\} \subset X^{\leq m}$, $\alpha \in [1, \infty)$, $\beta \in [1, \infty)$, $\delta \in (0, 1)$, and let $T' \subseteq T$ be an arbitrary subset such that $|T'| \geq |T| \cdot \beta^{-1}$. Suppose that there is an algorithm **Candidates** that given as input*

$T, \alpha, \beta, \delta$ outputs $C \subset X^{\leq \ell}$ such that with probability at least $1 - \delta$, $C$ contains a point sequence $c$ such that $\mathrm{cost}_p^q(T', c) \leq \alpha \cdot \mathrm{cost}_p^q(T', c^*)$, where $c^*$ is a restricted $p$-mean of $T'$.

Then, there is an algorithm $k\text{-}clustering$ that given as input $(T, \emptyset, k, \beta, \delta)$, where $\beta \in (2k, \infty)$, $\delta \in (0, 1)$, $p, q \in [1, \infty)$, returns with probability at least $1 - \delta$ a set $C = \{c_1, \ldots, c_k\} \subset X^{\leq \ell}$ with $\mathrm{cost}_p^q(T, C) \leq \left(1 + \frac{4k}{\beta - 2k}\right) \cdot \alpha \cdot \mathrm{cost}_p^q(T, C^*)$, where $C^*$ is an optimal solution to the $(k, \ell, p, q)$-clustering problem with input $T$.

**Theorem 10 (Theorem 7.3 [11]).** Let $T_1(n, \alpha, \beta, \delta)$ denote the worst-case running time of $Candidates$ for an arbitrary input-set $T \subset X^{\leq m}$ with $|T| = n$ and let $C(n, \alpha, \beta, \delta)$ denote the maximum number of candidates it returns. If $T_1$ and $C$ are non-decreasing in $n$, then $k\text{-}clustering$ has running time in

$$O\left(C(n, \alpha, \beta, \delta)^{k+2} \cdot nm \cdot T_\rho + C(n, \alpha, \beta, \delta)^{k+1} \cdot T_1(n, \alpha, \beta, \delta)\right),$$

where $T_\rho$ denotes the worst-case running time needed to compute the distance between two points in $X$.

The algorithm $k\text{-}clustering$ is a general recursive scheme that collects $k$ medians by repeatedly calling the algorithm $Candidates$. We now adapt our algorithms from Sects. 2.2 and 2.3 such that they can serve as $Candidates$. Algorithm 1 can be easily modified to return the set of candidates, instead of returning the best among them. We show that setting parameters appropriately yields an algorithm that satisfies the properties required by the above-mentioned framework and leads to a randomized algorithm for the $(k, \ell, p, p)$-clustering problem with approximation factor in $O(2^p)$, probability of success $1 - \delta$ and running time in $O\left((2^p km \ln(\ell/\delta))^{\ell(k+2)} \cdot nm\right)$, assuming that the time needed to compute the distance between two points is constant. Similarly, the random sampling method implied by Theorem 7 can be used to produce a sufficiently large sample of candidates, which leads to a randomized algorithm for the $(k, \ell, p, 1)$-clustering problem with approximation factor in $O(m^{1/p} \ell^{1/p})$, probability of success $1 - \delta$ and running time in $O\left((k \log(1/\delta))^{k+2} \cdot nm + m^4 (k \log(1/\delta))^{k+2}\right)$, assuming again constant time for distance computations of points.

## 3.1 $(k, \ell, p, p)$-Clustering

In this section, we apply the result of Sect. 2.2 to design a randomized algorithm for the $(k, \ell, p, p)$-clustering problem. The following algorithm is an adaptation of Algorithm 1.

---

**Algorithm 5.** $(1, \ell, p, p)$-clustering approximate candidates

1: **procedure** $\mathrm{CAND1}(T = \{(\tau_{1,1}, \ldots, \tau_{1,|\tau_1|}), \ldots, (\tau_{n,1}, \ldots, \tau_{n,|\tau_n|})\}, \beta, \delta, \varepsilon, p)$
2:     $P \leftarrow \bigcup_{i=1}^n \bigcup_{j=1}^{|\tau_i|} \{\tau_{i,j}\}$
3:     $S \leftarrow$ sample $\lceil (2^p \varepsilon^{-1} + 1)\beta m \ln(\ell/\delta) \rceil$ points from $P$
        uniformly and independently at random with replacement
4:     **return** $S^{\leq \ell}$

**Lemma 12.** *Let $T \subset X^{\leq m}$, $\beta > 1$, $\delta \in (0,1)$, $\varepsilon > 0$, $p \geq 1$. Let $T' \subseteq T$ such that $|T'| \geq |T| \cdot \beta^{-1}$ and let $C$ be a set obtained by running Algorithm 5 with input $(T, \beta, p, \delta)$. Let $c \in X^{\ell'}$, $\ell' \leq \ell$, be an optimal $p$-mean of $T'$. With probability at least $1 - \delta$, there exists $\tau' \in C$ such that*

$$\mathrm{cost}_p^p(T', \tau') \leq (2^p + \varepsilon) \cdot \mathrm{cost}_p^p(T', c).$$

*Proof.* By Lemma 1 applied on $T'$, we have that there exist sets $B_1, \ldots, B_{\ell'} \subseteq P$, each of cardinality at least $\left(\frac{\varepsilon}{2^p + \varepsilon}\right) \cdot |T'|$ such that any point sequence $c' = (c'_1, \ldots, c'_{\ell'})$ with $\forall i \in [\ell] : c'_i \in B_i$, is a $(2^p + \varepsilon)$-approximate restricted $p$-mean of $T'$. We upper bound the probability that $S$ does not contain any point from a fixed $B_i$:

$$\Pr\left[\forall x \in B_i : x \notin S\right] \leq \left(\frac{|P| - \left(\frac{\varepsilon}{2^p + \varepsilon}\right) \cdot |T'|}{|P|}\right)^{|S|} \leq \left(1 - \left(\frac{\varepsilon}{(2^p + \varepsilon)\beta m}\right)\right)^{|S|}$$

$$\leq \frac{\delta}{\ell}.$$

Then, by a union bound we have that the probability that there exists $i \in [\ell']$ such that $\forall x \in B_i : x \notin S$, is at most $\delta$. Hence, with probability at least $1 - \delta$, there is a point sequence $c' \in S^{\leq \ell}$ which is a $(2^p + \varepsilon)$-approximate restricted $p$-mean of $T'$. ∎

**Theorem 11.** *There is an algorithm that given a set $T \subset X^{\leq m}$ of $n$ point sequences, $p \in [1, \infty)$, $\beta \in (2k, \infty)$ and $\delta \in (0,1)$, returns with probability at least $1 - \delta$ a set $C = \{c_1, \ldots, c_k\}$ with $\mathrm{cost}_p^p(T, C) \leq \left(1 + \frac{4k}{\beta - 2k}\right) \cdot (2^p + \varepsilon) \cdot \mathrm{cost}_p^p(T, C^*)$, where $C^*$ is an optimal set of $k$-medians of $T$, under $d_{\mathrm{DTW}p}$. The algorithm has running time in*

$$O\left(((2^p \varepsilon^{-1} + 1)\beta m \ln(\ell/\delta))^{\ell(k+2)} \cdot nm\right),$$

*assuming that the time needed to compute the distance between two points of $X$ is constant.*

*Proof.* We plug Algorithm 5 into Theorems 9 and 10. By Lemma 12, for any $T' \subseteq T$ with $|T'| \geq |T|\beta^{-1}$ Algorithm 5 returns a set of point sequences which contains a $(2^p + \varepsilon)$-approximate restricted $p$-mean of $T'$. Therefore, by Theorem 9 the clustering algorithm is correct. The running time of Algorithm 5 is upper bounded by $O(|S|^\ell + nm)$. The running time then follows by Theorem 10. ∎

## 3.2　$k$-medians Under $p$-DTW

In this section, we apply the random sampling bound developed in Sect. 2.3 to design a randomized algorithm for the $(k, \ell, p, 1)$-clustering problem, that is the problem of computing $k$-medians of complexity at most $\ell$, under $\mathrm{d}_{\mathrm{DTW}p}$. We achieve an approximation factor in $O(m^{1/p}\ell^{1/p})$.

The main idea is that one can use random sampling and approximate simplifications, to obtain a simple algorithm for computing a set of 1-median candidates. Those candidates are guaranteed, up to some user-defined probability, to contain a point sequence which is an approximate 1-median for a fixed but unknown subset of the input.

---

**Algorithm 6.** $(1, \ell, p, 1)$-clustering approximate candidates

---

1: **procedure** CAND2$(T = \{\tau_1, \ldots, \tau_n\}, \beta, p, \delta)$
2:    $S \leftarrow$ sample $\lceil 2\beta \cdot \log(2/\delta) \rceil$ point sequences from $T$
        uniformly and independently at random with replacement
3:    $C \leftarrow \emptyset$
4:    **for each** $\tau \in S$ **do**
5:        $\tau' \leftarrow (2, \ell)$-simplification of $\tau$, under $\mathrm{d}_{\mathrm{DTW}p}$
6:        $C \leftarrow C \cup \{\tau'\}$
7:    **return** $C$

---

**Lemma 13.** *Let* $T \subset X^{\leq m}$, $\beta > 1$, $p \geq 1$, $\delta \in (0, 1)$. *Let* $T' \subseteq T$ *such that* $|T'| \geq |T| \cdot \beta^{-1}$ *and let* $C$ *be a set obtained by running Algorithm 6 with input* $(T, \beta, p, \delta)$. *Let* $c$ *be an optimal restricted* $(p, 1)$-*mean of* $T'$. *With probability at least* $1 - \delta$, *there exists* $\tau' \in C$ *such that*

$$\mathrm{cost}_p^1(T', \tau') \leq 8 \cdot m^{1/p}\ell^{1/p} \cdot \mathrm{cost}_p^1(T', c).$$

*Proof.* We use a standard Chernoff bound (see [23, Theorem 4.5]) to upper bound the probability that $|S \cap T'| \leq |S|/(2\beta)$. Notice that $\mathrm{E}\left[|S \cap T'|\right] \geq |S| \cdot \beta^{-1}$. Hence,

$$\Pr\left[|S \cap T'| \leq \frac{|S|}{2\beta}\right] \leq \exp\left(-\frac{|S|}{8\beta}\right) \leq \frac{\delta}{2}. \tag{6}$$

Let $\mathcal{E}_{T'}$ be the event that $|S \cap T'| > \frac{|S|}{2\beta}$. We condition the rest of the proof on the event $\mathcal{E}_{T'}$. Let $\tau'$ be a $(2, \ell)$-simplification of any point sequence $\tau \in S \cap T'$. Then, by Theorem 7,

$$\mathrm{E}_\tau\left[\mathrm{cost}_p^1(T', \tau') \mid \mathcal{E}_{T'}\right] \leq 4m^{1/p}\ell^{1/p} \cdot \mathrm{cost}_p^1(T', c).$$

By Markov's inequality, $\Pr\left[\mathrm{cost}_p^1(T', \tau') \geq 8m^{1/p}\ell^{1/p} \cdot \mathrm{cost}_p^1(T', c) \mid \mathcal{E}_{T'}\right] \leq \frac{1}{2}$. Hence, by independence of the random sampling,

$$\Pr\left[\forall \tau \in S \cap T' : \mathrm{cost}_p^1(T', \tau') \geq 8m^{1/p}\ell^{1/p} \cdot \mathrm{cost}_p^1(T', c) \mid \mathcal{E}_{T'}\right] \leq \frac{1}{2^{|S|/(2\beta)}}$$

$$\leq \frac{\delta}{2}. \tag{7}$$

A union bound using inequalities (6), (7) completes the proof.

**Theorem 12.** *There is an algorithm that given a set $T \subset X^{\leq m}$ of $n$ point sequences, $p \in [1, \infty)$, $\beta \in (2k, \infty)$ and $\delta \in (0, 1)$, returns with probability at least $1 - \delta$ a set $C = \{c_1, \ldots, c_k\}$ with $\mathrm{cost}_p^1(T, C) \leq \left(1 + \frac{4k}{\beta - 2k}\right) \cdot (8m^{1/p}\ell^{1/p}) \cdot \mathrm{cost}_p^1(T, C^*)$, where $C^*$ is an optimal set of $k$-medians of $T$, under $\mathrm{d}_{\mathrm{DTW}p}$. The algorithm has running time in*

$$O\left((\beta \log(1/\delta))^{k+2} \cdot nm + (\beta \log(1/\delta))^{k+1} \cdot \beta m^4 \log(1/\delta)\right),$$

*assuming that the distance between two points of $X$ can be computed in constant time.*

*Proof.* We plug Algorithm 6 into Theorems 9 and 10. Lemma 13 guarantees that with probability at least $1 - \delta$, there exists a point sequence in the set $C$, returned by Algorithm 6, which is an $(8m^{1/p}\ell^{1/p})$-approximate 1-median to an arbitrary subset $T' \subset T$, as required by Theorem 9. Let $T_\rho$ be time needed to compute the distance between two elements of $X$. Using Theorem 6 to compute simplifications, Algorithm 6 needs $O(T_\rho \cdot \beta m^4 \ell \log(1/\delta)))$ time to compute $C$. Taking into account the time needed to read the input, and assuming that $d, \ell$ are constants, the total running time of Algorithm 6 is in $O(nm + T_\rho \cdot \beta m^4 \log(1/\delta))$. Therefore, by Theorem 9, there is an algorithm that returns an $\left(1 + \frac{4k}{\beta - 2k}\right) \cdot (8m^{1/p}\ell^{1/p})$-approximate solution to the $k$-medians problem, and by Theorem 10 the algorithm has running time in

$$O\left((\beta \log(1/\delta))^{k+2} \cdot nm \cdot T_\rho + (\beta \log(1/\delta))^{k+1} \cdot (nm + T_\rho \cdot \beta m^4 \log(1/\delta))\right).$$

## 4    Conclusions

We have studied mean problems for point sequences under the $p$-DTW distance and devised exact and approximation algorithms for several relevant problem variants where the complexity of the mean is restricted by a parameter $\ell$. Our exact algorithm runs in polynomial time for constant $\ell$ and $d$. The running times of our approximation algorithms depend only linearly on the number of input sequences. The dependency on the length of the input sequences, however, is high; the dependency on the parameter $\ell$ is even exponential. We hope that the algorithmic techniques developed in this paper will inspire further work on the topic. In particular, we think the weak triangle inequality and the simplification algorithm could be of great use. In contrast, a proof of hardness of approximation for the central problem studied in this paper is not in sight. We leave this as an open problem.

# References

1. Aach, J., Church, G.M.: Aligning gene expression time series with time warping algorithms. Bioinformatics **17**(6), 495–508 (2001)
2. Abboud, A., Backurs, A., Williams, V.V.: Tight hardness results for LCS and other sequence similarity measures. In: Guruswami, V. (ed.) IEEE 56th Annual Symposium on Foundations of Computer Science, FOCS 2015, Berkeley, CA, USA, 17–20 October 2015, pp. 59–78. IEEE Computer Society (2015)
3. Abdulla, W.H., Chow, D., Sin, G.: Cross-words reference template for DTW-based speech recognition systems. In: TENCON 2003. Conference on Convergent Technologies for Asia-Pacific Region, vol. 4, pp. 1576–1579 (2003)
4. Berndt, D.J., Clifford, J.: Using dynamic time warping to find patterns in time series. In: Fayyad, U.M., Uthurusamy, R. (eds.) Knowledge Discovery in Databases: Papers from the 1994 AAAI Workshop, Seattle, Washington, USA, July 1994. Technical report WS-94-03, pp. 359–370. AAAI Press (1994)
5. Brankovic, M., Buchin, K., Klaren, K., Nusser, A., Popov, A., Wong, S.: (k, l)-medians clustering of trajectories using continuous dynamic time warping. In: Lu, C., Wang, F., Trajcevski, G., Huang, Y., Newsam, S.D., Xiong, L. (eds.) SIGSPATIAL 2020: 28th International Conference on Advances in Geographic Information Systems, Seattle, WA, USA, 3–6 November 2020, pp. 99–110. ACM (2020)
6. Brill, M., Fluschnik, T., Froese, V., Jain, B.J., Niedermeier, R., Schultz, D.: Exact mean computation in dynamic time warping spaces. In: Proceedings of the 2018 SIAM International Conference on Data Mining, SDM 2018, 3–5 May 2018, San Diego Marriott Mission Valley, San Diego, CA, USA, pp. 540–548 (2018)
7. Brill, M., Fluschnik, T., Froese, V., Jain, B., Niedermeier, R., Schultz, D.: Exact mean computation in dynamic time warping spaces. Data Min. Knowl. Disc. **33**(1), 252–291 (2018). https://doi.org/10.1007/s10618-018-0604-8
8. Bringmann, K., Künnemann, M.: Quadratic conditional lower bounds for string problems and dynamic time warping. In: Guruswami, V. (ed.) IEEE 56th Annual Symposium on Foundations of Computer Science, FOCS 2015, Berkeley, CA, USA, 17–20 October 2015, pp. 79–97. IEEE Computer Society (2015)
9. Brönnimann, H., Chazelle, B., Matousek, J.: Product range spaces, sensitive sampling, and derandomization. SIAM J. Comput. **28**(5), 1552–1575 (1999)
10. Buchin, K., Driemel, A., Struijs, M.: On the hardness of computing an average curve. In: 17th Scandinavian Symposium and Workshops on Algorithm Theory, SWAT 2020, 22–24 June 2020, Tórshavn, Faroe Islands, pp. 19:1–19:19 (2020)
11. Buchin, M., Driemel, A., Rohde, D.: Approximating (k, ℓ)-median clustering for polygonal curves. In: Marx, D. (ed.) Proceedings of the 2021 ACM-SIAM Symposium on Discrete Algorithms, SODA 2021, Virtual Conference, 10–13 January 2021, pp. 2697–2717. SIAM (2021)
12. Bulteau, L., Froese, V., Niedermeier, R.: Tight hardness results for consensus problems on circular strings and time series. SIAM J. Discret. Math. **34**(3), 1854–1883 (2020)
13. Caiani, E.G., et al.: Warped-average template technique to track on a cycle-by-cycle basis the cardiac filling phases on left ventricular volume. In: Computers in Cardiology 1998, vol. 25 (Cat. No.98CH36292), pp. 73–76 (1998)
14. Chazelle, B., Edelsbrunner, H., Guibas, L.J., Sharir, M.: A singly exponential stratification scheme for real semi-algebraic varieties and its applications. Theoret. Comput. Sci. **84**(1), 77–105 (1991)

15. Datta, S., Karmakar, C.K., Palaniswami, M.: Averaging methods using dynamic time warping for time series classification. In: 2020 IEEE Symposium Series on Computational Intelligence (SSCI), pp. 2794–2798 (2020)

16. Edelsbrunner, H., O'Rourke, J., Seidel, R.: Constructing arrangements of lines and hyperplanes with applications. SIAM J. Comput. 15(2), 341–363 (1986)

17. Har-peled, S.: Geometric Approximation Algorithms. American Mathematical Society, USA (2011)

18. Hautamäki, V., Nykanen, P., Franti, P.: Time-series clustering by approximate prototypes. In: 2008 19th International Conference on Pattern Recognition, pp. 1–4 (2008)

19. Lemire, D.: Faster retrieval with a two-pass dynamic-time-warping lower bound. Pattern Recogn. 42(9), 2169–2180 (2009)

20. Liu, Y.T., Zhang, Y., Zeng, M.: Adaptive global time sequence averaging method using dynamic time warping. IEEE Trans. Signal Process. 67, 2129–2142 (2019)

21. Luca, A.D., Hang, A., Brudy, F., Lindner, C., Hussmann, H.: Touch me once and i know it's you!: implicit authentication based on touch screen patterns. In: Konstan, J.A., Chi, E.H., Höök, K. (eds.) CHI Conference on Human Factors in Computing Systems, CHI 2012, Austin, TX, USA, 05–10 May 2012, pp. 987–996. ACM (2012)

22. Matousek, J.: Lectures on Discrete Geometry. Graduate Texts in Mathematics, vol. 212. Springer, New York (2002). https://doi.org/10.1007/978-1-4613-0039-7

23. Mitzenmacher, M., Upfal, E.: Probability and Computing: Randomization and Probabilistic Techniques in Algorithms and Data Analysis, 2nd edn. Cambridge University Press, Cambridge (2017)

24. Morel, M., Achard, C., Kulpa, R., Dubuisson, S.: Time-series averaging using constrained dynamic time warping with tolerance. Pattern Recogn. 74, 77–89 (2018)

25. Muda, L., Begam, M., Elamvazuthi, I.: Voice recognition algorithms using MEL frequency cepstral coefficient (MFCC) and dynamic time warping (DTW) techniques (2010)

26. Munich, M.E., Perona, P.: Continuous dynamic time warping for translation-invariant curve alignment with applications to signature verification. In: Proceedings of the International Conference on Computer Vision, Kerkyra, Corfu, Greece, 20–25 September 1999, pp. 108–115. IEEE Computer Society (1999)

27. Okawa, M.: Time-series averaging and local stability-weighted dynamic time warping for online signature verification. Pattern Recogn. 112, 107699 (2021)

28. Petitjean, F., Ketterlin, A., Gançarski, P.: A global averaging method for dynamic time warping, with applications to clustering. Pattern Recognit. 44(3), 678–693 (2011)

29. Rabiner, L., Wilpon, J.: Considerations in applying clustering techniques to speaker independent word recognition. In: ICASSP 1979. IEEE International Conference on Acoustics, Speech, and Signal Processing, vol. 4, pp. 578–581 (1979)

30. Sakoe, H., Chiba, S.: Dynamic programming algorithm optimization for spoken word recognition. IEEE Trans. Acoust. Speech Signal Process. 26(1), 43–49 (1978)

31. Sauer, N.: On the density of families of sets. J. Comb. Theory Ser. A 13, 145–147 (1972)

32. Schaar, N., Froese, V., Niedermeier, R.: Faster binary mean computation under dynamic time warping. In: 31st Annual Symposium on Combinatorial Pattern Matching, CPM 2020, 17–19 June 2020, Copenhagen, Denmark, pp. 28:1–28:13 (2020)

33. Schultz, D., Jain, B.: Nonsmooth analysis and subgradient methods for averaging in dynamic time warping spaces. Pattern Recogn. 74, 340–358 (2018)

34. Shelah, S.: A combinatorial problem; stability and order for models and theories in infinitary languages. Pac. J. Math. **41**(1), 247–261 (1972)
35. Vapnik, V., Chervonenkis, A.: On the uniform convergence of relative frequencies of events to their probabilities. Theory Probab. Appl **16**, 264–280 (1971)
36. Whittaker, E.T., Watson, G.N.: A Course of Modern Analysis. Cambridge Mathematical Library, 4th edn. Cambridge University Press, Cambridge (1996)
37. Zhu, Y., Shasha, D.E.: Warping indexes with envelope transforms for query by humming. In: Halevy, A.Y., Ives, Z.G., Doan, A. (eds.) Proceedings of the 2003 ACM SIGMOD International Conference on Management of Data, San Diego, California, USA, 9–12 June 2003, pp. 181–192. ACM (2003)

# Author Index

Printed in the United States
by Baker & Taylor Publisher Services

Printed in the United States
by Baker & Taylor Publisher Services